New Insights into Endometrial Cancer 2022

New Insights into Endometrial Cancer 2022

Editor

Laura Paleari

MDPI • Basel • Beijing • Wuhan • Barcelona • Belgrade • Manchester • Tokyo • Cluj • Tianjin

Editor
Laura Paleari
Research, Innovation and HTA Unit
Liguria Health Authority A.Li.Sa.
Genova
Italy

Editorial Office
MDPI
St. Alban-Anlage 66
4052 Basel, Switzerland

This is a reprint of articles from the Special Issue published online in the open access journal *International Journal of Molecular Sciences* (ISSN 1422-0067) (available at: www.mdpi.com/journal/ijms/special_issues/Endometrial_Cancer_2021).

For citation purposes, cite each article independently as indicated on the article page online and as indicated below:

LastName, A.A.; LastName, B.B.; LastName, C.C. Article Title. *Journal Name* **Year**, *Volume Number*, Page Range.

ISBN 978-3-0365-7915-3 (Hbk)
ISBN 978-3-0365-7914-6 (PDF)

Contents

Preface to "New Insights into Endometrial Cancer 2022"

Endometrial cancer is a hormone-dependent cancer with an increasing incidence that is estimated to continue to grow over the next several years. It is typically treated with surgery and/or chemo-/radiation therapy. The clinical benefit of hormone therapies for advanced and recurrent endometrial cancers underlines the need to examine their characteristics (particularly, steroid hormone receptor expression and functions) to assess how they can be used better. Furthermore, a critical phase that can drive clinicians' therapeutic choices is histopathological and molecular classification. Thus, a current challenge is to integrate IHC markers with molecular tests to identify prognostic groups. Moreover, the observations of the immunosuppressive nature of the endometrial cancer environment are leading to studies that assess therapies with the aim to boost the immune response, which might represent significant potential in the treatment of the disease. For this reason, current and ongoing studies are trying to improve clinical responses through immunotherapy strategies, either alone or combined with classic treatments.

Laura Paleari

Editor

International Journal of *Molecular Sciences*

Editorial

New Strategies for Endometrial Cancer Detection and Management

Laura Paleari

Research, Innovation and HTA Unit, A.Li.Sa., Liguria Health Authority, 16121 Genoa, Italy; laura.paleari@alisa.liguria.it; Tel.: +39-010-5484243

With 400,000 new cases and over 80,000 deaths a year worldwide, endometrial cancer (EC) holds a rather unfortunate record, namely, that of the tumour with the highest increase in incidence, a unique trend among gynaecological cancers [1]. EC is more common in high-income countries where obesity rates are high, and this cancer is the fourth most common malignancy in women. It mainly affects postmenopausal women, where patients' mean age at diagnosis is 61 years, and most cases are diagnosed in women aged 45 to 74 years [1]. The prognosis is worse in older patients with higher-grade and more prevalent tumours, with a median 5-year survival rates of 95% for stage I/II and of 60% for stage III/IV. Overall, 63% of patients are tumour free $\geq$5 years after therapy [1].

To date, the standard management of early-stage cancer includes surgery, radiotherapy and/or chemotherapy. Thus, there is a clear unmet medical need for disease management, and it is essential to find a molecular target to better classify ECs in terms of prognosis and to drive the therapeutical choice. The intention of this Special Issue is to increase our understanding of EC development, which may lead to the discovery of new molecular diagnostic technologies and targeted therapeutics. For instance, although EC is a hormone-dependent neoplasm, there are no recommendations for the determination of steroid hormone receptors [2]. Currently, only a few studies have assessed the impact hormonal therapy (HT) has in EC treatment, suggesting how HT is a useful treatment modality to prolong progression-free survival (PFS) and overall survival (OS) [3]. Recently, several studies have assessed diagnostic and therapeutic markers. Bioinformatics analyses indicated that NRAS gene deregulation affected the OS rate of patients with EC, leading to prognostic significance [4]. Celsi and colleagues with a gel-based proteomic assay showed that suprabasin (SBSN), an oncogene previously associated with poor prognosis in different cancers, could be a potential novel biomarker in EC [5]. Interestingly, the study of Benbrook et al. analysed serum from EC patients (with or without lymph node metastases, and compared with that of benign gynaecological surgery patients) to determine if it is possible to predict the presence of positive lymph nodes, which represents a critical factor guiding treatment decisions [6]. The results suggest that pathways implicated in metastases included the loss of PTEN, PI3K, AKT and PKA activation, leading to calcium signalling, oxidative phosphorylation and oestrogen receptor-induced transcription, platelet activation, the epithelial-to-mesenchymal transition, and senescence. Upstream activators implicated in these events included neurostimulation and inflammation, the activation of G-protein-coupled receptor $G\beta\gamma$, the loss of HER-2 activation and the upregulation of the insulin receptor [6].

Moreover, advancements have newly been made in EC pathological classification through the new molecular classification that allows for a more accurate categorisation of the pathology, which is useful for making a therapeutic choice. In fact, for decades, EC risk stratification has been based on histopathological features, such as tumour grade and histotype, depth of myometrial invasion, and cervical and adnexal involvement. The Cancer Genome Atlas (TGCA) has proposed a new and disruptive molecular classification of ECs which identifies, based on molecular characteristics and expression, four distinct subgroups.

Citation: Paleari, L. New Strategies for Endometrial Cancer Detection and Management. *Int. J. Mol. Sci.* **2023**, *24*, 6462. https://doi.org/10.3390/ijms24076462

Received: 20 March 2023
Accepted: 27 March 2023
Published: 30 March 2023

This approach has revolutionized diagnostics, allowing data to be refined from a precision medicine point of view through the definition of a real therapeutic algorithm [7,8].

Furthermore, in recent years, vibrational biospectrometry methods for the screening and diagnosis of EC have gained ground. Vibrational biospectroscopy is a noninvasive objective technique that uses the interaction of light with tissue to obtain detailed information about the chemical composition of biological samples. Current diagnostic strategies are time-consuming, invasive and have limited accuracy, and there is no population-level screening. Treatment depends on the patients' health, type of disease and prevalence at the time of diagnosis. Most women will be offered surgery; however, the role of lymph node dissection in the early stage of the disease remains controversial. There is, therefore, a need for a test that is objective, accurate, and capable of detecting pre-cancer and early cancer, as well as able to identify metastatic lymph node involvement, so that lymph node excision is performed only when necessary. However, while there has been significant progress made in vibrational spectroscopy development, the techniques are still at the trial phase and have not yet been translated into clinical practice [9].

Moreover, for over 20 years, the availability of a new therapy capable of affecting the prognosis of a tumour which, although not considered particularly aggressive, had a proportion of recurrences that were difficult to treat, had not been verified. In the last year, interesting advances have taken place with the introduction of immunotherapy for the treatment of ECs. The drug dostarlimab is an antibody that blocks the PD-1 receptor, and is indicated for adult patients with advanced or recurrent EC with a mismatch repair/microsatellite instability system defect. This immunotherapy drug, in fact, recently had accelerated approval by the US FDA, and the European Commission itself gave the green light to its conditional marketing. This approval is based on results obtained through the GARNET multi-cohort study [10,11]. The data collected showed that the dostarlimab treatment resulted in an objective response rate (ORR) of 43.5% in 71 patients with a DNA mismatch repair deficiency, including a complete response rate (CRR) of 12.7% and a partial control rate (PCR) of 29.6%. A duration of the response at six months was seen in more than 93.3% of the respondents [10,11].

The main aim of this Special Issue is to provide an open-source sharing platform for significant works in the field of molecular oncology that can increase our understanding of EC development, and which may lead to the discovery of new molecular diagnostic technologies and targeted therapeutics.

Acknowledgments: Many thanks to Violetto, who supports my research.

Conflicts of Interest: The author declares no conflict of interest.

References

1. American Cancer Society: Key Statistics for Endometrial Cancer. Available online: https://www.cancer.org/cancer/endometrial-cancer/about/key-statistics.html (accessed on 10 March 2023).
2. Available online: https://www.nccn.org/professionals/physician_gls/pdf/uterine.pdf (accessed on 10 March 2023).
3. Paleari, L.; Rutigliani, M.; Siri, G.; Provinciali, N.; Colombo, N.; DeCensi, A. Aromatase Inhibitors as Adjuvant Treatment for ER/PgR Positive Stage I Endometrial Carcinoma: A Retrospective Cohort Study. *Int. J. Mol. Sci.* **2020**, *21*, 2227. [CrossRef] [PubMed]
4. Alessandro, L.; Low, K.-J.E.; Abushelaibi, A.; Lim, S.-H.E.; Cheng, W.-H.; Chang, S.-K.; Lai, K.-S.; Sum, Y.W.; Maran, S. Identification of NRAS Diagnostic Biomarkers and Drug Targets for Endometrial Cancer—An Integrated in Silico Approach. *Int. J. Mol. Sci.* **2022**, *23*, 14285. [CrossRef] [PubMed]
5. Celsi, F.; Monasta, L.; Arrigoni, G.; Battisti, I.; Licastro, D.; Aloisio, M.; Di Lorenzo, G.; Romano, F.; Ricci, G.; Ura, B. Gel-Based Proteomic Identification of Suprabasin as a Potential New Candidate Biomarker in Endometrial Cancer. *Int. J. Mol. Sci.* **2022**, *23*, 2076. [CrossRef] [PubMed]
6. Benbrook, D.M.; Hocker, J.R.S.; Moxley, K.M.; Hanas, J.S. Sera Protein Signatures of Endometrial Cancer Lymph Node Metastases. *Int. J. Mol. Sci.* **2022**, *23*, 3277. [CrossRef] [PubMed]
7. Arciuolo, D.; Travaglino, A.; Raffone, A.; Raimondo, D.; Santoro, A.; Russo, D.; Varricchio, S.; Casadio, P.; Inzani, F.; Seracchioli, R.; et al. TCGA Molecular Prognostic Groups of Endometrial Carcinoma: Current Knowledge and Future Perspectives. *Int. J. Mol. Sci.* **2022**, *23*, 11684. [CrossRef] [PubMed]

8. Paleari, L.; Pesce, S.; Rutigliani, M.; Greppi, M.; Obino, V.; Gorlero, F.; Vellone, V.G.; Marcenaro, E. New Insights into Endometrial Cancer. *Cancers* **2021**, *13*, 1496. [CrossRef] [PubMed]
9. Schiemer, R.; Furniss, D.; Phang, S.; Seddon, A.B.; Atiomo, W.; Gajjar, K.B. Vibrational Biospectroscopy: An Alternative Approach to Endometrial Cancer Diagnosis and Screening. *Int. J. Mol. Sci.* **2022**, *23*, 4859. [CrossRef] [PubMed]
10. Oaknin, A.; Gilbert, L.; Tinker, A.V.; Brown, J.; Mathews, C.; Press, J.; Sabatier, R.; O'Malley, D.M.; Samouelian, V.; Boni, V.; et al. Safety and antitumor activity of dostarlimab in patients with advanced or recurrent DNA mismatch repair deficient/microsatellite instability-high (dMMR/MSI-H) or proficient/stable (MMRp/MSS) endometrial cancer: Interim results from GARNET-a phase I, single-arm study. *J. Immunother. Cancer* **2022**, *10*, e003777. [CrossRef] [PubMed]
11. Oaknin, A.; Tinker, A.V.; Gilbert, L.; Samouëlian, V.; Mathews, C.; Brown, J.; Barretina-Ginesta, M.P.; Moreno, V.; Gravina, A.; Abdeddaim, C.; et al. Clinical activity and safety of the anti-PD-1 monoclonal antibody dostarlimab for patients with recurrent or advanced dMMR endometrial cancer. *Future Oncol.* **2021**, *17*, 3781–3785. [CrossRef] [PubMed]

International Journal of *Molecular Sciences*

MDPI

Article

Combination of EphA2- and Wee1-Targeted Therapies in Endometrial Cancer

Santosh K. Dasari [1,2], Robiya Joseph [1], Sujanitha Umamaheswaran [1,3], Lingegowda S. Mangala [1], Emine Bayraktar [1], Cristian Rodriguez-Aguayo [4], Yutuan Wu [1], Nghi Nguyen [5], Reid T. Powell [5], Mary Sobieski [5], Yuan Liu [1], Mamur A. Chowdhury [1], Paola Amero [4], Clifford Stephan [5], Gabriel Lopez-Berestein [4], Shannon N. Westin [1] and Anil K. Sood [1,*]

1 Department of Gynecologic Oncology and Reproductive Medicine, The University of Texas MD Anderson Cancer Center, Houston, TX 77030, USA
2 National Institute of Animal Biotechnology, Hyderabad 500029, India
3 UTHealth Houston Graduate School of Biomedical Sciences, The University of Texas MD Anderson Cancer Center, Houston, TX 77030, USA
4 Department of Experimental Therapeutics, The University of Texas MD Anderson Cancer Center, Houston, TX 77030, USA
5 High-Throughput Research and Screening Center, Center for Translational Cancer Research, Texas A&M Health Science Center, Institute of Biosciences and Technology, Houston, TX 77030, USA
* Correspondence: asood@mdanderson.org; Tel.: +1-713-745-5266

Abstract: EphA2 tyrosine kinase is upregulated in many cancers and correlated with poor survival of patients, including those with endometrial cancer. EphA2-targeted drugs have shown modest clinical benefit. To improve the therapeutic response to such drugs, we performed a high-throughput chemical screen to discover novel synergistic partners for EphA2-targeted therapeutics. Our screen identified the Wee1 kinase inhibitor, MK1775, as a synergistic partner to EphA2, and this finding was confirmed using both in vitro and in vivo experiments. We hypothesized that Wee1 inhibition would sensitize cells to EphA2-targeted therapy. Combination treatment decreased cell viability, induced apoptosis, and reduced clonogenic potential in endometrial cancer cell lines. In vivo Hec1A and Ishikawa-Luc orthotopic mouse models of endometrial cancer showed greater anti-tumor responses to combination treatment than to either monotherapy. RNASeq analysis highlighted reduced cell proliferation and defective DNA damage response pathways as potential mediators of the combination's effects. In conclusion, our preclinical findings indicate that Wee1 inhibition can enhance the response to EphA2-targeted therapeutics in endometrial cancer; this strategy thus warrants further development.

Keywords: endometrial cancer; EphA2; Wee1

Citation: Dasari, S.K.; Joseph, R.; Umamaheswaran, S.; Mangala, L.S.; Bayraktar, E.; Rodriguez-Aguayo, C.; Wu, Y.; Nguyen, N.; Powell, R.T.; Sobieski, M.; et al. Combination of EphA2- and Wee1-Targeted Therapies in Endometrial Cancer. *Int. J. Mol. Sci.* **2023**, *24*, 3915. https://doi.org/10.3390/ijms24043915

Academic Editor: Laura Paleari

Received: 6 December 2022
Revised: 7 February 2023
Accepted: 8 February 2023
Published: 15 February 2023

1. Introduction

EphA2 is a receptor tyrosine kinase that has multiple roles in facilitating malignant progression. Although EphA2 was first studied in the context of neuronal migration during embryogenesis, it has since been shown to regulate cancer cell growth, migration, invasion, and angiogenesis [1]. In addition, EphA2 is overexpressed in various cancers, including breast cancer [2], esophageal cancer [3], melanoma [4], lung cancer [5], prostate cancer [6], ovarian cancer [7,8], and endometrial cancer [9,10]. Over the years, many therapeutic strategies have been developed to target EphA2, including tyrosine kinase inhibitors, monoclonal antibodies, immunoconjugates, aptamers, and short-interfering RNA (siRNA) [11,12]. We have previously demonstrated that delivery of EphA2 siRNA through 1,2-dioleoyl-*sn*-glycero-3-phosphatidylcholine (DOPC) neutral liposome nanoparticles (EPHARNA) showed highly efficient in vivo delivery to the tumor, resulting in decreased tumor burden in mouse ovarian cancer models [13]. In addition, EPHARNA combined well with paclitaxel and significantly reduced tumor growth in these preclinical models [13]. A phase 1 trial of EPHARNA in patients with solid cancers is ongoing. However, rational combinations with EphA2-targeted therapy are not yet known; therefore, we aimed to identify novel therapeutic combinations through high-throughput chemical screens for use in endometrial cancer.

In the present study, we identified the Wee1 kinase inhibitor, MK1775, as a synergistic partner to EphA2-targeted therapy in endometrial cancer cells. To test our hypothesis that Wee1 inhibition sensitizes cells to EphA2-targeted therapy, we examined the anti-tumor effects of both agents in endometrial cancer mouse models and evaluated potential mechanisms of synergy. Our findings supported our hypothesis, justifying further investigation of this combination.

2. Results

2.1. High-Throughput Drug Screening Identifies Rational Combinations to EphA2 Inhibition

Given the oncogenic function of EphA2 in endometrial cancer, we performed a systematic high-throughput drug screen on an endometrial cancer cell line (Hec1A) transfected with either control siRNA or EphA2 siRNA with two drug libraries containing a total of 1510 drugs comprising FDA-approved drugs, clinical candidates, active pharmaceutical ingredients, and chemotherapeutic agents. We hypothesized that candidate small molecule drugs from these libraries would overcome resistance to anti-EphA2 therapeutics such as EPHARNA and thereby enhance their therapeutic benefit when given as combined therapy. Among the top hits, the biggest class represented was targeted kinase inhibitors (MK-1775, TAK-632, and BML-277), followed by microtubule poisons (darinaparsin and CYT997) and GPCR and G protein inhibitors (azilsartan, medoxomil, and matrine) (Figure 1A). Among the three kinase inhibitors in the hit list, the Wee1 inhibitor, MK1775, demonstrated the greatest synergistic interaction score (Bliss synergy score of 1.14) when combined with the EphA2 inhibitor, ALW (Figure 1C,D). Furthermore, EphA2 knockdown increased Wee1 activity as evidenced by elevated phospho-cdc2 levels (Figure 1B), hinting that Wee1 activation may be a compensatory mechanism to adapt to the loss of EphA2 and maintain cell viability.

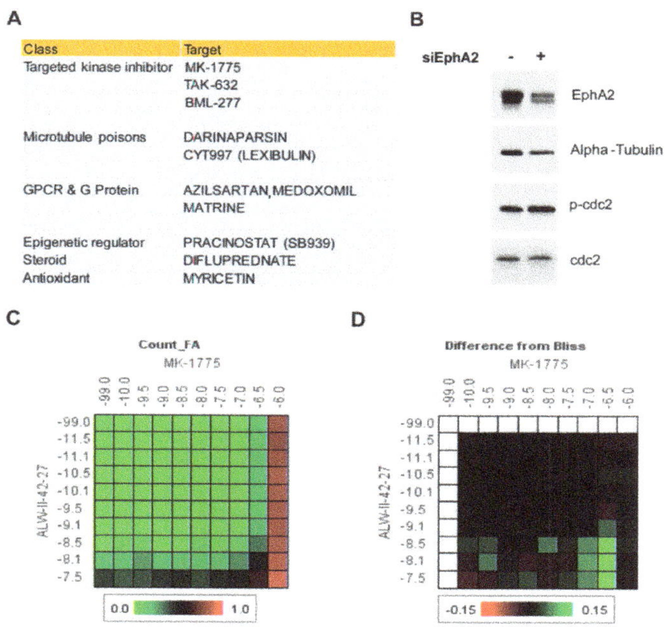

Figure 1. *Cont.*

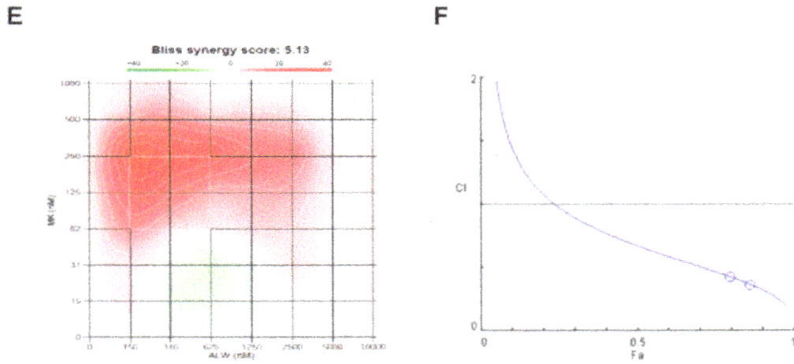

Figure 1. High–throughput screen for synergistic partners for EphA2–targeted therapy and verification of Wee1 as a target. (**A**) Top 10 hits from the chemical screen. (**B**) Effect of EphA2 silencing on Wee1 activity, as seen by phosphorylation of cdc2, in Hec1A cells. (**C**) Dose–response matrix for combinatorial analysis of the EphA2 inhibitor ALW and Wee1 inhibitor MK1775. (**D**) Difference from predicted Bliss independence surface model for the combination of ALW and MK1775. (**E**) Visualization of the calculated 2D synergy maps for the MTT cell viability assay from the SynergyFinder Bliss independence model combinatorial analysis. Red regions represent synergy, and green regions represent antagonism. (**F**) Plots showing fraction of cells affected (Fa) and combination index (CI) values for ALW and MK1775. The white circles representing the CI values of the ALW:MK1775 combination (1 μM:0.5 μM and 1 μM:0.75 μM concentrations, respectively) were less than one, indicating synergy.

Next, we analyzed the synergism of EphA2 and Wee1 inhibition in two endometrial cancer cell lines (Hec1A and Ishikawa) known to have high expression levels of EphA2. To determine the effect of combination therapy of EphA2 and Wee1 inhibitors (ALW and MK1775, respectively), cancer cells were treated with either each single drug or both in combination for a period of 72 h, followed by cell viability analysis using an MTT assay. We observed a significant decrease in cell viability in a dose-dependent manner in both Hec1A and Ishikawa cells; at each dose tested, cell viability was lower after combination therapy than after individual drug treatments (Figure S1A,B). Furthermore, we used SynergyFinder and CompuSyn software to test the drug–drug interactions using the median effect equation to derive combination index values, and we observed that the drug combination produced a synergistic effect (Figure 1E,F).

2.2. *MK1775 Sensitizes Endometrial Cancer Cells to EphA2 Inhibition*

Upon morphological analyses, single-drug-treated wells showed a lower number of cells in comparison to control cells, and the cell numbers were even lower with the combination treatment (Figure 2A). Of interest here is the appearance of many round cells for the ALW and MK1775 combination treatment at both the 24 h and 48 h time points, which could be either cells undergoing mitosis or cells preparing to die by apoptosis. To understand this, we performed cell cycle analysis and observed that there was a decrease in number of cells in the G1 phase and a concomitant increase in number of cells in the S phase for the combination treatment at the 48 h time point (Figures 2B and S1C,D). In addition, there was a statistically significant increase in the sub-G1 population in cells given combination treatment compared with the control cells (Figure 2C). Consistent with the associated changes in the cell cycle profile (Figures 2D and S2A), we observed elevated levels of the mitotic marker phospho H3 (Ser10) in cells treated with the drug combination compared with cells treated with single drugs or the control cells (Figure 2E). This result indicates that the cells moved into mitosis prematurely, overriding the G2/M checkpoint (Figure 2D) and thereby triggering mitotic arrest and leading to mitotic-catastrophe-mediated apoptotic cell death.

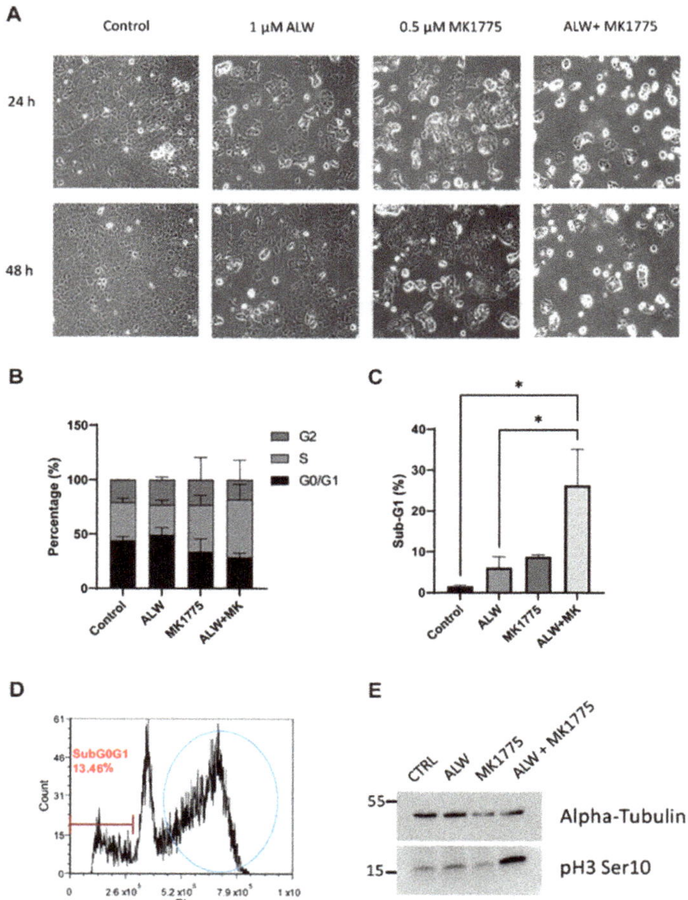

Figure 2. Effects of combination therapy on cell proliferation. (**A**) Cell morphology images of untreated (control) and ALW- and MK1775-treated Hec1A cells at 24 h and 48 h timepoints. (**B**,**C**) Flow cytometry analysis of cell cycle phases in Hec1A cells that were untreated or treated for 48 h. (**D**) Cell cycle profile of ALW- and MK1775-treated cells showing cells arrested between the S and G2 phases. (**E**) Western blot analysis for phosphorylated histone H3, an indicator of cells arrested in mitosis. Statistics were performed using one-way ANOVA with the Tukey's post hoc test for multiple comparisons for more than two groups. * $p < 0.05$.

The synergistic effect of the combination treatment was further evaluated using an Annexin V/PI-based apoptosis assay. Higher levels of apoptosis were seen for combination treatment with ALW and MK1775 than for individual drug treatments in both cell lines tested (Figure 3A,B). This result was confirmed using PARP cleavage as a marker of apoptotic cell death; as expected, we saw higher levels of cleaved PARP in the combination treatment (Figure 3C,D). In addition to caspase-3 activation (Figure S2B) and PARP cleavage, one of the hallmarks of apoptosis is double-strand DNA break formation, which is marked by γH2AX phosphorylation at S139. We observed increased levels of γH2AX phosphorylation after combination treatment, indicating that extensive double-strand breaks occurred during cell death (Figure S2C). Furthermore, we performed a colony formation assay to assess the effect of therapy on clonogenic survival, and in comparison to control cells, cells treated with the drug combination had significantly suppressed colony-forming activity in Hec1A ($p < 0.001$) and Ishikawa ($p < 0.01$) cells (Figure 3E,F).

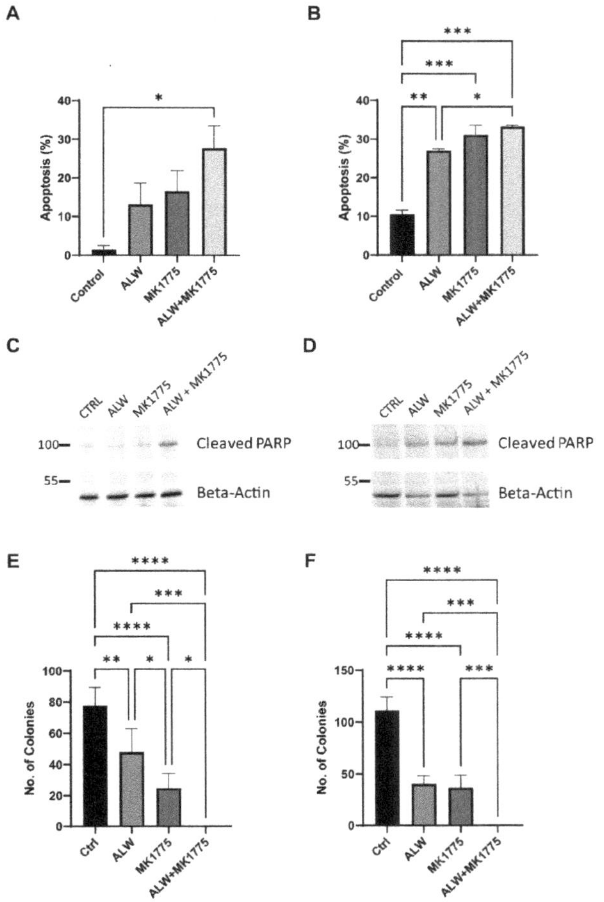

Figure 3. Apoptotic and clonogenic effects of combination therapy. (**A**,**B**) Flow cytometry analysis of apoptosis in Hec1A (**A**) and Ishikawa (**B**) cells that were untreated or treated with ALW, MK1775, or both drugs. (**C**,**D**) Expression of apoptosis marker cleaved PARP in untreated or drug-treated Hec1A (**C**) and Ishikawa (**D**) cells. (**E**,**F**) Clonogenic colony formation assay in untreated or drug-treated Hec1A (**E**) and Ishikawa (**F**) cells. Statistics were performed using one-way ANOVA with the Tukey's post hoc test for multiple comparisons for more than two groups. ns, nonsignificant; * $p < 0.05$; ** $p < 0.01$; *** $p < 0.001$; **** $p < 0.0001$.

2.3. EPHARNA and MK1775 Suppress Tumor Growth in an Endometrial Cancer Xenograft Model

To test the anti-tumor effects of EPHARNA and MK1775 in vivo, we next used the Hec1A and Ishikawa-Luc orthotopic endometrial cancer models. Tumor nodules were localized primarily in the uterus with few metastases to the intestine, stomach, and peritoneal wall (Figure 4A,E). At the end of the experiment, mice treated with EPHARNA and MK1775 monotherapies showed lower tumor weight (Figure 4B) and fewer tumor nodules (Figure 4C) compared with the control group, although the differences did not reach statistical significance. In contrast, EPHARNA and MK1775 combination therapy led to significantly lower tumor weight ($p < 0.05$) and fewer tumor nodules compared with the control group (Figure 4B,C). There were no significant differences in mouse body weight across all four groups (Figure 4D).

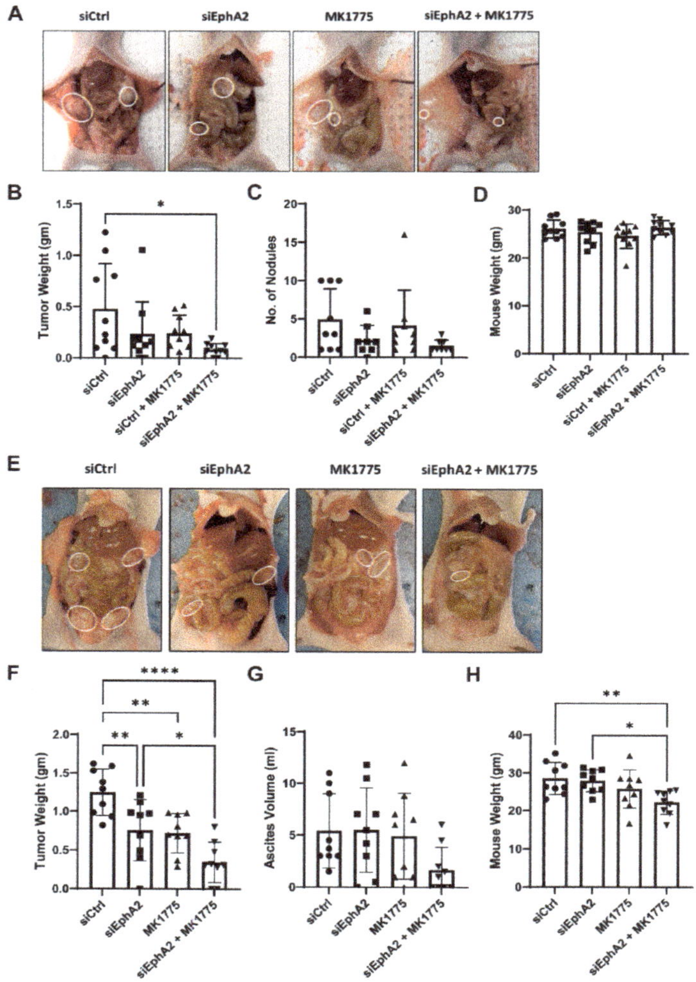

Figure 4. Anti-tumor effects of EphA2- and Wee1-targeted therapy in endometrial cancer xenograft models. Representative images of tumor burden in mice with Hec1A (**A**) and Ishikawa-Luc (**E**) tumors with siRNA nanoparticles and/or MK1775 therapy. (**B**,**C**) Tumor weights (**B**) and number of nodules (**C**) after therapy in Hec1A model. (**F**,**G**) Tumor weights (**F**) and ascites (**G**) in Ishikawa-Luc model. Mouse body weights at the end of the experiment in Hec1A (**D**) and Ishikawa-Luc (**H**) models. Statistics were performed using one-way ANOVA with the Tukey's post hoc test for multiple comparisons for more than two groups. * $p < 0.05$; ** $p < 0.01$; **** $p < 0.0001$.

In the Ishikawa-Luc model, treatment with EPHARNA and MK1775 combination therapy resulted in a significant decrease in tumor burden in comparison to siControl or EPHARNA monotherapy (Figure 4E,F). However, we observed higher ascites in siControl-, EPHARNA-, and MK1775-treated animals, which is partly attributed to the increased body weight in siControl- and EPHARNA-treated groups in comparison to combination treatment (Figure 4G,H).

2.4. Inhibition of EphA2 and Wee1 Leads to Suppression of DNA Damage Response Repair Pathways and Downregulates Cell Survival Pathways

To identify potential downstream mechanisms associated with the synergistic interaction of EphA2 and Wee1 inhibition, we performed RNA-Seq analysis followed by IPA analysis. To better analyze the observed synergy between the drugs, we performed a comparative analysis to identify enriched canonical pathways that were high under EphA2 inhibition alone and low under combination therapy. This analysis identified therapeutic compensatory cell survival pathways activated under EphA2 monotherapy that were lost under combination therapy. Comparative analysis identified downregulation of cell cycle checkpoint regulation, DNA damage response, and cell survival pathways (Figure 5A).

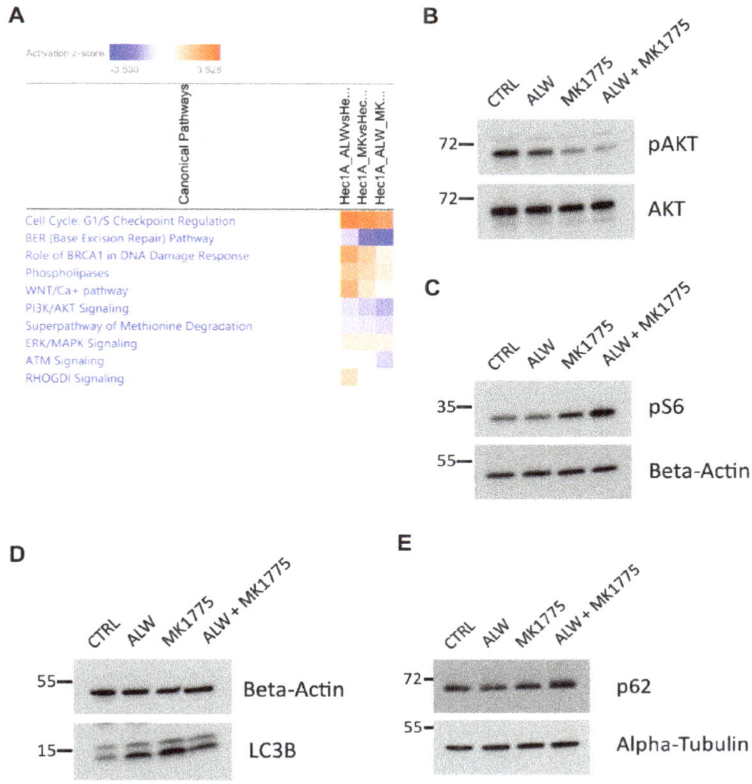

Figure 5. IPA analysis for identification of molecular mechanisms associated with synergy of EphA2- and Wee1–targeted therapy. (**A**) Comparative analysis of pathways downregulated in combination therapy. (**B**,**C**) Validation of RNA–Seq analysis of AKT pathway repression (**B**) and mTOR pathway activation (**C**). (**D**,**E**) Validation of RNA–Seq analysis with autophagy marker LC3B (**D**) and autophagy substrate p62 (**E**).

Further investigation was performed to better understand the molecular mechanisms associated with the synergism of Epha2 and Wee1 inhibition. In comparison to monotherapy, pAKT levels were lower with the combination therapy, hinting at reduced cell survival capacity (Figure 5B). In addition, an increase in phosphorylated S6 levels served as a surrogate for mTOR activation status (Figure 5C). Because mTOR was activated with the combination treatment, we examined the status of autophagy in these cells and observed that the combination treatment led to a block in the autophagic flux, as evidenced by the accumulation of LC3B-II and p62 levels (Figure 5D,E). These results suggest that presence of a block in autophagic flux, along with additional defects in cell cycle progression and DNA damage responses, pushes the cells into apoptotic cell death.

3. Discussion

In this study, we sought to identify novel agents for combination with EphA2-targeted therapy in endometrial cancer. One of the top hits identified in our high-throughput screen was the Wee1 kinase inhibitor, MK1775. We found EphA2- and Wee1-targeted therapies had synergistic effects in vitro, and the combination therapy led to enhanced anti-tumor efficacy in vivo.

Over two decades of preclinical research have identified EphA2 as a promising target for clinical translation. Previous studies have shown that EphA2 silencing was effective in reducing tumor burden in combination with chemotherapeutic agents such as paclitaxel and docetaxel [13]. Here, we showed that silencing EphA2 expression with siRNA increases Wee1 activity, as evidenced by increased cdc2 phosphorylation, which regulates the G2/M cell cycle block and allows cells to evade cell death by inhibiting premature entry into mitosis. Therefore, increased Wee1 activity may serve as an adaptive cell survival mechanism to evade cell death upon EphA2 inhibition. Blocking such secondary cell survival loops shows promise in enhancing the therapeutic efficacy of EphA2 inhibition.

In vitro experiments confirmed the synergy between the EphA2- and Wee1-targeted drugs, as seen through an increase in apoptosis and the inhibition of colony-forming efficiency. Furthermore, RNA-Seq analysis confirmed the synergy was partly mediated by suppression of cell proliferation and of DNA damage response pathways. EphA2 has been shown to be highly expressed in many cancers and has been shown to regulate the PI3K-AKT signaling pathway. However, it has been shown to have opposing effects in different cancers. For example, in pancreatic cancer and hepatocellular carcinoma, ligand-induced EphA2 signaling activates AKT signaling by enhancing its phosphorylation [14,15]. In contrast, in glioblastoma, EphA2 signaling decreases AKT function by reduced phosphorylation [16]. Here, we observed that EphA2 inhibition decreased AKT signaling modestly, with further reduction when EphA2 inhibition was combined with the Wee1 inhibitor MK1775. This suppression of AKT signaling with the combination treatment enhanced mTOR activity, leading to inhibition of cell survival autophagy and ultimately leading to cell death by apoptosis in the absence of efficient DNA damage response activation.

At present, several EphA2-targeted therapeutics are being evaluated in clinical trials in various cancers in which EphA2 has an established oncogenic function [10,12]. Our findings provide further support to explore additional combination therapies that may synergize with EphA2-targeted therapeutics. In addition, recent ADAGIO phase 2 trial results of the Wee1 inhibitor adavosertib (AZD1775) demonstrating response rates of 30% in uterine serous carcinoma are promising and warrant further investigation [17]. Furthermore, since EphA2 is overexpressed in ovarian serous carcinoma and is associated with poor clinical outcomes [18], it is possible that the combination therapy with Epha2 inhibition may be a viable option for further improving clinical outcomes of patients with other histological subtypes such as serous carcinoma.

4. Materials and Methods

4.1. Cell Culture

Hec1A (RRID:CVCL_0293) and Ishikawa cells (RRID:CVCL_2529) were procured from ATCC and The University of Texas MD Anderson Cancer Center Characterized Cell Line Core, respectively. Cell lines were validated by short tandem repeat fingerprinting in the core facility. Cells were routinely screened for mycoplasma. Hec1A cells were grown in McCoy's 5A medium (HyClone, Logan, UT, USA), and Ishikawa cells were grown in Dulbecco's modified Eagle's medium (HyClone, Logan, UT, USA), supplemented with 10% fetal bovine serum (Sigma-Aldrich, St. Louis, MO, USA) and 0.1% gentamicin sulfate (Gemini Bioproducts, West Sacramento, CA, USA). Cells were incubated in a humidified atmosphere containing 5% CO_2 at 37 °C. All experiments were conducted with cells at 70%

to 80% confluence and cultured for fewer than 20 passages for in vitro work and for fewer than 10 passages for in vivo experiments.

4.2. siRNA Transfection

Cells were plated in six-well plates so that the cells could reach 60% to 70% confluence by the next day. For each well, 1.3 μg of siRNA was added to 150 μL of reduced serum medium (Opti-MEM, Thermo Fisher Scientific, Waltham, MA, USA), and in a separate tube, 8 μL of Lipofectamine RNAiMAX transfection reagent (Thermo Fisher Scientific, Waltham, MA, USA) was incubated in 150 μL of Opti-MEM for 5 min. The siRNA/media mixture was added dropwise to the transfection reagent mixture, vortexed and then incubated for 15 to 20 min at room temperature. Wells to be transfected were washed once with PBS, and then 900 μL of Opti-MEM and 300 μL of siRNA mixture were added dropwise to each well. The plates were gently swirled and placed in the incubator for 4 to 6 h. The transfection medium was replaced with complete media. For the high-throughput screening, after 24 h of transfection, cells were trypsinized, counted, and seeded in clear-bottom 384-well plates. For Western blot analysis, cells were collected 48 to 72 h after transfection.

4.3. High-Throughput Screening

High-throughput chemical screens were performed by the Gulf Coast Consortia's Combinatorial Drug Discovery Program at the Institute of Biosciences and Technology in Texas A&M Health Science Center, Houston, TX, USA. Hec1A endometrial cancer cells transfected with either siControl or siEphA2 were screened against two drug library collections: The Broad Collection-Informer Set (358 compounds) and Selleck Bioactives Collection (1150 compounds) libraries. A brief description of the contents of the libraries can be found at the following link (https://ibt.tamu.edu/cores/high-throughput/core-libraries/approved-drugs.html, accessed on 12 July 2019). For screening assays, a total of 800 cells per well were suspended in 50 μL of media and seeded into black 384-well μClear plates (Greiner Bio-One International, Monroe, NC, USA) using a Multidrop Combi liquid dispenser (Thermo Fisher Scientific, Waltham, MA, USA). After cell seeding, the plates were kept at room temperature for 40 to 60 min before being moved into a cell culture incubator. The cells were grown overnight at 37 °C in a humidified chamber (>95% relative humidity) with 5% CO_2. The following day, 50 nl of drugs were transferred into each well using an Echo 550 acoustic dispensing platform (Labcyte, San Jose, CA, USA). A non-treated plate was immediately fixed with 4% paraformaldehyde and followed by nuclei staining with 4′,6-diamidino-2-phenylindole (DAPI) at the start of drug treatment (day 0) to estimate the number of cells present at the start of treatment.

In the primary screen, the drug libraries were tested at three concentrations (1 μM, 0.1 μM, and 0.01 μM) with a fixed volume of dimethyl sulfoxide (DMSO) (0.1% v/v) and two biological replicates. Each assay plate contained a fixed concentration of the drugs in addition to a negative control (0.1% DMSO) and two positive controls (etoposide and dasatinib). After 72 h of incubation, plates were fixed with 0.4% paraformaldehyde and nuclei stained with DAPI using an integrated HydroSpeed plate washer (Tecan Life Sciences, Männedorf, Switzerland) and Multidrop Combi dispenser. Plates were imaged on an IN Cell Analyzer 6000 laser-based confocal imaging platform (GE Healthcare Bio-Sciences, Marlborough, MA, USA), and nuclei were counted using IN Cell Developer Toolbox software (version 1.6). To evaluate the cells' response to the drug screen, we performed curve fitting followed by a calculation of area under the curve values.

4.4. Orthogonal Cell Viability Assay

To evaluate the cytotoxicity of ALW-II-41-27 (ALW; ApexBio Technology, Houston, TX, USA) and MK1775 (ApexBio Technology, Houston, TX, USA) both alone and in combination, cells were plated in a 96-well plate at a starting density of 2000 cells per well for Hec1A cells and 1000 cells per well for Ishikawa cells. After 24 h, the medium was aspirated, and 100 μL of fresh medium containing serial dilutions of individual drugs was placed over the cells. After 72 h of incubation, the medium was aspirated, and cells were incubated with 0.05% MTT solution for 1 h. The supernatant was removed, and the formazan crystals were dissolved in 100 μL DMSO. The plates were read at 570 nm by a uQuant microplate spectrophotometer (BioTek, Winooski, VT, USA). Triplicate biological experiments were performed. Dose–response curves were plotted using Prism 8.0.0 (GraphPad Software, San Diego, CA, USA), the combination index was determined by CompuSyn software [19] (ComboSyn, combosyn.com, accessed on 12 July 2019), and synergy assessment was performed using the Bliss model in the SynergyFinder web application [20] (https://synergyfinder.fimm.fi, accessed on 29 July 2019). The Bliss synergy scores in this platform indicate synergy if they are greater than 10, additivity if they are between −10 and 10, and antagonism if they are less than −10. Based on the cell viability results for all future experiments, Hec1A cells were treated with DMSO, 1 μM ALW, 0.5 μM MK1775, and the combination of 1 μM ALW and 0.5 μM MK1775. Ishikawa cells were treated with DMSO, 0.5 μM ALW, 0.25 μM MK1775, and the combination of 0.5 μM ALW and 0.25 μM MK1775.

4.5. Cell Cycle and Apoptosis Analysis

For the cell cycle assay, control and drug-treated cells were trypsinized, washed with PBS twice, fixed in ice-cold 70% ethanol, and stored at –20 °C. On the day of analysis, cells were washed twice with PBS and then incubated in 50 μg/mL of propidium iodide (PI) solution containing 0.5 μg/mL RNase A for 4 h in the dark and analyzed by flow cytometry. For the apoptosis assay, the cell supernatant as well as trypsinized cells were mixed and pelleted and then washed with PBS. The apoptosis assay was performed using the FITC Annexin V Apoptosis Detection Kit I (BD Biosciences, Franklin Lakes, NJ, USA). After Annexin V–fluorescein isothiocyanate (FITC) and PI staining, cells were analyzed by flow cytometry.

4.6. Colony Formation Assay

Cells were seeded at a density of 500 to 1000 cells per well in a 12-well plate, and the cells were left to grow in the incubator for 10 to 14 days with the respective drug combinations for both Hec1A and Ishikawa cells. After visible colonies containing more than 50 cells appeared, the plates were fixed with a solution containing glutaraldehyde (6.0%, *v*/*v*) and crystal violet (0.5%, *w*/*v*) for 15 to 20 min at room temperature. After that, the crystal-violet-fixing solution was decanted, and the plates were washed in water three to five times and then left to dry at room temperature. The plates were imaged, and the number of colonies was counted.

4.7. Western Blotting

Harvested cells were spun down at 2000 rpm for 5 min, washed with ice-cold PBS, and then pelleted at 3000 rpm for 3 min. Cell pellets were lysed in RIPA buffer supplemented with protease and phosphatase inhibitors and quantified using a Pierce BCA protein assay kit (Thermo Fisher Scientific, Waltham, MA, USA). Equal amounts of protein (20 μg) were boiled at 95 °C for 10 min, run on an SDS-PAGE gel (8–12%), transferred onto a nitrocellulose membrane, incubated in 5% milk (in Tris buffered saline-Tween 20 [TBS-T]) for 1 h, and then incubated overnight in the appropriate primary antibodies (listed below). Blots were washed with TBS-T thrice for 5 min each and then incubated with corresponding secondary antibodies (1:2500 dilution, GE Healthcare, Chicago, IL, USA) for 1 h. Enhanced chemiluminescence substrate (ECL; Thermo Fisher Scientific, Waltham, MA, USA) was then added to the blots for 1 min, and immunoblot images were captured using an Azure Biosystems imaging machine (Azure Biosystems, Dublin, CA, USA).

The following antibodies were used: anti-EphA2 (1:1000 dilution; Cell Signaling Technology, Danvers, MA, USA), anti-phosphorylated (phospho) cdc2 (1:1000 dilution; Cell Signaling Technology, Danvers, MA, USA), anti-cdc2 (1:1000 dilution; Cell Signaling Technology, Danvers, MA, USA), Cell Cycle and Apoptosis WB Cocktail (pCdk/pHH3/Actin/cleaved PARP) (1:250 dilution; Abcam, Cambridge, UK); Apoptosis and DNA damage WB Cocktail (pH2A.X/GAPDH/cleaved PARP) (1:250 dilution; Abcam, Cambridge, UK); anti-phospho S6 (1:1000 dilution; Cell Signaling Technology, Danvers, MA, USA), anti-S6 (1:1000 dilution; Cell Signaling Technology, Danvers, MA, USA), anti-AKT (1:1000 dilution; Cell Signaling Technology, Danvers, MA, USA); anti-pAKT (1:1000 dilution; Cell Signaling Technology, Danvers, MA, USA); anti-cleaved caspase 3 (1:1000 dilution; Cell Signaling Technology, Danvers, MA, USA); anti-P62 (1:3000; BD Biosciences, Franklin Lakes, NJ, USA); anti-GAPDH (1:5000; Thermo Fisher Scientific, Waltham, MA, USA), anti-LC3B (1:1000 dilution; Cell Signaling Technology, Danvers, MA, USA), anti-alpha-tubulin (1:1000 dilution; Cell Signaling Technology, Danvers, MA, USA), and antibeta-actin (1:3000; Sigma-Aldrich, St. Louis, MO, USA).

4.8. Liposomal Nanoparticle Preparation

For in vivo delivery, siRNAs were incorporated into DOPC liposomes as described earlier [13]. In brief, DOPC and siRNA were mixed in a ratio of 1:10 (*w*/*w*) siRNA:DOPC in the presence of excess tertiary butanol. Tween 20 was added to the siRNA/DOPC mixture in a ratio of 1:19 (Tween-20:siRNA/DOPC). The mixture was vortexed, frozen in an acetone/dry-ice bath, and lyophilized. Before in vivo administration, this preparation was hydrated with magnesium- and calcium-free PBS to achieve a desired concentration of 5 μg of siRNA in 200 μL volume per dose per mouse.

4.9. In Vivo Model of Endometrial Cancer

Female nude mice aged 4–8 weeks were purchased from Taconic Biosciences, USA. All mice were housed at The University of Texas MD Anderson Cancer Center animal facility under specific pathogen-free conditions. All animal-related experiments were approved by the Institutional Animal Care and Use Committee of MD Anderson Cancer Center. The right uterine horns of 8-week-old female athymic nude mice were injected with five million Hec1A cells in 100 μL of Hank's Balanced Salt Solution (HyClone, Logan, UT, USA) to the uterine horn. For the second model, one million Ishikawa-Luc cells were injected into the peritoneal cavity of 6–8-week-old female mice. After eight days, mice were randomized to four groups (10 mice per group): siControl-DOPC nanoparticles (NPs), siEphA2-DOPC NPs, siControl-DOPC NPs with MK1775, and siEphA2-DOPC NPs with MK1775. The siRNA-DOPC NPs were given to mice twice a week intraperitoneally, and MK1775 (30 mg/kg) was administered daily by oral gavage. Once mice from any group became moribund, all mice were euthanized; mouse weight, tumor weight, ascites volume, and number of nodules were recorded.

4.10. RNA-Seq Analysis

Hec1A cells were plated in six-well plates at a density of 100,000 cells per well, in triplicate, and incubated overnight. After 24 h, cells were treated with DMSO, 1 μM ALW, 0.5 μM MK1775, or the combination of 1 μM ALW and 0.5 μM MK1775 for 8 h, after which the RNA was extracted using Direct-zol RNA Miniprep Plus kit (ZYMO Research, Irvine, CA, USA). RNA quality was determined by RNA integrity number using a Bioanalyzer (Agilent Technologies, Santa Clara, CA, USA), and the samples were shipped to Novogene (Sacramento, CA, USA) for RNA-Seq analysis on the Illumina NovaSeq 6000 platform. Downstream analysis was performed using a combination of programs, including hisat2, DEseq2, and ClusterProfiler software. Pathway analysis was performed using Ingenuity Pathway Analysis (IPA) (Qiagen, Hilden, Germany).

4.11. Statistical Analysis

Statistics were performed using unpaired t-tests for comparisons between two groups and one-way ANOVAs with the Tukey post hoc test for multiple comparisons between more than two groups (Prism). Statistical significance was defined as a p value of <0.05.

5. Conclusions

In conclusion, EphA2- and Wee1-targeted therapies show synergistic interaction in endometrial cancer in both in vitro and in vivo experiments. The combination of cell cycle checkpoint inhibitors and EphA2-targeted therapy may have utility in the treatment of endometrial cancer and warrants further investigation.

Supplementary Materials: The following supporting information can be downloaded at: https://www.mdpi.com/article/10.3390/ijms24043915/s1.

Author Contributions: Conceptualization, S.K.D. and A.K.S.; methodology, S.K.D., N.N., R.T.P. and L.S.M.; validation, S.K.D., R.J. and L.S.M.; formal analysis, S.K.D., R.J., L.S.M., R.T.P., C.S. and A.K.S.; investigation, S.K.D., R.J., S.U., E.B., C.R.-A., L.S.M., Y.W., M.S., N.N., R.T.P., Y.L., M.A.C. and P.A.; resources, G.L.-B., S.N.W., C.S. and A.K.S.; data curation, S.K.D., E.B. and L.S.M.; writing—original draft preparation, S.K.D. and A.K.S.; writing—review and editing, S.K.D. and A.K.S.; visualization, S.K.D., N.N., R.T.P. and R.J.; supervision, A.K.S.; funding acquisition, G.L.-B., C.S., S.N.W. and A.K.S. All authors have read and agreed to the published version of the manuscript.

Funding: This work is supported, in part, by NIH grants CA098258, R35CA209904, the American Cancer Society Research Professor Award, the Frank McGraw Memorial Chair in Cancer Research, the Dunwoody Fund, the Gordon Fund, and NIH-NCI grant U01 CA213759. C.S. is supported by the CPRIT-funded Combinatorial Drug Discovery Program (RP200668 and RP150578).

Institutional Review Board Statement: The animal study protocol was approved by Institutional Animal Care & Use Committee of MDACC (00001029-RN02; aarm0005614).

Informed Consent Statement: Not applicable.

Data Availability Statement: The datasets generated during the current study are available from the corresponding author on reasonable request.

Acknowledgments: This work was supported by the National Institutes of Health/National Cancer Institute under award number P30CA016672. We thank Madison Semro and Sunita Patterson (Scientific Publications, Research Medical Library, MD Anderson Cancer Center) for editorial work.

Conflicts of Interest: S.N.W.: Consultant (AstraZeneca (Gaithersburg, MD, USA), Bayer (Whippany, NJ, USA), Caris (New York, NY, USA), Clovis Oncology (Boulder, CO, USA), Eisai (Nutley, NJ, USA), EQRX (Cambridge, MA, USA), GSK (Waltham, MA, USA), ImmunoGen (Waltham, MA, USA), Lilly (Indianapolis, IN, USA), Merck (Rahway, NJ, USA), Mereo (San Jose, CA, USA), Mersana (Cambridge, MA, USA), NGM Bio (South San Francisco, CA, USA), Nuvectis (Fort Lee, NJ, USA), Roche/Genentech (South San Francisco, CA, USA), SeaGen (Bothell, Washington, USA), Verastem (Boston, MA, USA), Vincerx (Palo Alto, CA, USA) and Zentalis (New York, NY, USA); research support to institution (AstraZeneca (Gaithersburg, MD, USA), AvengeBio (Cambridge, MA, USA), Bayer (Whippany, NJ, USA), Bio-Path (Bellaire, TX, USA), Clovis Oncology (Boulder, CO, USA), GSK (Waltham, MA, USA), Mereo (San Jose, CA, USA), Novartis (Cambridge, MA, USA), Roche/Genentech (South San Francisco, CA, USA), and Zentalis (New York, NY, USA). A.K.S.: Consultant [Merck (Rahway, NJ, USA) Iylon (Houston, TX, USA), Onxeo (New York, NY, USA), GSK (Waltham, MA, USA), ImmunoGen (Waltham, MA, USA)and Kiyatec (Greenville, SC, USA)]; shareholder (BioPath, Bellaire, TX, USA). All remaining authors have declared no conflicts of interest.

References

1. Dodelet, V.C.; Pasquale, E.B. Eph receptors and ephrin ligands: Embryogenesis to tumorigenesis. *Oncogene* **2000**, *19*, 5614–5619. [CrossRef] [PubMed]
2. Zelinski, D.P.; Zantek, N.; Stewart, J.C.; Irizarry, A.R.; Kinch, M. EphA2 overexpression causes tumorigenesis of mammary epithelial cells. *Cancer Res.* **2001**, *61*, 2301–2306. [PubMed]
3. Miyazaki, T.; Kato, H.; Fukuchi, M.; Nakajima, M.; Kuwano, H. EphA2 overexpression correlates with poor prognosis in esophageal squamous cell carcinoma. *Int. J. Cancer* **2003**, *103*, 657–663. [CrossRef] [PubMed]
4. Easty, D.J.; A Guthrie, B.; Maung, K.; Farr, C.J.; A Lindberg, R.; Toso, R.J.; Herlyn, M.; Bennett, D. Protein B61 as a new growth factor: Expression of B61 and up-regulation of its receptor epithelial cell kinase during melanoma progression. *Cancer Res.* **1995**, *55*, 2528–2532. [PubMed]
5. Kinch, M.S.; Moore, M.-B.; Harpole, D.H., Jr. Predictive value of the EphA2 receptor tyrosine kinase in lung cancer recurrence and survival. *Clin. Cancer Res.* **2003**, *9*, 613–618. [PubMed]

6. Walker-Daniels, J.; Coffman, K.; Azimi, M.; Rhim, J.; Bostwick, D.; Snyder, P.; Kerns, B.; Waters, D.; Kinch, M. Overexpression of the EphA2 tyrosine kinase in prostate cancer. *Prostate* **1999**, *41*, 275–280. [CrossRef]
7. Moyano-Galceran, L.; Pietilä, E.A.; Turunen, S.P.; Corvigno, S.; Hjerpe, E.; Bulanova, D.; Joneborg, U.; Alkasalias, T.; Miki, Y.; Yashiro, M.; et al. Adaptive RSK-EphA2-GPRC5A signaling switch triggers chemotherapy resistance in ovarian cancer. *EMBO Mol. Med.* **2020**, *12*, e11177. [CrossRef] [PubMed]
8. Takahashi, Y.; Hamasaki, M.; Aoki, M.; Koga, K.; Koshikawa, N.; Miyamoto, S.; Nabeshima, K. Activated EphA2 Processing by MT1-MMP Is Involved in Malignant Transformation of Ovarian Tumours In Vivo. *Anticancer. Res.* **2018**, *38*, 4257–4266. [CrossRef] [PubMed]
9. Hudecek, R.; Kohlova, B.; Siskova, I.; Piskacek, M.; Knight, A. Blocking of EphA2 on Endometrial Tumor Cells Reduces Susceptibility to Vδ1 Gamma-Delta T-Cell-Mediated Killing. *Front. Immunol.* **2021**, *12*, 752646. [CrossRef] [PubMed]
10. Xiao, T.; Xiao, Y.; Wang, W.; Tang, Y.Y.; Xiao, Z.; Su, M. Targeting EphA2 in cancer. *J. Hematol. Oncol.* **2020**, *13*, 114. [CrossRef] [PubMed]
11. Tandon, M.; Vemula, S.V.; Mittal, S.K. Emerging strategies for EphA2 receptor targeting for cancer therapeutics. *Expert Opin. Ther. Targets* **2011**, *15*, 31–51. [CrossRef] [PubMed]
12. Wilson, K.; Shiuan, E.; Brantley-Sieders, D.M. Oncogenic functions and therapeutic targeting of EphA2 in cancer. *Oncogene* **2021**, *40*, 2483–2495. [CrossRef] [PubMed]
13. Landen, C.N.; Chavez-Reyes, A.; Bucana, C.; Schmandt, R.; Deavers, M.T.; Lopez-Berestein, G.; Sood, A.K. Therapeutic *EphA2* Gene Targeting *In vivo* Using Neutral Liposomal Small Interfering RNA Delivery. *Cancer Res.* **2005**, *65*, 6910–6918. [CrossRef] [PubMed]
14. Chang, Q.; Jorgensen, C.; Pawson, T.; Hedley, D.W. Effects of dasatinib on EphA2 receptor tyrosine kinase activity and downstream signalling in pancreatic cancer. *Br. J. Cancer* **2008**, *99*, 1074–1082. [CrossRef] [PubMed]
15. Wang, H.; Hou, W.; Perera, A.; Bettler, C.; Beach, J.R.; Ding, X.; Li, J.; Denning, M.F.; Dhanarajan, A.; Cotler, S.J.; et al. Targeting EphA2 suppresses hepatocellular carcinoma initiation and progression by dual inhibition of JAK1/STAT3 and AKT signaling. *Cell Rep.* **2021**, *34*, 108765. [CrossRef] [PubMed]
16. Miao, H.; Li, D.-Q.; Mukherjee, A.; Guo, H.; Petty, A.; Cutter, J.; Basilion, J.P.; Sedor, J.; Wu, J.; Danielpour, D.; et al. EphA2 Mediates Ligand-Dependent Inhibition and Ligand-Independent Promotion of Cell Migration and Invasion via a Reciprocal Regulatory Loop with Akt. *Cancer Cell* **2009**, *16*, 9–20. [CrossRef] [PubMed]
17. Liu, J.; Oza, A.M.; Colombo, N.; Oaknin, A. ADAGIO: A phase IIb international study of the Wee1 inhibitor adavosertib in women with recurrent or persistent uterine serous carcinoma. *Int. J. Gynecol. Cancer* **2021**, *32*, 89–92. [CrossRef] [PubMed]
18. Han, L.-P.; Liu, J.-F.; Liu, X.-R.; Dong, Z.-M.; Suo, Z.-H. Expression and prognostic significance of EphA2 and EphrinA-1 in ovarian serous carcinomas. *Sichuan Da Xue Xue Bao Yi Xue Ban J. Sichuan Univ. Med Sci.* **2011**, *42*, 179–184.
19. Chou, T.C.; Martin, N. *CompuSyn for Drug Combinations: Pc software and User's Guide: A Computer Program for Quantitation of Synergism and Antagonism in Drug Combinations, and the Determination of IC50 and ED50 and LD50 Values*; ComboSyn: Paramus, NJ, USA, 2005.
20. Ianevski, A.; He, L.; Aittokallio, T.; Tang, J. SynergyFinder: A web application for analyzing drug combination dose-response matrix data. *Bioinformatics* **2017**, *33*, 2413–2415. [CrossRef] [PubMed]

International Journal of *Molecular Sciences*

MDPI

Article

A Hypoxia Molecular Signature-Based Prognostic Model for Endometrial Cancer Patients

Yang Jiao [1,†], Rui Geng [1,†], Zihang Zhong [1], Senmiao Ni [1], Wen Liu [1], Zhiqiang He [1], Shilin Gan [1], Qinghao Huang [1], Jinhui Liu [2,*] and Jianling Bai [1,*]

[1] Department of Biostatistics, School of Public Heath, Nanjing Medical University, 101 Longmian Avenue, Jiangning District, Nanjing 211166, China

[2] Department of Gynecology, Nanjing Medical University, 101 Longmian Avenue, Jiangning District, Nanjing 211166, China

* Correspondence: jinhuiliu@njmu.edu.cn (J.L.); baijianling@njmu.edu.cn (J.B.)

† These authors contributed equally to this work.

Abstract: Endometrial cancer has the highest incidence of uterine corpus cancer, the sixth most typical cancer in women until 2020. High recurrence rate and frequent adverse events were reported in either standard chemotherapy or combined therapy. Hence, developing precise diagnostic and prognostic approaches for endometrial cancer was on demand. Four hypoxia-related genes were screened for the EC prognostic model by the univariate, LASSO, and multivariate Cox regression analysis from the TCGA dataset. QT-PCR and functional annotation analysis were performed. Associations between predicted risk and immunotherapy and chemotherapy responses were investigated by evaluating expressions of immune checkpoint inhibitors, infiltrated immune cells, m6a regulators, and drug sensitivity. The ROC curve and calibration plot indicated a fair predictability of our prognostic nomogram model. NR3C1 amplification, along with IL-6 and SRPX suppressions, were detected in tumor. High stromal score and enriched infiltrated aDCs and B cells in the high-risk group supported the hypothesis of immune-deserted tumor. Hypoxia-related molecular subtypes of EC were then identified via the gene signature. Cluster 2 patients showed a significant sensitivity to Vinblastine. In summary, our hypoxia signature model accurately predicted the survival outcome of EC patients and assessed translational and transcriptional dysregulations to explore targets for precise medical treatment.

Keywords: endometrial cancer; hypoxia; tumor microenvironment (TME); prognosis; risk model; immune cells; chemotherapy; targeted treatment

Citation: Jiao, Y.; Geng, R.; Zhong, Z.; Ni, S.; Liu, W.; He, Z.; Gan, S.; Huang, Q.; Liu, J.; Bai, J. A Hypoxia Molecular Signature-Based Prognostic Model for Endometrial Cancer Patients. *Int. J. Mol. Sci.* **2023**, *24*, 1675. https://doi.org/10.3390/ijms24021675

Academic Editor: Laura Paleari

Received: 21 November 2022
Revised: 28 December 2022
Accepted: 4 January 2023
Published: 14 January 2023

1. Introduction

Endometrial cancer (EC) has a high incidence rate among the subclassification of uterine corpus cancer until 2020 [1]. Unfortunately, early screening mainly focused on abnormal bleeding and might require additional evaluations like the pipelle method with high accuracy but sampling difficulty [2–4]. Patients diagnosed with stage III or IV endometrial cancer achieved a locoregional recurrence rate of 20% while treated with standard chemotherapy (doxorubicin-cisplatin (AP)) [5]. Additionally, In a randomized phase 3 trial, 58% of cases were reported with adverse events in chemo-radiotherapy and 63% of cases were reported in chemotherapy-only [6]. Hence, there is a growing need for developing endometrial cancer diagnostic and prognostic approaches.

While targeted therapies for endometrial cancer regarding glucose metabolism and the PI3K/Akt/mTOR pathway have been developed, there are rising concerns about a synchronous disturbance on other biological pathways of drugs [7,8]. Therefore, a high recurrence rate of endometrial cancer after radiotherapy and chemotherapy can be a result of tumor cell proliferation and angiogenesis [9]. According to previous studies, endometrial cancer was stratified into copy-number high, DNA-polymerase epsilon, microsatellite instability hyper mutated, and copy number low [10,11]. Moreover, the sequencing-based classification shows a potential association between molecular characteristics under hypoxia and adjuvant treatment for patients with high-grade tumors.

The hypoxic tumor microenvironment always leads to poorer clinical results as tumor cells adapt to conditions of low oxygen and nutrition and become resistant to radiation and chemotherapy [12,13]. Recently, some studies indicated that hypoxia influences tumor cells in metabolism and immunity, thus resulting in immune infiltration and acidosis [14,15]. Therefore, novel molecular subtypes capable of distinguishing patients with similar histologic characteristics under hypoxic conditions is needed for customized treatments [16,17].

Specifically, targeting hypoxia in tumor cells gives rise to the development of immunotherapy via controlling immunosuppressive cells and effector T cells [18]. However, whether the genes related to the hypoxic tumor microenvironment can systematically contribute to the increased risk of endometrial cancer is unclear.

Our study obtained the gene and clinical-relevant data of endometrial cancer patients from the Cancer Genome Atlas dataset. Several bioinformatic programs and packages were used in analyses, including linear models for microarray data (limma), clusterProfiler r packages and the Cell type identification by estimating relative subsets of RNA transcripts (CIBERSORT) algorithm. Limma has recently become famous for identifying differentially expressed mRNAs with thresholds of fold changes in an unsupervised clustering of samples [19,20]. In the annotation function analysis, the clusterProfiler package provides a comprehensive way to compare essential biological pathways among the classified gene set [21]. We established a hypoxia gene signature to calculate risk scores for patients and identified particular molecular subtypes of endometrial cancer from the samples. The nomogram consisted of risk scores, and several clinical characteristics were finally built for prognosis. To look into the potential benefits of existing therapies and the new targets of the treatment, we evaluated the immune cell infiltration by CIBERSORT algorithm, immune checkpoints by Estimation of stromal and immune cells in malignant tumors using expression data (ESTIMATE) algorithm, and the semi-inhibitory concentration (IC50) in drug sensitivity of chemotherapy. CIBERSORT has great power over computing infiltrating immune fractions with 22 immune cell types by the deconvolution of genetic microarray expression profiles and defining the immune phenotypes with signature genes from the TCGA samples. In Wang et al., the authors investigated the tumor-infiltrated immune cell levels and characteristics of tumor microenvironment for the constructed circRNA signature via CIBERSORT and ESTIMATE algorithms [22]. Practically, phenotype-genotype-dependent subtyping of EC provided an insight into the proper selection of suitable patients and their follow-ups into personalized therapies.

2. Results

2.1. Differential Expression Profile and Gene Enrichment Analysis of Hypoxia-Related DEGs in EC

In the TCGA-EC cohort, 29 hypoxia-related differentially expressed genes (DEGs) with FDR < 0.001 were retained for further analysis (Figure 1A, Table S1), including 12 genes up-regulated and 17 down-regulated (Figure S1A). As explicitly shown in the heatmap, differential expression profiles of the 29 DEGs related to hypoxia were exhibited in normal or endometrial tumor cell types (Figure 1B). Several hypoxia-related DEGs showed a highly correlated relationship to their expression levels in tumor samples, such as FOS and DUSP-1 (Figure 1C).

Function analysis results indicated that these DEGs could have immune-related roles (Figure 1D). From Figure S1B, one of the most significant functions was the response to steroid hormone, which involves seven DEGs. KEGG enrichment analysis results found that one out of two EC-related signaling pathways were significantly enriched by these DEGs (Figure S1C, Insulin resistance p-value = 0.008, Human T-cell leukemia virus 1 infection $p = 0.061$).

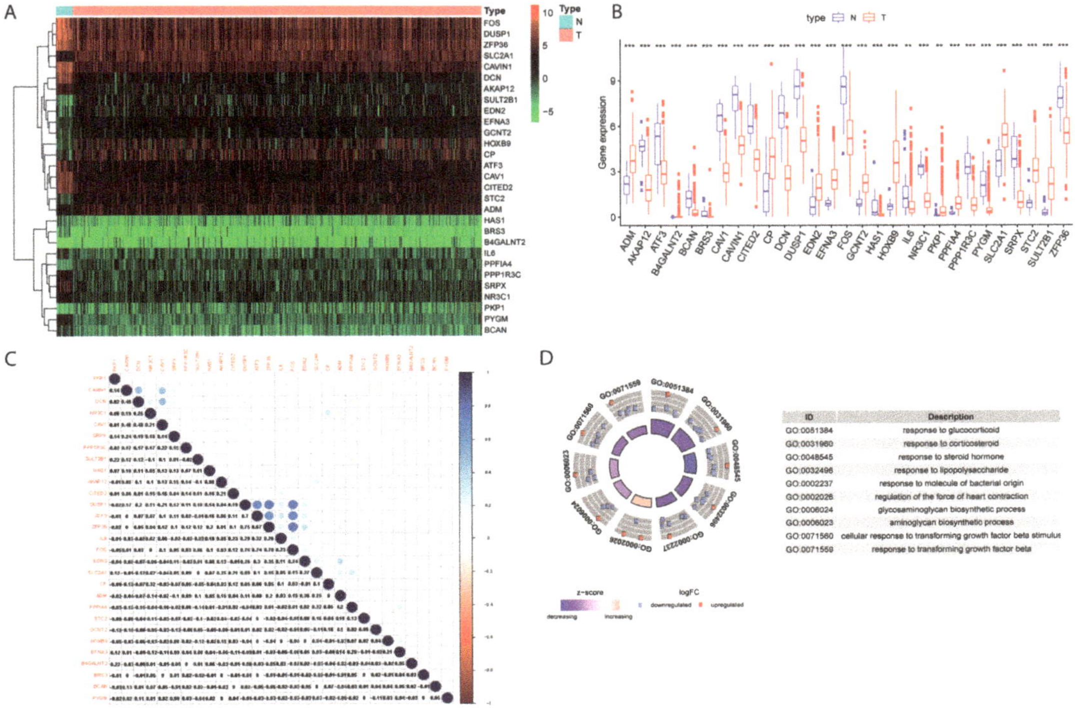

Figure 1. Differential expression profile of hypoxia-related genes in endometrial carcinoma. (**A**) heatmap of differentially expressed hypoxia-related genes clustered in N (Normal) and T (Tumor) cell types. (**B**) Differential expression of hypoxia related genes between N (Normal) and T (Tumor) cell types. ** means $p < 0.01$, *** means $p < 0.001$. (**C**) Correlation matrix plot of hypoxia-related differential expressed genes. (**D**) Gene Ontology (GO) Functional Annotation analysis of 29 hypoxia-related DEGs.

2.2. Construction of a Prognostic Four-Gene Model for EC

Through the Univariate Cox regression and least absolute shrinkage and selection operation (LASSO) analysis, four out of twenty-nine prognostic significant DEGs were obtained for constructing the hypoxia gene signature of EC (Figure S2). From the TCGA-EC cohort, 256 samples were defined as the training cohort, and 256 samples were for model testing. The four-gene signature prognostic model was constructed via Multivariate Cox regression analysis: Risk score = (0.062 × expression level of SRPX) + (0.016 × expression level of IL6) + (0.006 × expression level of HOXB9) + (0.155 × expression level of NR3C1) (Table S2). Based on the calculated risk scores, samples were divided into high- or low-risk groups. Principal components analysis (PCA) analysis for the testing and entire sets illustrated a fit of the model (Figure S3A. center, right). From the boxplots, four genes were all expressed differentially in two risk groups (Figure 2A, $p < 0.001$).

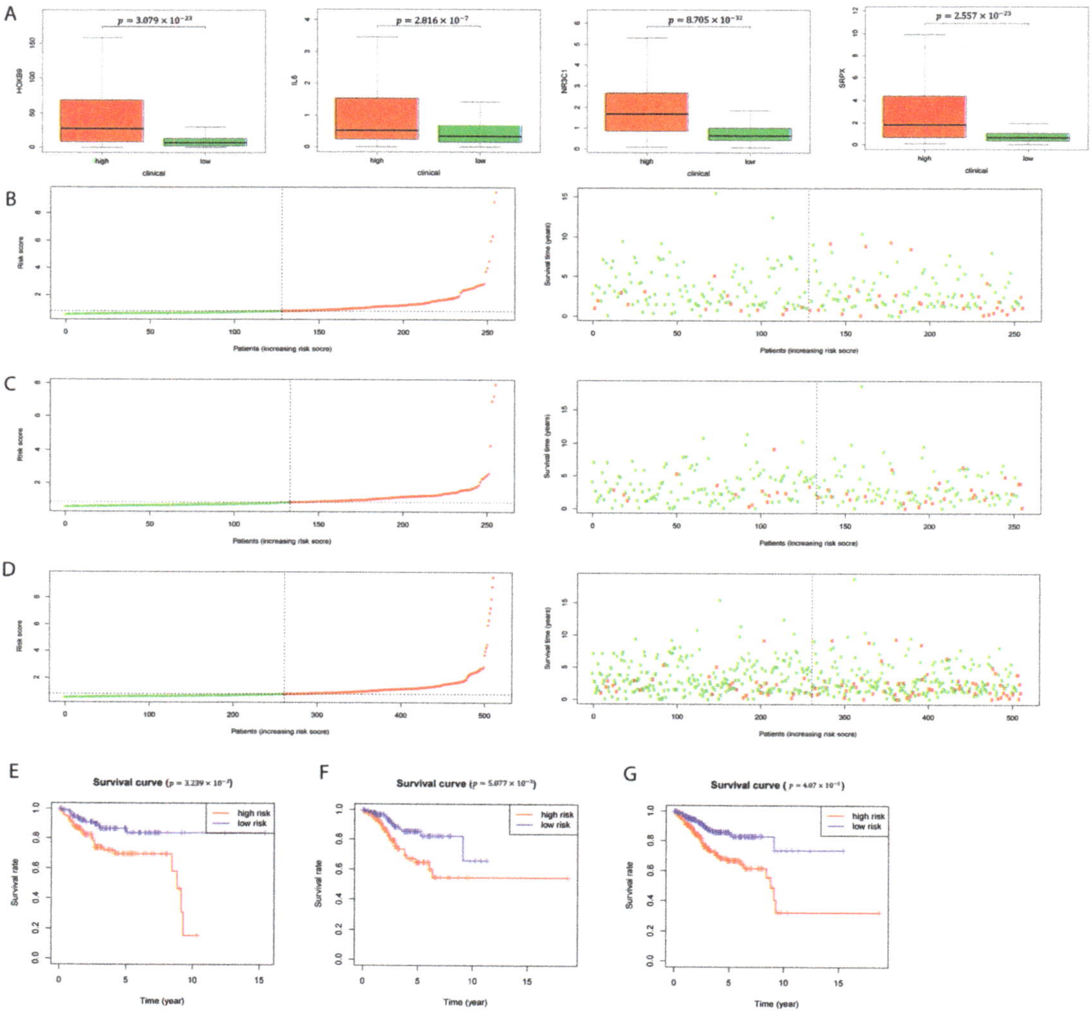

Figure 2. Development of prognostic model combining hypoxia-related gene signature and clinical factors. (**A**) mRNA Expression of gene HOXB9, IL6, NR3C1, and SRPX compared between low and high-risk groups. (**B–D**) The risk score rank (**left**) and distribution of survival status (**right**) of the four genes in the training set, testing set, and entire set. Green for alive and red for dead in high-risk groups. (**E–G**) Kaplan-Meier OS for high-risk group and low-risk group in the training set, testing set, and entire set (**from left to right**).

Patients with increasing risk scores had an observed possibility of death status (Figure 2B–D). Moreover, the K-M survival curves of the training, testing and entire sets implicated the high-risk group's lower OS rate (Figure 2E–G, *p*-value < 0.001). To note, a drop was found between the 9th and 10th year of the high risk group, which may due to the randomized division of samples or a batch effect while TCGA collecting patients' survival information. We also computed the survival probability of subdivisions in clinical factors in two risk groups and uncovered distinct patterns within the subdivisions of each clinical factor (Figure S3B).

The risk score estimated from the four-gene signature was then incorporated with the clinical characteristics for the further multivariate cox regression analysis. Moreover, the univariate and multivariate Cox model of training dataset corroborated the adequate predictability of the model with an independent variable, the stage factor (Table 1, $p < 0.001$). This result could suggest a complex relationship among patients' age, histological type, grade, and hypoxia-related risk scores in EC. Consequently, the trained model was validated by the entire set, which included three factors: Stage, grade, and risk score ($p < 0.05$).

Table 1. Univariate and multivariate Cox model result of the training set, testing set, and entire set.

Variable	Univariate Cox Model				Multivariate Cox Model			
	HR	HR.95L	HR.95H	*p* Value	HR	HR.95L	HR.95H	*p* Value
Training set								
age	2.3010	1.1898	4.4499	0.0133	2.2176	1.1146	4.4121	0.0233
stage	3.6435	2.0219	6.5656	0.0000	3.3310	1.7492	6.3431	0.0003
histological_type	2.0604	1.1423	3.7165	0.0163	0.9318	0.4581	1.8952	0.8454
grade	2.1739	1.1239	4.2049	0.0211	1.3027	0.6030	2.8143	0.5010
riskScore	1.3662	1.1800	1.5817	0.0000	1.2244	1.0440	1.4360	0.0128
Testing set								
age	1.3332	0.6821	2.6061	0.4003	0.9553	0.4598	1.9848	0.9025
stage	4.7100	2.5689	8.6357	0.0000	2.8626	1.5063	5.4402	0.0013
histological_type	4.6591	2.5439	8.5331	0.0000	2.0849	1.0231	4.2483	0.0431
grade	7.5966	2.7083	21.3077	0.0001	3.9697	1.3121	12.0107	0.0147
riskScore	1.3974	1.1728	1.6649	0.0002	1.1233	0.9240	1.3657	0.2433
Entire set								
age	1.7782	1.1121	2.8432	0.0162	1.5240	0.9300	2.4974	0.0946
stage	4.1162	2.7000	6.2754	0.0000	3.0942	1.9669	4.8676	0.0000
histological_type	3.0435	2.0032	4.6242	0.0000	1.3605	0.8292	2.2323	0.2231
grade	3.3973	1.9765	5.8397	0.0000	1.9294	1.0493	3.5478	0.0345
riskScore	1.3956	1.2489	1.5596	0.0000	1.1991	1.0610	1.3552	0.0036

In the receiver operating characteristic curves (ROC) analysis, the one-, three-, and five-year AUC were shown in Figure S3C. It is recommended to use the risk score model for facilitating molecular subtype-based diagnosis.

2.3. *Evaluations of Immune Cells and Highlighted mRNA Modifications between Risk Groups of EC*

In order to study the related immune cells or pathways in EC, gene set variation analysis (GSVA) was performed to calculate the enrichment scores of low and high-risk groups of the EC patients in the TCGA cohort. In the gene enrichment of EC high-risk groups, human-activated dendritic cells (aDCs) and B cells were significantly differentiated among all measured immune cells (up-regulated) (Figure 3A). In contrast, human immature dendritic cells (iDCs) were highly upregulated in tissues of low-risk patients. Two immune pathways achieved elevated enrichment scores in EC high-risk groups: Parainflammation and Type I IFN response (Figure 3B). Recent studies showed that B cells could be a good indicator for the prolonged survival of high-grade EC patients. Besides, IgA regulation mediated by plgR in the EC tumor cells enhances the activation of inflammatory pathways involving IFN signaling and the hindrance to DNA repairing [23].

By the ESTIMATE algorithm, 22 types of immune cells were evaluated and eight were found significantly associated with the hypoxia gene signature (Figure S4A, $p < 0.05$). Specifically, gamma delta T cells and memory B cells were upregulated in the high-risk group with a high correlation (Figure 3C: $p < 0.001$, $p = 0.001$, Figure 3D: $p = 0.00028$, $p = 0.0012$), and Macrophages M1 was upregulated in the high-risk group with a correlation of $p < 0.005$ (Figure 3C: $p = 3.47 \times 10^{-5}$). Gene SRPX was highly expressed in the T cells gamma delta and Macrophages M1, which was consistent with the result of the GSVA (Figure 3D). Then, we calculated the correlation between the DEG expression profile in the signature and the immune cells in TCGA (Figure 3E).

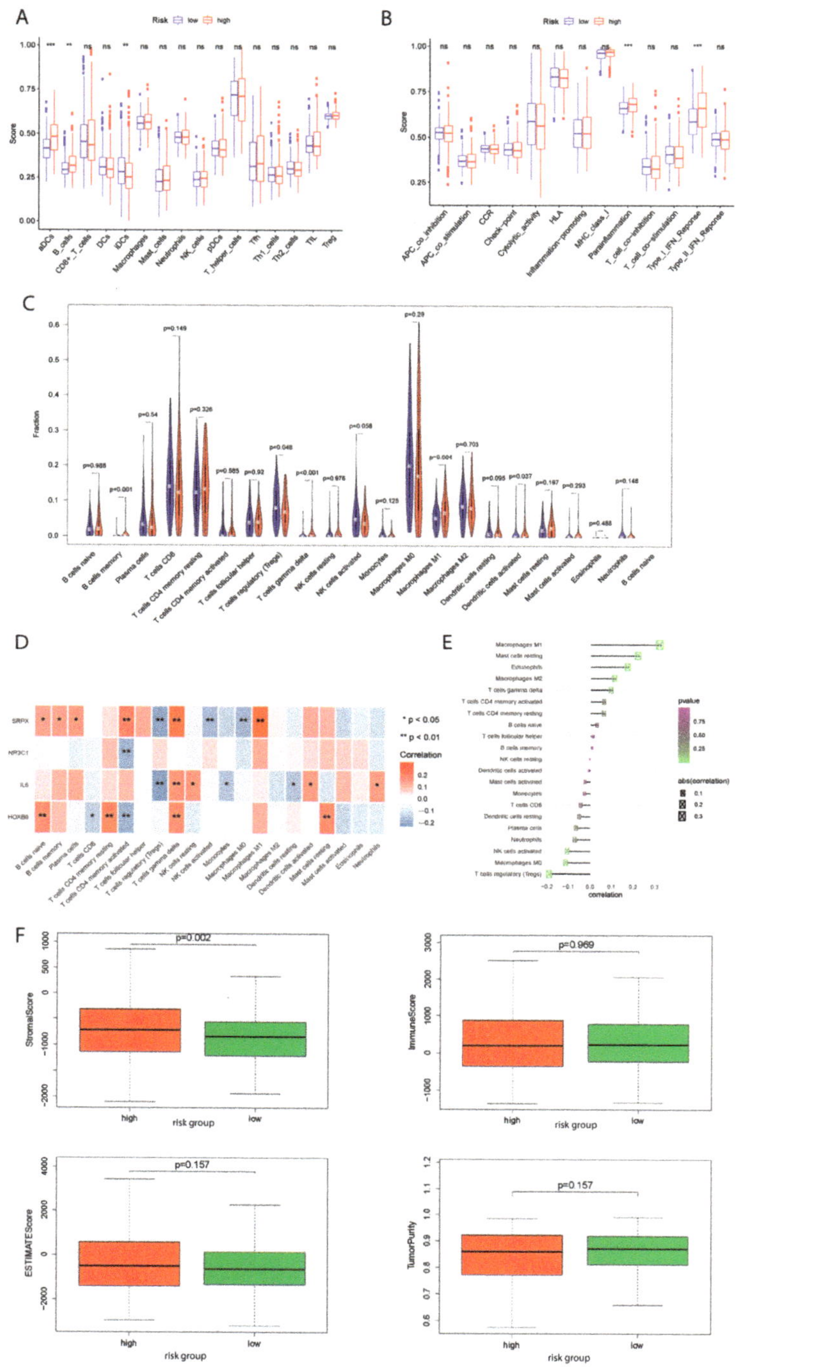

Figure 3. Infiltrating immune cells and tumor cell mRNA modifications correlated with hypoxia-related DEGs in EC patients. (**A**) The function scores of twenty-two immune cell types in EC samples

between two risk groups. ** means $p < 0.01$; *** means $p < 0.001$. (**B**) The differences in thirteen immune pathways between two risk groups. (**C**) Abundance of 22 infiltrating immune cell types between two risk groups. (**D**) Correlation plot of 22 infiltrating immune cell types with four hypoxia-related DEGs in TCGA-EC cohort. (**E**) Correlation plot of 22 infiltrating immune cell types with four genes from the prognostic signature in TCGA-EC cohort. (**F**) Estimated Patterns of stromal cell scores (**up-left**), immune cell scores (**up-right**), ESTIMATE scores (**down-left**), and tumor purity (**down-right**) based on risk scores.

The high-risk group in EC has a moderately higher occupancy of stromal cells (Figure 3F(up-left), $p < 0.005$). However, neither high nor low-risk group showed significant differences between the immune cell fractions (Figure 3F(up-right)). Moreover, the undifferentiated estimate score and estimated tumor purity can be explained by the genetic heterogeneity of endometroid type or serous tumor (Figure 3F(down)). Besides, the subtype-specific immune cell expression clustered in levels of stromal, immune, and estimate scores, plus tumor purity, was presented (Figure S4B). To depict stemness featured in risk groups in EC samples, mRNA expression-based stemness index (mRNAsi) and epigenetically regulated mRNAsi (EREG-mRNAsi) were evaluated, and no significant differences were shown (Figure S4C). It suggested that the risk progression of the tumor was neither characterized by undifferentiating expression of cells nor co-expression regulations related to immune invasion but rather by simple stromal invasions resulting in the ectopic endometrial-like epithelium and stroma [24]. As a result, molecular mechanisms of stromal invasion in tumor tissues during pathogenesis are worth investigation for EC patients.

We later examine the expression level of the N6-methyladenosine (m6a) regulators compared between two risk groups. There were five m6a regulators significantly expressed in tumor tissues of high-risk patients, indicating possible epigenetic modifications or transcriptional dysregulations during EC tumorigenesis (Figure S4D).

2.4. Assessment of Tumor Microenvironment in Different Risk Groups of EC Samples

Among 17 immune checkpoints, IDO1, ICOS, PD-L2, B7-H3, CD40, LAG3, CD86, PD-L1, and CD270 were differentially expressed in the low and high-risk groups of EC samples (Table 2, $p < 0.05$; Figure S5). Since PD-L1 and PD-L2 were reported to be promising candidates for immunotherapy [25,26], we therefore investigated expression profiles of immune checkpoint inhibitors (ICIs), PD-L1 and PD-L2 on patients in high and low risk-group from the TCGA cohort. The expression of PD-L1 and PD-L2 were positively correlated with patients' risk scores with $p < 0.005$ (Figure 4A,B). Additionally, as indicated in the boxplots, the high-risk group achieved a relatively predominant expression of PD-L1 and PD-L2 (Figure 4C,D).

Table 2. The expression difference of immune checkpoints between low and high-risk groups.

Gene	*p*-Value
IDO1	0.0326
CD27	0.2540
CD58	0.5882
CTLA4	0.3644
ICOS	0.0129
PD-L2	0.0000
B7-H3	0.0153
B7-H4	0.9876
TIGIT	0.1388
PD-1	0.3375
CD40	0.0000
LAG3	0.0061
TIM-3	0.1870
CD86	0.0214

Table 2. *Cont.*

Gene	*p*-Value
PD-L1	0.0023
CD70	0.1660
CD270	0.0239

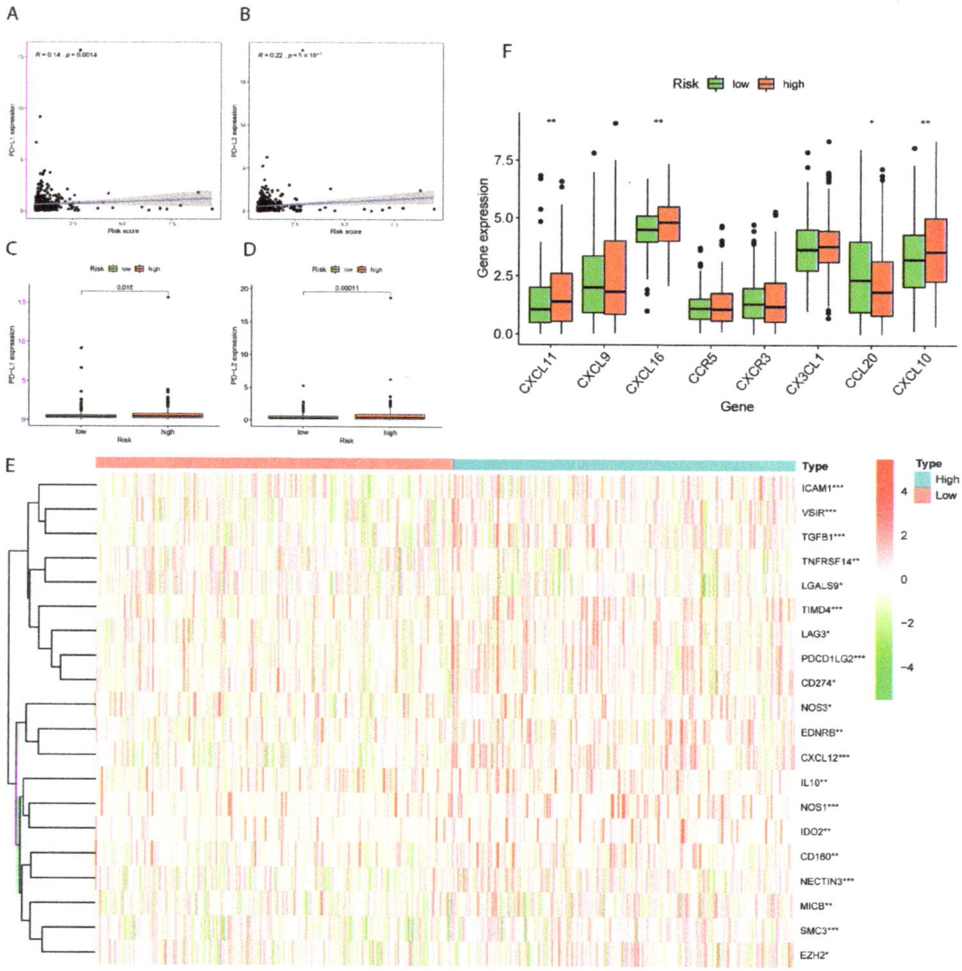

Figure 4. Immune checkpoint and immunosuppressive cytokine gene expression levels in high- and low-risk groups. (**A**) Scattered plot illustrating a correlation between the expression level of Immune checkpoint PD-L1 and risk scores in TCGA cohort. (**B**) Scattered plot illustrating a correlation between the expression level of Immune checkpoint PD-L2 and risk scores in TCGA cohort. (**C**) Expression level of PD-L1 compared between low and high-risk groups. (**D**) Expression level of PD-L2 compared between low and high-risk groups. (**E**) Heatmap of distinct immune checkpoint inhibitor marker expressed between low and high-risk groups. (**F**) Differential gene expression of the immunosuppressive cytokines in two risk groups from the TCGA-EC samples. * means $p < 0.05$; ** means $p < 0.01$; *** means $p < 0.001$.

The following heatmap illustrating the risk-specific expression of immune checkpoint suppressors showed the complex modification of immuno-pathways, which can contribute to the pathological process and tumorigenesis (Figure 4E).

Cytokines functioned in inflammatory pathways and were reported to actively participate in EC pathogenesis. The high-risk group of EC samples disclosed substantial enrichment of CXCL11, CXCL16 and CCL20, while CXCL10 were upregulated in the low-risk group (Figure 4F).

2.5. TMB Evaluation and Chemotherapeutic Sensitivity Analysis in Prognostic Risk Groups

In the TCGA-EC cohort, 241 samples were sorted as high-risk, whereas 253 samples were as low-risk (Figure 5A). Eight genes were found for highly differential mutation frequencies between high and low-risk groups ($p > 0.05$). Tumor mutation burden (TMB) is significantly higher in the low-risk group indicating better prognostic immunotherapy benefits for EC cases in the low-risk group (Figure 5B). The correlation line with a *p*-value of 0.08999 (Figure S6, correlation index = −0.076). Thus, the prognostic model using only the risk score is recommended as a diagnostic tool and it is worthwhile to consider incorporating the clinical factors into the model.

Specifically, 97.51% of samples in the high-risk group had gene alterations, including missense mutation, such as mutations at gene TP53, while 94.47% of low-risk samples had gene alterations (Figure 5C,D and Figure S7).

Previous reports discovered influenced sensitivity of drugs Bleomycin, cisplatin, doxorubicin, and doxorubicin when either in a hypoxic condition or acidic conditions associated with hypoxia [27–29]. After patients' responses to the chemotherapy were tested in terms of hypoxia-related genes signature, Bleomycin, Docetaxel, and Vinblastine displayed relatively higher sensitivity in high-risk EC samples (Figure 5E,G,I, $p < 0.05$). Treatments with Cisplatin and Doxorubicin displayed a lower sensitivity in high-risk samples (Figure 5F,H, $p < 0.05$). According to an article by Deschoemaeker et al., cisplatin resistance can be inverted with the removal of acidic stress, which resembles reoxygenation [30]. However, when compared with a normoxic condition, EC cells under hypoxic condition showed reduced sensitivity [31]. Therefore, while the majority of the chemotherapy results were validated, other sensitivity results need more studies related to hypoxic conditions to confirm in the endometrial cancer cases.

2.6. Definition of Hypoxia Molecular Subtypes in EC for Diagnosis

Accounting for the expression levels of four hypoxia-differentiated DEGs, EC samples were grouped from TCGA via The ConsensusClusterPlus package in R software. The consistent cumulative distribution function (CDF) graph and the delta region graph decided the optimal value of k, which is the cluster number (Figure 6A,B). When starting from $k = 2$, the consensus CDF curve is stable enough, corresponding to the insignificant delta area changes. Figure S8A explicitly illustrated the samples in TCGA allocated into CLUSTER 1 and CLUSTER 2 subtypes of EC and other possible clustering. Therefore, the heatmap in Figure 6C exhibited the consensus matrix when $k = 2$ in simplicity. The PCA and t-distributed stochastic neighbor embedding (t-SNE) results also agreed with our clustering outcome of two subtypes (Figure S8B,C). Further analysis of the tumor microenvironment and targets of adjuvant therapies featured with the hypoxia-related molecular subtypes will serve as suggestions for diagnosis use under the complex hypoxic TME. The alluvial diagram demonstrates the distribution overlap of the EC samples between risk score and molecular subtype (Figure S8D). CLUSTER 1 achieved a close ratio of two risk groups in the samples, while the distribution of CLUSTER 2 subtype samples was mainly confined to high risk scores.

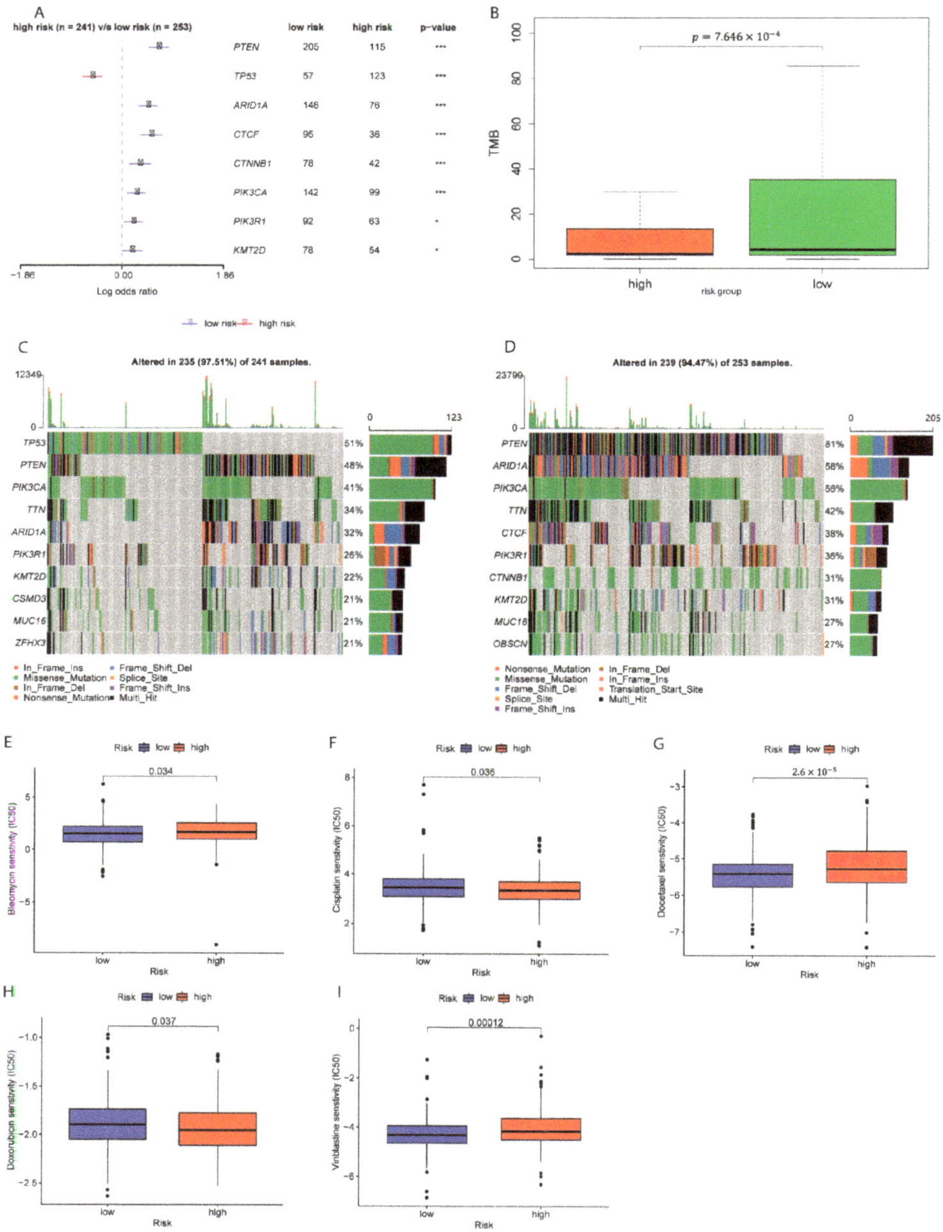

Figure 5. Gene mutation and drug sensitivity analysis in EC prognostic risk groups (**A**) Differential Expression of Top eight altered genes in risk groups. * means $p < 0.05$; *** means $p < 0.001$. (**B**) TMB for low and high-risk groups in the TCGA cohort. (**C**) Mutation profile of the high-risk group. (**D**) Mutation profile of the low-risk group. (**E–I**) Box plot of the estimated IC50 values for Bleomycin, Cisplatin, Docetaxel, Doxorubicin, and Vinblastine with two risk groups.

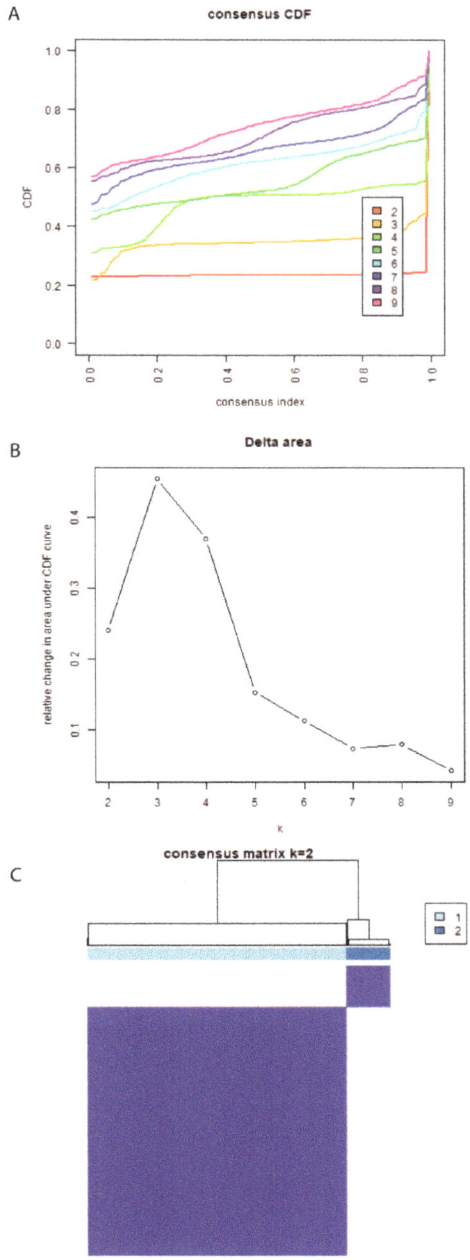

Figure 6. Consensus clustering of molecular subgroups in EC based on hypoxia-related DEGs (**A**) Cumulative distribution function (CDF) curve from k = 2 to k = 9. (**B**) CDF Delta area curve. The horizontal axis represents the number k and the vertical axis represents the relative change in the area under the CDF curve. (**C**) Consensus matrix heatmap of two clusters (k = 2) and the correlation area.

2.7. Expression Profile of Prognostic DEGs Clustered by Subtype and Clinical Factors

To investigate the relative gene expressions of the four prognostic DEGs, the forward and reverse primer sequence of genes SRPX, IL6, HOXB9, and NR3C1 was shown as follows (Table 3). The mRNAs for the four genes were measured by real-time quantitative polymerase chain reaction (RT-qPCR). Endometrial tissues were treated with TRIzol reagent (Invitrogen, Waltham, MA, USA) for total-RNA extraction. Therefore, we explored the gene expressions of these four hypoxia-related genes in the subdivisions of the clinical categories as well as the total risk scores calculated compared between the subdivisions (Figure S9A). Expression of the prognostic genes were also compared between paratumor tissues and tumor tissues (Figure S9B). Summaries of differential expression of each prognostic DEG classified by age, histological type, grade, and stage was illustrated in the heatmap (Figure S10).

Table 3. PCR results of four hypoxia-related genes in the EC prognostic model.

Primer	Primer Sequence (5′ to 3′)	Base Pairs
SRPX F	ATCAAGGTGAAGTATGGGGATGT	23
SRPX R	GTTTGACTGGCAGATCAGTAGG	22
IL6 F	ACTCACCTCTTCAGAACGAATTG	23
IL6 R	CCATCTTTGGAAGGTTCAGGTTG	23
HOXB9 F	CCATTTCTGGGACGCTTAGCA	21
HOXB9 R	TGTAAGGGTGGTAGACGGACG	21
NR3C1 F	ACAGCATCCCTTTCTCAACAG	21
NR3C1 R	AGATCCTTGGCACCTATTCCAAT	23

Furthermore, the EC samples obtained from the training cohort were classified into four typical clinical categories under each subtype (Figure 7A). The K-M OS curves of the two subtypes showed a significant prognostic difference in the TCGA-EC cohort that CLUSTER 1 has a higher survival probability compared to CLUSTER 2 ($p = 0.004$; Figure 7B). However, we observed no stage differences between CLUSTER 1 and CLUSTER 2, while these two molecular subtypes can be distinguished among survival probability patterns of age, histological type, and grade (Figure 7C).

2.8. Identification of Potential Targets for Immunotherapy and Chemotherapy in EC Molecular Subtypes

We compared the amount of non-epithelial cells in each TCGA sample tissue between two molecular subtypes via the ESTIMATE algorithm. Patterns of Immune score, stromal score, and estimate score were increased in CLUSTER 1 with a decreasing Tumor purity (Figure 8A–D, $p < 0.05$). High immune and stromal scores indicated an enriched level of immune-reactive as well as mesenchymal expression [24]. Furthermore, low tumor purity indicates better prognostic outcomes. Several methods were used to compare the substantial differences in immune cell expressions in the two clusters (Figure S11).

To explore potential immunotherapy responses of patients under hypoxia, immune checkpoint expressions in different molecular subtypes were assessed. Classic differentially expressed immune checkpoints like CD27, CD70, CTLA4, and PDCD1 were included (Figure 8E).

In terms of the chemotherapy, two drugs traditionally used in adjuvant chemotherapy for EC were evaluated for their subtype-specific sensitivity, respectively. A higher IC50 score of Doxorubicin was related to C1 (Figure 8F, $p = 0.028$). A lower IC 50 score of Vinblastine was related to Cluster 1 (Figure 8G, $p = 0.0057$).

2.9. A Nomogram Predicting Overall Survival for EC Patients by Subtype-Specific Signature and Clinical Factors

A nomogram was developed to further accurately predict the clinical outcomes by integrating the four-gene signature with two selected clinical characteristic variables (grade and stage) (Figure 9A). The prediction accuracy of prognostic models using risk score only and two independent clinical factors only or models using two clinical factors together and risk score with two clinical factors were compared by the multi-ROC analysis (Figure 9B). Additionally, calibration of the nomogram further validated a high consistency between the predicted survival probabilities of one-, three- and five-year OS and the observed data (Figure 9C). For this reason, the nomogram developed from our prognostic model should improve the prognostic result's predictive power for EC patients compared to the previous signature model.

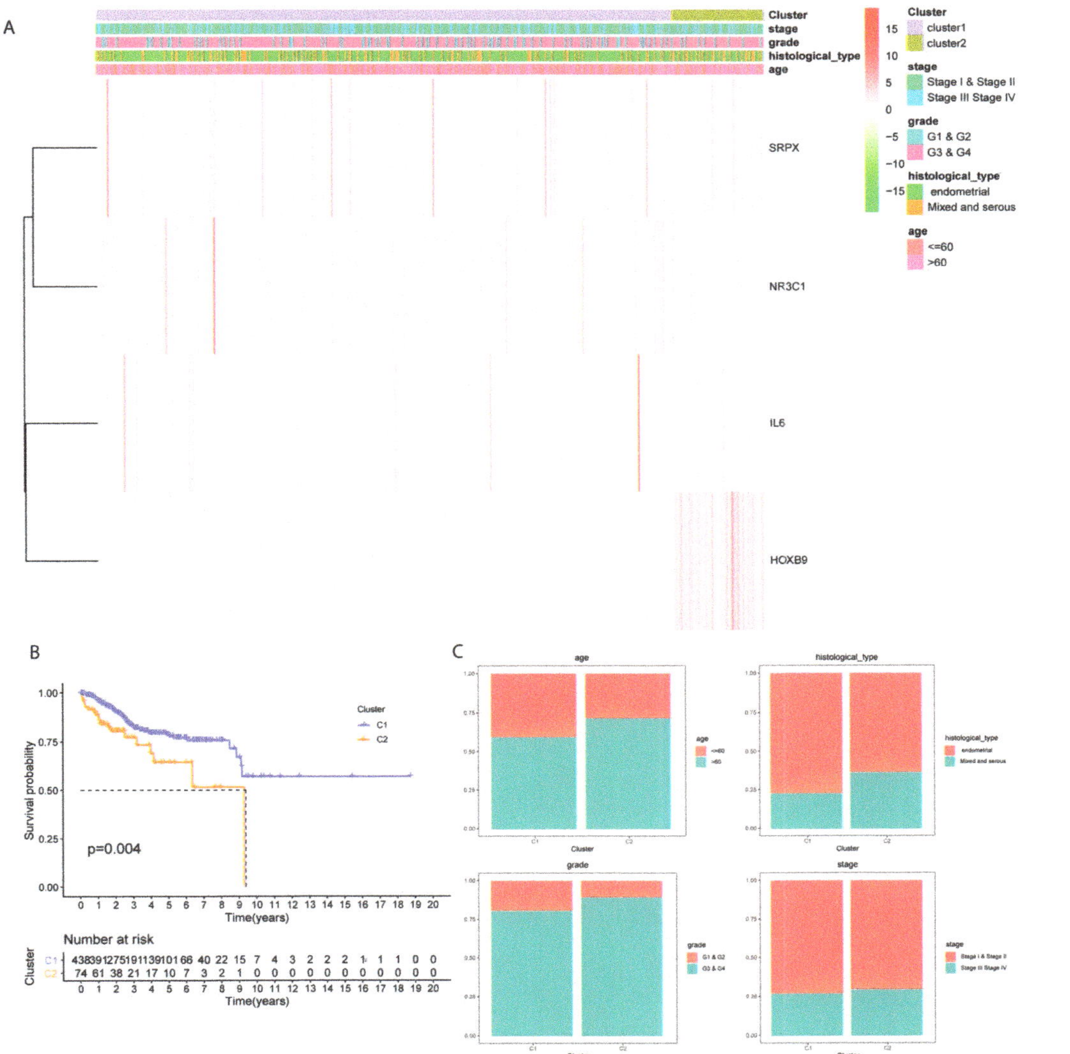

Figure 7. Main Clinical factors clustered into EC molecular subtypes. (**A**) Gene expression of four prognostic DEGs clustered by subtype and clinical factors. (**B**) Survival analysis of patients grouped by subtypes. (**C**) Barplots of survival probability of CLUSTER 1 and CLUSTER 2 in four main clinical factors (age, histological type, grade, and stage).

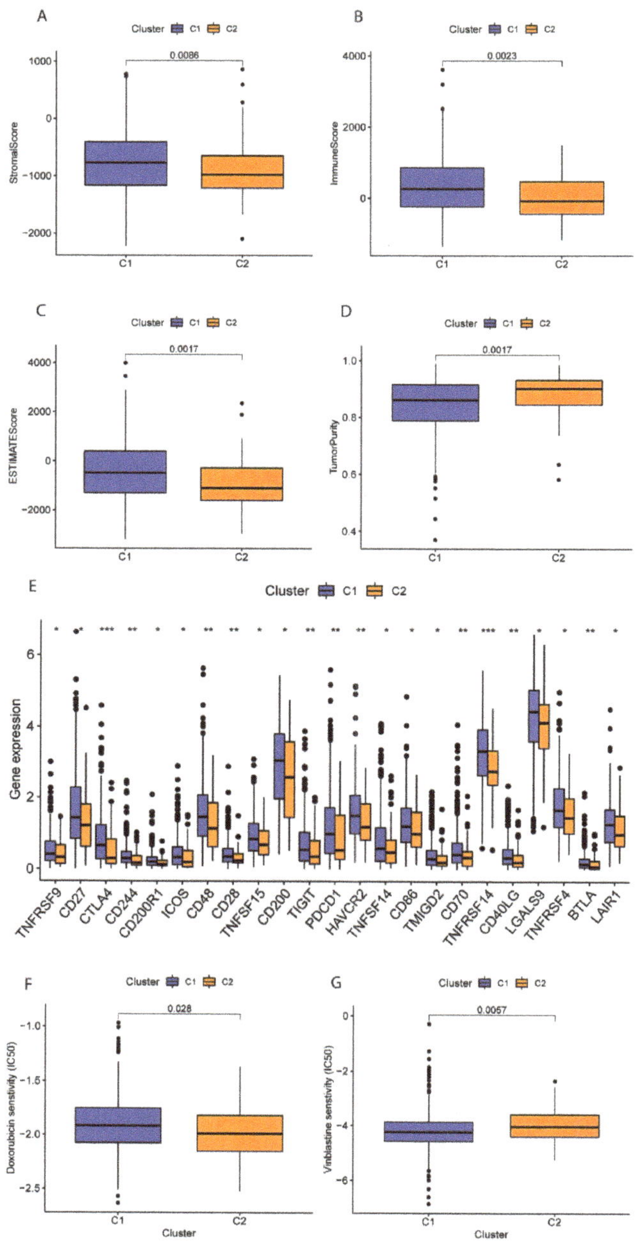

Figure 8. Tumor immune microenvironment analysis and drug sensitivity test for EC molecular subtypes. (**A–D**) Patterns of stromal cell scores, immune cell scores, ESTIMATE scores, and tumor purity between subtypes C1 and C2 of EC. (**E**) Expression profiles of 23 immune checkpoints between EC molecular subtypes. * means $p < 0.05$, ** means $p < 0.01$, *** means $p < 0.001$. (**F**) Box plot displaying the estimated IC50 values for Doxorubicin from the two molecular subtypes. (**G**) Box plot displaying the estimated IC50 values for Vinblastine from the two molecular subtypes.

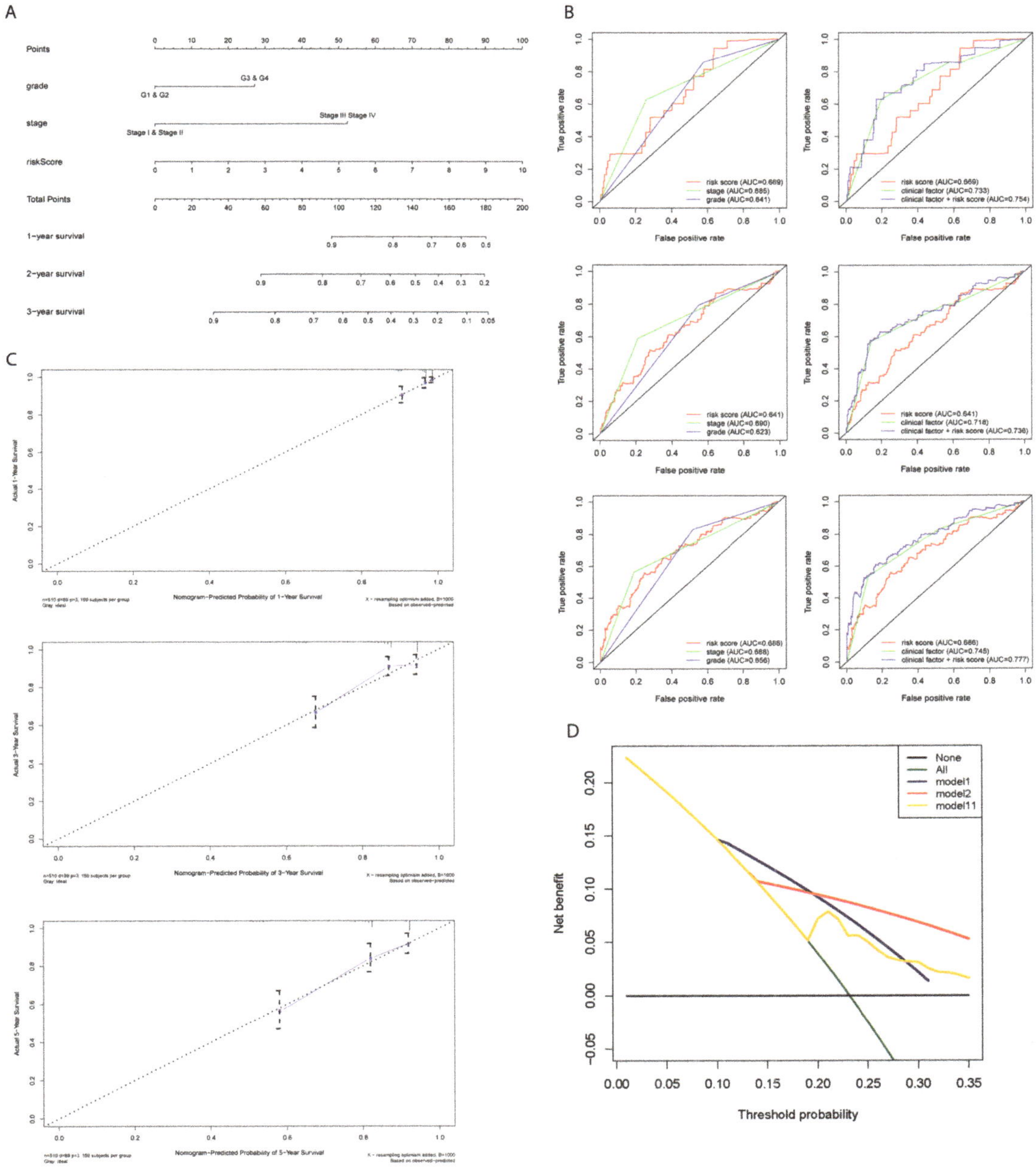

Figure 9. Predictive significance of signature verified in the nomogram model incorporating hypoxia-related gene signature and clinical characteristics. (**A**) Construction of a nomogram combining the four-gene signature and clinical features for the prediction of OS. (**B**) 1-,3-,5-year multi-ROC analysis for the final decision of the prognostic models. (**C**) Calibration plots displayed the actual and nomogram-predicted probability of one- (up), three- (middle), and five-year OS (down). (**D**) Decision curve analysis (DCA) curves of the nomogram for OS in HCC.

Importantly, the DCA plot showed no significantly higher net benefit between the risk score model, the clinical factor model, and the combined prognostic model (Figure 9D). Therefore, three models can be utilized under consideration of different applications.

3. Discussion

While histo-pathological tumor characteristics have been widely utilized for making clinical decisions in the past decade, the molecular subtype-based diagnostic and prognostic approaches are developing rapidly and have achieved advances in decision-making on targeted adjuvant therapies [32]. Some studies have pointed out that hypoxia commonly occurred during tumorigenesis and caused therapy resistance, which may influence the differentially expressed gene regulation in metabolic and immune systems [15,33]. In this study, we built a hypoxia gene signature to investigate the hypoxic TME's association with tumor recurrence and to provide suggestions to both risk-based prognosis of EC patients and subtype-based diagnoses of EC progression.

Four DEGs (HOXB9, IL6, NR3C1, and SRPX) were selected as the predictor variables in the prognostic gene signature for an estimation of risk scores for the EC patients from the TCGA. According to our hypoxia-related gene signature, somatic gene alterations, especially PTEN mutation, were highly active regardless of a high or low risk score. During the development of EC in low grade, PTEN gene mutation is one of the most frequent mutations that tend to co-occur with PIK3CA and PIK3R1 gene mutations. Patients with PTEN mutation were susceptible to developing cancers like breast cancer, kidney cancer, and skin cancer [34]. Besides, loss of PTEN tended to cooperatively happen with CTNNB1 missense mutation and PIK3CA activation to boost myometrial invasion and thus form EC [35]. Another prominent mutation in our result was found at ARID1a, usually known as the tumor suppressor gene. It is reported that loss of ARID1a up-regulated PTEN in terms of the tumor cell proliferation in endometrial glands [36–38]. TP53 mutation is often associated with ECs in higher grades. However, the mutual occurrence of TP53 mutation and PTEN mutation is unique in USCs, which is closely related to our high-risk group [39–41]. Therefore, EC tumor cells in the hypoxic microenvironment gained specific somatic mutations including TP53 in high-risk patients, which promoted cell proliferation and lymph node metastasis. Nevertheless, the molecular mechanisms behind it were unclear and require further studies aided by animal models.

To look into the abnormalities at the transcriptional level, we explored the mRNA modification over the low and high-risk groups in EC samples. High level modifications in protein translation during the development of EC tumors were due to abnormal higher expression of m6A "readers" and "writers" [42,43]. Expressions of three m6A "readers", YTHDF1, YTHDF3, and FMR1, as well as two m6A "writers," KIAA1429 and WTAP, were enhanced in EC with predicted higher risks.

Gene ontology analysis disclosed a relationship between hypoxia-related DEGs and immune response, especially inflammatory, via regulations over macrophages or other cytokine receptors. One of the most active GO pathways in EC tissues was the response to the steroid hormone, which explained the effects of sex hormones interacting with insulin-like growth factors on EC tumorigenesis [44].

From the PCR results of the risk-predictor genes, we speculated high expression levels of NR3C1 in tumor tissues and expressions of IL-6 and SRPX converged in paratumor tissues. Previously known, amplification of HOXB9 in EC was correlated with poorer overall survival and EC progression [45]. IL-6, identified as a pro-inflammatory cytokine during inflammation and tumorigenesis, is related to microbial communities involved in immune responses [46]. NR3C1 was involved in the activation and enrichment of infiltrated immune cells including naive B cells, M1 macrophages, neutrophils, CD4 memory resting T cells, follicular helper T cells, gamma delta T cells, and regulatory T cells (Tregs) [47,48]. All four DEGs' expressions were enhanced in high-risk EC patients.

Consistent with the previous studies, we also found that three out of four DEGs in hypoxia signature, SRPX, IL6, and HOXB 9 were positively associated with naive B cells,

CD4 memory resting T cells, gamma delta T cells, M1 Macrophages, and resting Mast cells. Except for functions in innate immunity and autoimmunity, the immune cells mentioned were reported to have a role in modulating viral and bacterial [49,50]. CD4 memory resting T cells engaged in the secretion of C-X-C motif chemokine ligand (CXCL)10 during viral infection [49]. This interaction is consistent with a significantly higher expression of (CXCL)10 in high-risk EC patients. Aside from this, SRPX and IL6 were negatively correlated with regulatory T cells (Tregs). Tregs were recognized as a subset of T cells that suppresses immunity in sterilization and anti-tumor [51], thus SRPX and IL6 can be a novel target for recovering immune response in EC tumor cells. Moreover, Ryan, R et al. reasoned that SRPX also plays a part in the cell metabolism relating to the glucocorticoid GO pathway [52]. Consequently, our prognostic signature model could precisely predict the potential risk of EC patients by detecting aberrancies in the genome and proteomes.

Surprisingly, gathering results of ESTIMATE scores, high profiles of stromal cells and unique neuroendocrine-like immune cells bridged our hypoxia signature to an immune-desert molecular subtype of ovarian cancer determined by quantitative immune phenotypes [53]. "Cold" tumor subtype was defined as scarce CD8+T cells and identified in pancreatic, gastric, and ovarian cancer [54–56]. In details, a smaller number of T cells driven by CD8 were associated with higher risk in EC patients. TME of desert subtype explored in pancreatic cancer was proved to have enriched B cell expression [54], which was also supported by infiltrated neuroendocrine-like immune cell including human-activated dendritic cell (aDC) and B cell enrichments in the EC high-risk group. In Zhang B et al. study, the desert subtype was found consistent with the m6a modification patterns, which confirmed that desert sub-TMEs has a close relationship with immune ignorance and loss of T cells [55]. KIAA1429, among all m6a regulators, were positively associated with sub TMEs of desert or non-infiltrated subtype in gastric cancer, which is consistent with the high-risk group predicted with hypoxia signature. Noticeably, either desert-dominant or co-occurred subtype was correlated with a poorer survival outcomes [55]. However, further identification including the definitive measurement and spatial distribution of CD8+ T-cell versus stromal cells were required.

Standard treatments on EC are confined to excision surgeries on the lesion [54,57]. Along with the deepened insight into EC's staging and histological classifications, patients diagnosed with higher grades will be treated with systemic therapies mixed with radiotherapy, and low-grade will be treated with chemotherapy and targeted therapies for trials. Several chemo-drugs were previously studied regarding EC treatments [58]. Gebbia V et al. tested a combination including Cisplatin and Vinorelbine and insisted that the regimen was preferable to the classical Anthracycline [59]. Another drug, Docetaxel was reported to have considerable efficacy and a bearable range of toxicity in two phase II trials [51,60]. The significant chemotherapeutic sensitivity of Vinorelbine and Docetaxel in the high-risk group proved their fit for patients with high risk even encountering the hypoxia-raised therapy resistance.

Emerging targeted medications include ICI anti-PD-1/PD-L1 and chemotherapy [35,61]. PD-L1 and PD-L2 displayed positive relationships with hypoxia DEGs prognostic risks, which indicated anti-PD-1/PD-L1 and anti-PD-1/PD-L1 treatment should be a good choice explicitly targeting the predicted high-risk group. Present studies further supplemented that increasing number of mutations gave rise to a growing class of neo-antigens, which could be targets for T-cell attack [62,63]. Tumor mutation burden was therefore considered an index of patients' response to immunotherapy [64,65]. We figured out a higher TMB for low risk group indicating higher immunotherapy benefit by targeting various checkpoints in terms of advanced prevention, although PD-1/PD-L1 immunotherapy was effective in the high risk group.

Two molecular subtypes were characterized upon the hypoxia gene expressions. An overlap between distributions of molecular subtype and risk groups was discovered in TCGA dataset. CLUSTER 2 was majorly high-risk, while CLUSTER 1 had a subtly higher ratio distributed in low-risk than high-risk. Significantly higher immune score, stromal score and estimate score calculated on CLUSTER 1 subtype explained the abundance of immune cells and stromal cells in tumor tissues. CLUSTER 1 was also estimated with a lower tumor purity, which suggested a less complex tumor microenvironment and better clinical result. Our chemotherapeutic sensitivity results also revealed relatively significant effects of Vinorelbine over CLUSTER 1, compatible with results of high-risk patients. As a result, subtype-based diagnosis recommended surgery and targeted chemotherapy on patients in CLUSTER 1, while more personalized medications on patients in CLUSTER 2.

The hypoxia gene signature was eventually coupled with clinical factors, which were selected by the univariate and multivariate hazard analysis, to build a prognostic model. We constructed a nomogram for the model, which was validated for the improved predictive power and thus was recommended for EC prognosis.

Due to the research subjects' particularity, and ethical reasons, our study still had some limitations. Firstly, although a large sample size and quality control were obtained, data resources were limited to TCGA, and more evidence of hypoxia signature from other databases can be sought for confirmation. Secondly, as a cross-sectional retrospective study without longitudinal follow-up data, we cannot confirm whether TME changes occurred after the recorded histology and stages. However, our model showed a superiority over other prognostic models solely comprising gene signature. Additionally, we have provided some auxiliary analysis regarding m6A regulators and cancer cell stemness, which can be supportive evidence for cancers progressions associated with metabolic reprogramming.

Recent studies have found several gene mutations, including PTEN, TP53, PIK3CA, and classified four major genomic classes of endometrial cancer. However, the scope of prominent mutations can be narrowed to identify the target with the most direct relationship with tumor cell growth and cancer progression. Besides, our study proposed a possible relationship with immune-desert subtype. Future investigation can be directed to the differential responses to chemotherapy and immune checkpoint inhibition for three immune subtypes with the consideration of hypoxic TME.

4. Methods and Materials

4.1. Data Collection and Preprocessing

There were 200 hypoxia genes chosen from The Molecular Signatures Database v7.2 (https://www.gsea-msigdb.org/gsea/msigdb, "HALLMARK_HYPOXIA", accessed on 3 May 2022), which were found to be up-regulated in hypoxic condition via Gene Set Enrichment Analysis v4.1.0 software [66].

We retrieved the data regarding somatic mutations and clinical factors (age, histological type, grade, stage, and survival information) for EC from the TCGA public database (http://cancergenome.nih.gov/, accessed on 6 May 2022). Therefore, 512 EC samples from TCGA were randomly divided into the training and testing cohort with an approximately 1:1 ratio. The hypoxia-related differentially expressed genes (DEGs) were filtered by the linear models for microarray data with FDR < 0.001 using the R limma package.

4.2. Functional Annotation Analysis

Kyoto Encyclopedia of Genes and Genomes (KEGG) pathway and Gene Ontology (GO) enrichment analyses were performed on hypoxia-associated DEGs between high and low-risk cohorts by the R clusterProfiler package [21]. Gene enrichment analysis includes three GO terms, biological process (BP), molecular function (MF), and cellular component. GO terms and KEGG pathways were considered statistically significant with $p < 0.05$.

4.3. Establishment of a Hypoxia Gene Signature

To explore the prognostic significance of the 29 DEGs relating to hypoxia in EC, a univariate Cox regression analysis was performed. Five prognosis-related DEGs with $p < 0.05$ were screened in the training dataset. After the LASSO Cox regression analysis by the glmnet R package [67], four out of five hypoxia-related DEGs were identified

(Figure S2). A multivariate Cox regression was performed to formulate a hypoxia-related gene combination (Table S2).

4.4. Formulation and Validation of a Nomogram for the Prognostic Model

We further integrated the risk score evaluated by the hypoxia gene signature and the clinical factors into a nomogram facilitating the OS prediction by the R rms package [68]. Based on the nomogram model, the total survival probability of each patient can be calculated by summarizing the corresponding points of all variables. Calibration plots of the nomogram were used to illustrate the fitness of the predicted 3-, 5-, and 10-year survival compared to the observed value. The decision curve analysis (DCA) was used to check the predictive power.

4.5. Quantitative Real-Time Polymerase Chain Reaction PCR after the RNA Isolation

The research was approved by the First Affiliated Hospital of Nanjing Medical University Ethics Committee. The participants entered the research cohort strictly provided us with written informed consent. RNA was extracted from 15 EC and 15 normal sample tissues with TRIzol reagent (Thermo Fisher Scientific, Waltham, MA, USA), and complementary DNA (cDNA) was synthesized using the total RNA via the high-capacity reverse transcription kits (TaKaRa, Shiga, Japan) (Table S3). Assays were used to perform the RT-qPCR based on SYBR Green PCR Kit (Thermo Fisher Scientific, Waltham, MA, USA). The $2^{-\Delta\Delta CT}$ method was applied on Light Cycler 480 (Roche, Basel, Switzerland). The forward and reverse primer sequences used in qRT-PCR are listed in Table 3.

4.6. Genomic Alteration Analysis

We analyzed the gene variations from the Genomic Identification of Significant Targets in Cancer v2.0 by the software, genePattern. Specifically, frequencies of somatic mutations were calculated using the MutSigCV algorithm [69]. Moreover, we plotted the TMB score for EC patients from the TCGA dataset to predict the immunotherapeutic impacts for patients with varying risk scores. To determine the disparities in TMB levels, the Wilcoxon rank sum test was employed.

4.7. Identification of Immune Cell Types and Assessment of Significant Immune Checkpoint Inhibitors

The infiltration levels of twenty-two kinds of immune cells were calculated utilizing the Cell type identification by estimating relative subsets of RNA transcripts (CIBERSORT) algorithm. We calculated immune infiltration variations between two molecular subtypes through the Wilcoxon rank-sum test [70]. Besides, the correlation of the immune cell types with hypoxia gene signature was evaluated with absolute value and $p < 0.05$.

4.8. Estimation of Immune and Stromal Cells in EC

Using expression profiles from TCGA samples, we evaluated infiltrating stromal and immune cell levels in EC diseased cells by the Estimation of stromal and immune cells in malignant tumors using the expression data (ESTIMATE) algorithm. By pooling stromal and immune scores, the ESTIMATE score was subsequently evaluated. The tumor purity of samples from each TCGA patient was then determined for the corresponding ESTIMATE scores [24].

4.9. Expression Analysis of m6A RNA Methylation Regulators

According to recent papers, we chose twenty m6A RNA methylation regulators (METTL3, HNRNPC, YTHDC1, ZC3H13, YTHDF2, FTO, YTHDF1, YTHDF3, YTHDC2, METTL14, RBM15, WTAP, KIAA1429, FMR1, METTL16, HNRNPA2B1, and ALKBH5) for further research. Using the CIBERSORT algorithm, the risk differences in expression profiles of m6A regulators were compared for the EC tissues in TCGA samples with $p < 0.001$.

4.10. Chemotherapy Sensitivity Test

To explore substantially different responses of chemotherapeutic drugs in the molecular subtypes of EC and risk groups predicted by the hypoxia gene signature model, we calculated the IC50 of drugs typically applied in regiments of ECs using the R pRRophetic package. The sensitivity response of each patient in chemotherapy was predicted by database extracted from the Genomics of Drug Sensitivity in Cancer (GDSC; https://www.cancerrxgene.org/, accessed on 7 May 2022) [71].

4.11. Clustering Analysis

The consistent clustering identified molecular subtypes of EC samples from TCGA via the ConsensusClusterPlus package in R software [72]. In the clustering analysis, we used transcriptomic profiling data of four hypoxia-related signature genes, survival time, survival status, predicted risk score, and risk groups as dimensions of each sample. Then, we calculated the Euclidean squared distance metric and the K-means clustering algorithm from $k = 2$ to $k = 9$. Besides, we performed the principal components analysis (PCA) and t-distributed stochastic neighbor embedding (t-SNE) to perform multiples tests on clustering results established on the transcriptome expression profile of the above hypoxia-related genes.

4.12. Statistical Analysis

We predominantly performed data analysis with the aid of the R language v4.0.2 software throughout the study (https://www.r-project.org/, accessed on 7 May 2022). Different hypoxia subtypes were compared by the Kruskal–Wallis test. The differential survival time was figured out using the log-rank test with $p < 0.05$ and we applied Kaplan Meier curves to illustrate the striking distinctions in survival time.

5. Conclusions

We portrayed two hypoxia-related molecular subtypes of EC based on the four screened DEGs signature, which integrated with clinical factors to serve as a predictive model for EC patients. The assessments on infiltrating immune cell types, immune checkpoint inhibitors, and chemotherapy responses can be referred to as some insights into the hypoxic impacts of the genome, methylome, and transcriptome on EC progression in the future.

Supplementary Materials: The following supporting information can be downloaded at: https://www.mdpi.com/article/10.3390/ijms24021675/s1.

Author Contributions: Conceptualization, funding acquisition, and Resources, J.L. and J.B.; Methodology, S.N.; Investigation and software, Z.Z.; Formal analysis, W.L.; Project administration and visualization, Z.H.; Validation, S.G. and Q.H.; Writing–original draft, review & editing, Y.J. and R.G. All authors have read and agreed to the published version of the manuscript.

Funding: This work was funded by the Jiangsu Province Nature Science Foundation (Grant No. BK20220729) and the National Natural Science Foundation of China (Grant No. 82273738).

Institutional Review Board Statement: All subjects have their informed consent for inclusion before they participated in the study. The study was conducted in accordance with the Helsinki, and the protocol was approved by the Ethics Committee (2022-SR-740, 14 September 2022).

Informed Consent Statement: The research was approval by the First Affiliated Hospital of Nanjing Medical University ethics committee.

Data Availability Statement: The datasets supporting the conclusions of this article are included within the article. Public datasets in this study were available for additional analysis. Datasets needed for reproducibility can be retrieved from "HALLMARK_HYPOXIA" repository, the TCGA website [http://cancergenome.nih.gov/</b&gt] (accessed on 3 May 2022). Data analysis was performed using the R language v4.0.2 software throughout the study (https://www.r-project.org/, accessed on 3 May 2022).

Conflicts of Interest: The authors declare that they have no competing interest.

Abbreviations

aDCs	human-activated dendritic cells
CIBERSORT	Cell type identification by estimating relative subsets of RNA transcripts
DEGs	differentially expressed genes
EC	endometrial carcinoma
ESTIMATE	Estimation of stromal and immune cells in malignant tumors using expression data
GDSC	Genomics of Drug Sensitivity in Cancer
GSVA	Gene set variation analysis
GO	Gene Ontology
HR	hazard ratio
ICI	immune checkpoint inhibitors
IC50	the semi-inhibitory concentration
iDCs	immature dendritic cells
KEGG	Kyoto Encyclopedia of Genes and Genomes
LASSO	least absolute shrinkage and selection operation
Limma	linear models for microarray data
m6a	N^6-methyladenosine
OS	overall survival
PCA	principal components analysis
ROC	receiver operating characteristic curves
TCGA	The Cancer Genome Atlas
TMB	tumor mutation burden
Tregs	regulatory T cells
t-SNE	t-distributed stochastic neighbor embedding

References

1. Gu, B.; Shang, X.; Yan, M.; Li, X.; Wang, W.; Wang, Q.; Zhang, C. Variations in incidence and mortality rates of endometrial cancer at the global, regional, and national levels, 1990–2019. *Gynecol. Oncol.* **2021**, *161*, 573–580. [CrossRef] [PubMed]
2. Smith, R.A.; von Eschenbach, A.C.; Wender, R.; Levin, B.; Byers, T.; Rothenberger, D.; Brooks, D.; Creasman, W.; Cohen, C.; Runowicz, C.; et al. American Cancer Society guidelines for the early detection of cancer: Update of early detection guidelines for prostate, colorectal, and endometrial cancers. Also: Update 2001—Testing for early lung cancer detection. *CA Cancer J. Clin.* **2001**, *51*, 150. [CrossRef] [PubMed]
3. Sorosky, J.I. Endometrial cancer. *Obstet. Gynecol.* **2012**, *120*, 383–397. [CrossRef] [PubMed]
4. American College of Obstetricians and Gynecologists. ACOG Committee Opinion No. 440: The Role of Transvaginal Ultrasonography in the Evaluation of Postmenopausal Bleeding. *Obstet. Gynecol.* **2009**, *114*, 409–411. [CrossRef] [PubMed]
5. Randall, M.E.; Filiaci, V.L.; Muss, H.; Spirtos, N.M.; Mannel, R.S.; Fowler, J.; Thigpen, J.T.; Benda, J.A. Randomized phase III trial of whole-abdominal irradiation versus doxorubicin and cisplatin chemotherapy in advanced endometrial carcinoma: A Gynecologic Oncology Group Study. *J. Clin. Oncol.* **2006**, *24*, 36–44. [CrossRef]
6. Matei, D.; Filiaci, V.; Randall, M.E.; Mutch, D.; Steinhoff, M.M.; DiSilvestro, P.A.; Moxley, K.M.; Kim, Y.M.; Powell, M.A.; O'Malley, D.M.; et al. Adjuvant Chemotherapy plus Radiation for Locally Advanced Endometrial Cancer. *N. Engl. J. Med.* **2019**, *380*, 2317–2326. [CrossRef]
7. Simpkins, F.; Drake, R.; Escobar, P.F.; Nutter, B.; Rasool, N.; Rose, P.G. A phase II trial of paclitaxel, carboplatin, and bevacizumab in advanced and recurrent endometrial carcinoma (EMCA). *Gynecol. Oncol.* **2015**, *136*, 240–245. [CrossRef]
8. Oza, A.M.; Pignata, S.; Poveda, A.; McCormack, M.; Clamp, A.; Schwartz, B.; Cheng, J.; Li, X.; Campbell, K.; Dodion, P.; et al. Randomized Phase II Trial of Ridaforolimus in Advanced Endometrial Carcinoma. *J. Clin. Oncol.* **2015**, *33*, 3576–3582. [CrossRef]
9. Muz, B.; de la Puente, P.; Azab, F.; Azab, A.K. The role of hypoxia in cancer progression, angiogenesis, metastasis, and resistance to therapy. *Hypoxia (Auckl)* **2015**, *3*, 83–92. [CrossRef]
10. Kandoth, C.; Schultz, N.; Cherniack, A.D.; Akbani, R.; Liu, Y.; Shen, H.; Robertson, A.G.; Pashtan, I.; Shen, R.; Benz, C.C.; et al. Integrated genomic characterization of endometrial carcinoma. *Nature* **2013**, *497*, 67–73. [CrossRef]
11. Bellone, S.; Roque, D.M.; Siegel, E.R.; Buza, N.; Hui, P.; Bonazzoli, E.; Guglielmi, A.; Zammataro, L.; Nagarkatti, N.; Zaidi, S.; et al. A phase 2 evaluation of pembrolizumab for recurrent Lynch-like versus sporadic endometrial cancers with microsatellite instability. *Cancer* **2022**, *128*, 1206–1218. [CrossRef]
12. Samanta, D.; Semenza, G.L. Metabolic adaptation of cancer and immune cells mediated by hypoxia-inducible factors. *Biochim. Biophys. Acta Rev. Cancer* **2018**, *1870*, 15–22. [CrossRef] [PubMed]
13. Brizel, D.M.; Scully, S.P.; Harrelson, J.M.; Layfield, L.J.; Bean, J.M.; Prosnitz, L.R.; Dewhirst, M.W. Tumor oxygenation predicts for the likelihood of distant metastases in human soft tissue sarcoma. *Cancer Res.* **1996**, *56*, 941–943.

14. Palazon, A.; Goldrath, A.W.; Nizet, V.; Johnson, R.S. HIF transcription factors, inflammation, and immunity. *Immunity* **2014**, *41*, 518–528. [CrossRef] [PubMed]
15. De Palma, M.; Biziato, D.; Petrova, T.V. Microenvironmental regulation of tumour angiogenesis. *Nat. Rev. Cancer* **2017**, *17*, 457–474. [CrossRef] [PubMed]
16. Pocrnich, C.E.; Ramalingam, P.; Euscher, E.D.; Malpica, A. Neuroendocrine Carcinoma of the Endometrium: A Clinicopathologic Study of 25 Cases. *Am. J. Surg. Pathol.* **2016**, *40*, 577–586. [CrossRef]
17. Rizzo, A. Immune Checkpoint Inhibitors and Mismatch Repair Status in Advanced Endometrial Cancer: Elective Affinities. *J. Clin. Med.* **2022**, *11*, 3912. [CrossRef]
18. Damgaci, S.; Ibrahim-Hashim, A.; Enriquez-Navas, P.M.; Pilon-Thomas, S.; Guvenis, A.; Gillies, R.J. Hypoxia and acidosis: Immune suppressors and therapeutic targets. *Immunology* **2018**, *154*, 354–362. [CrossRef]
19. Wu, Z.; Lu, Z.; Li, L.; Ma, M.; Long, F.; Wu, R.; Huang, L.; Chou, J.; Yang, K.; Zhang, Y.; et al. Identification and Validation of Ferroptosis-Related LncRNA Signatures as a Novel Prognostic Model for Colon Cancer. *Front. Immunol.* **2021**, *12*, 783362. [CrossRef]
20. Ritchie, M.E.; Phipson, B.; Wu, D.; Hu, Y.; Law, C.W.; Shi, W.; Smyth, G.K. limma powers differential expression analyses for RNA-sequencing and microarray studies. *Nucleic Acids Res.* **2015**, *43*, e47. [CrossRef]
21. Yu, G.; Wang, L.-G.; Han, Y.; He, Q.-Y. clusterProfiler: An R package for comparing biological themes among gene clusters. *Omics J. Integr. Biol.* **2012**, *16*, 284–287. [CrossRef] [PubMed]
22. Wang, S.; Su, W.; Zhong, C.; Yang, T.; Chen, W.; Chen, G.; Liu, Z.; Wu, K.; Zhong, W.; Li, B.; et al. An Eight-CircRNA Assessment Model for Predicting Biochemical Recurrence in Prostate Cancer. *Front. Cell Dev. Biol.* **2020**, *8*, 599494. [CrossRef] [PubMed]
23. Mandal, G.; Biswas, S.; Anadon, C.M.; Yu, X.; Gatenbee, C.D.; Prabhakaran, S.; Payne, K.K.; Chaurio, R.A.; Martin, A.; Innamarato, P.; et al. IgA-Dominated Humoral Immune Responses Govern Patients' Outcome in Endometrial Cancer. *Cancer Res.* **2022**, *82*, 859–871. [CrossRef]
24. Yoshihara, K.; Shahmoradgoli, M.; Martínez, E.; Vegesna, R.; Kim, H.; Torres-Garcia, W.; Treviño, V.; Shen, H.; Laird, P.W.; Levine, D.A.; et al. Inferring tumour purity and stromal and immune cell admixture from expression data. *Nat. Commun.* **2013**, *4*, 2612. [CrossRef]
25. Lin, H.; Wei, S.; Hurt, E.M.; Green, M.D.; Zhao, L.; Vatan, L.; Szeliga, W.; Herbst, R.; Harms, P.W.; Fecher, L.A.; et al. Host expression of PD-L1 determines efficacy of PD-L1 pathway blockade-mediated tumor regression. *J. Clin. Investig.* **2018**, *128*, 805–815. [CrossRef]
26. Marinelli, O.; Annibali, D.; Morelli, M.B.; Zeppa, L.; Tuyaerts, S.; Aguzzi, C.; Amantini, C.; Maggi, F.; Ferretti, B.; Santoni, G.; et al. Biological Function of PD-L2 and Correlation With Overall Survival in Type II Endometrial Cancer. *Front. Oncol.* **2020**, *10*, 538064. [CrossRef] [PubMed]
27. Wozniak, A.J.; Glisson, B.S.; Hande, K.R.; Ross, W.E. Inhibition of etoposide-induced DNA damage and cytotoxicity in L1210 cells by dehydrogenase inhibitors and other agents. *Cancer Res.* **1984**, *44*, 626–632.
28. Rohwer, N.; Dame, C.; Haugstetter, A.; Wiedenmann, B.; Detjen, K.; Schmitt, C.A.; Cramer, T. Hypoxia-inducible factor 1alpha determines gastric cancer chemosensitivity via modulation of p53 and NF-kappaB. *PLoS ONE* **2010**, *5*, e12038. [CrossRef]
29. Li, H.; Sun, X.; Li, J.; Liu, W.; Pan, G.; Mao, A.; Liu, J.; Zhang, Q.; Rao, L.; Xie, X.; et al. Hypoxia induces docetaxel resistance in triple-negative breast cancer via the HIF-1α/miR-494/Survivin signaling pathway. *Neoplasia* **2022**, *32*, 100821. [CrossRef]
30. Deschoemaeker, S.; Di Conza, G.; Lilla, S.; Martín-Pérez, R.; Mennerich, D.; Boon, L.; Hendrikx, S.; Maddocks, O.D.K.; Marx, C.; Radhakrishnan, P.; et al. PHD1 regulates p53-mediated colorectal cancer chemoresistance. *EMBO Mol. Med.* **2015**, *7*, 1350–1365. [CrossRef]
31. Tosatto, A.; Sommaggio, R.; Kummerow, C.; Bentham, R.B.; Blacker, T.S.; Berecz, T.; Duchen, M.R.; Rosato, A.; Bogeski, I.; Szabadkai, G.; et al. The mitochondrial calcium uniporter regulates breast cancer progression via HIF-1α. *EMBO Mol. Med.* **2016**, *8*, 569–585. [CrossRef] [PubMed]
32. van den Heerik, A.S.V.M.; Horeweg, N.; de Boer, S.M.; Bosse, T.; Creutzberg, C.L. Adjuvant therapy for endometrial cancer in the era of molecular classification: Radiotherapy, chemoradiation and novel targets for therapy. *Int. J. Gynecol. Cancer* **2021**, *31*, 594–604. [CrossRef] [PubMed]
33. Mantovani, A.; Sica, A.; Allavena, P.; Garlanda, C.; Locati, M. Tumor-associated macrophages and the related myeloid-derived suppressor cells as a paradigm of the diversity of macrophage activation. *Hum. Immunol.* **2009**, *70*, 325–330. [CrossRef] [PubMed]
34. Ngeow, J.; Stanuch, K.; Mester, J.L.; Barnholtz-Sloan, J.S.; Eng, C. Second malignant neoplasms in patients with Cowden syndrome with underlying germline PTEN mutations. *J. Clin. Oncol.* **2014**, *32*, 1818–1824. [CrossRef] [PubMed]
35. Arend, R.C.; Jones, B.A.; Martinez, A.; Goodfellow, P. Endometrial cancer: Molecular markers and management of advanced stage disease. *Gynecol. Oncol.* **2018**, *150*, 569–580. [CrossRef]
36. Ayhan, A.; Mao, T.-L.; Suryo Rahmanto, Y.; Zeppernick, F.; Ogawa, H.; Wu, R.-C.; Wang, T.-L.; Shih, I.-M. Increased proliferation in atypical hyperplasia/endometrioid intraepithelial neoplasia of the endometrium with concurrent inactivation of ARID1A and PTEN tumour suppressors. *J. Pathol. Clin. Res.* **2015**, *1*, 186–193. [CrossRef]
37. Kim, T.H.; Yoo, J.-Y.; Wang, Z.; Lydon, J.P.; Khatri, S.; Hawkins, S.M.; Leach, R.E.; Fazleabas, A.T.; Young, S.L.; Lessey, B.A.; et al. ARID1A Is Essential for Endometrial Function during Early Pregnancy. *PLoS Genet.* **2015**, *11*, e1005537. [CrossRef]
38. Wang, X.; Khatri, S.; Broaddus, R.; Wang, Z.; Hawkins, S.M. Deletion of Arid1a in Reproductive Tract Mesenchymal Cells Reduces Fertility in Female Mice. *Biol. Reprod.* **2016**, *94*, 93. [CrossRef]

39. Jones, S.; Stransky, N.; McCord, C.L.; Cerami, E.; Lagowski, J.; Kelly, D.; Angiuoli, S.V.; Sausen, M.; Kann, L.; Shukla, M.; et al. Genomic analyses of gynaecologic carcinosarcomas reveal frequent mutations in chromatin remodelling genes. *Nat. Commun.* **2014**, *5*, 5006. [CrossRef]
40. McConechy, M.K.; Hoang, L.N.; Chui, M.H.; Senz, J.; Yang, W.; Rozenberg, N.; Mackenzie, R.; McAlpine, J.N.; Huntsman, D.G.; Clarke, B.A.; et al. In-depth molecular profiling of the biphasic components of uterine carcinosarcomas. *J. Pathol. Clin. Res.* **2015**, *1*, 173–185. [CrossRef]
41. Cherniack, A.D.; Shen, H.; Walter, V.; Stewart, C.; Murray, B.A.; Bowlby, R.; Hu, X.; Ling, S.; Soslow, R.A.; Broaddus, R.R.; et al. Integrated Molecular Characterization of Uterine Carcinosarcoma. *Cancer Cell* **2017**, *31*, 411–423. [CrossRef]
42. Li, A.; Chen, Y.-S.; Ping, X.-L.; Yang, X.; Xiao, W.; Yang, Y.; Sun, H.-Y.; Zhu, Q.; Baidya, P.; Wang, X.; et al. Cytoplasmic mA reader YTHDF3 promotes mRNA translation. *Cell Res.* **2017**, *27*, 444–447. [CrossRef]
43. Luo, J.; Liu, H.; Luan, S.; He, C.; Li, Z. Aberrant Regulation of mRNA m^6A Modification in Cancer Development. *Int. J. Mol. Sci.* **2018**, *19*, 2515. [CrossRef]
44. Mullee, A.; Dimou, N.; Allen, N.; O'Mara, T.; Gunter, M.J.; Murphy, N. Testosterone, sex hormone-binding globulin, insulin-like growth factor-1 and endometrial cancer risk: Observational and Mendelian randomization analyses. *Br. J. Cancer* **2021**, *125*, 1308–1317. [CrossRef] [PubMed]
45. Wan, J.; Liu, H.; Feng, Q.; Liu, J.; Ming, L. HOXB9 promotes endometrial cancer progression by targeting E2F3. *Cell Death Dis.* **2018**, *9*, 509. [CrossRef] [PubMed]
46. Lu, W.; He, F.; Lin, Z.; Liu, S.; Tang, L.; Huang, Y.; Hu, Z. Dysbiosis of the endometrial microbiota and its association with inflammatory cytokines in endometrial cancer. *Int. J. Cancer* **2021**, *148*, 1708–1716. [CrossRef] [PubMed]
47. Sun, R.; Liu, J.; Nie, S.; Li, S.; Yang, J.; Jiang, Y.; Cheng, W. Construction of miRNA-mRNA Regulatory Network and Prognostic Signature in Endometrial Cancer. *Onco Targets Ther.* **2021**, *14*, 2363–2378. [CrossRef] [PubMed]
48. Wang, A.; Guo, H.; Long, Z. Integrative Analysis of Differently Expressed Genes Reveals a 17-Gene Prognosis Signature for Endometrial Carcinoma. *BioMed Res. Int.* **2021**, *2021*, 4804694. [CrossRef] [PubMed]
49. Wang, Z.; Yin, X.; Ma, M.; Ge, H.; Lang, B.; Sun, H.; He, S.; Fu, Y.; Sun, Y.; Yu, X.; et al. IP-10 Promotes Latent HIV Infection in Resting Memory CD4 T Cells LIMK-Cofilin Pathway. *Front. Immunol.* **2021**, *12*, 656663. [CrossRef]
50. Paul, S.; Shilpi; Lal, G. Role of gamma-delta ($\gamma\delta$) T cells in autoimmunity. *J. Leukoc. Biol.* **2015**, *97*, 259–271. [CrossRef]
51. Munn, D.H.; Sharma, M.D.; Johnson, T.S. Treg Destabilization and Reprogramming: Implications for Cancer ImmunotherapyTreg Instability and Reprogramming. *Cancer Res.* **2018**, *78*, 5191–5199. [CrossRef]
52. Ryan, R.; Booth, S.; Price, S. Corticosteroid-use in primary and secondary brain tumour patients: A review. *J. Neuro-Oncol.* **2012**, *106*, 449–459. [CrossRef] [PubMed]
53. Desbois, M.; Udyavar, A.R.; Ryner, L.; Kozlowski, C.; Guan, Y.; Dürrbaum, M.; Lu, S.; Fortin, J.-P.; Koeppen, H.; Ziai, J.; et al. Integrated digital pathology and transcriptome analysis identifies molecular mediators of T-cell exclusion in ovarian cancer. *Nat. Commun.* **2020**, *11*, 5583. [CrossRef] [PubMed]
54. Grünwald, B.T.; Devisme, A.; Andrieux, G.; Vyas, F.; Aliar, K.; McCloskey, C.W.; Macklin, A.; Jang, G.H.; Denroche, R.; Romero, J.M.; et al. Spatially confined sub-tumor microenvironments in pancreatic cancer. *Cell* **2021**, *184*, 5577–5592. [CrossRef] [PubMed]
55. Zhang, B.; Wu, Q.; Li, B.; Wang, D.; Wang, L.; Zhou, Y.L. mA regulator-mediated methylation modification patterns and tumor microenvironment infiltration characterization in gastric cancer. *Mol. Cancer* **2020**, *19*, 53. [CrossRef] [PubMed]
56. Hornburg, M.; Desbois, M.; Lu, S.; Guan, Y.; Lo, A.A.; Kaufman, S.; Elrod, A.; Lotstein, A.; DesRochers, T.M.; Munoz-Rodriguez, J.L.; et al. Single-cell dissection of cellular components and interactions shaping the tumor immune phenotypes in ovarian cancer. *Cancer Cell* **2021**, *39*, 928–944. [CrossRef]
57. Zhu, P.; Shen, L.; Ren, Q.; Zeng, Q.; He, X. Prognostic and Clinicopathological Significance of Hypoxia-Inducible Factor-1α in Endometrial Cancer: A Meta-Analysis. *Front. Oncol.* **2020**, *10*, 587420. [CrossRef]
58. Brooks, R.A.; Fleming, G.F.; Lastra, R.R.; Lee, N.K.; Moroney, J.W.; Son, C.H.; Tatebe, K.; Veneris, J.L. Current recommendations and recent progress in endometrial cancer. *CA Cancer J. Clin.* **2019**, *69*, 258–279. [CrossRef]
59. Gebbia, V.; Testa, A.; Borsellino, N.; Ferrera, P.; Tirrito, M.; Palmeri, S. Cisplatin and vinorelbine in advanced and/or metastatic adenocarcinoma of the endometrium: A new highly active chemotherapeutic regimen. *Ann Oncol* **2001**, *12*, 767–772. [CrossRef]
60. Scribner, D.R.; Puls, L.E.; Gold, M.A. A phase II evaluation of docetaxel and carboplatin followed by tumor volume directed pelvic plus or minus paraaortic irradiation for stage III endometrial cancer. *Gynecol. Oncol.* **2012**, *125*, 388–393. [CrossRef]
61. Dong, D.; Lei, H.; Liu, D.; Bai, H.; Yang, Y.; Tang, B.; Li, K.; Liu, J.; Xu, G.; Xiao, X. POLE and Mismatch Repair Status, Checkpoint Proteins and Tumor-Infiltrating Lymphocytes in Combination, and Tumor Differentiation: Identify Endometrial Cancers for Immunotherapy. *Front. Oncol.* **2021**, *11*, 640018. [CrossRef]
62. Le, D.T.; Uram, J.N.; Wang, H.; Bartlett, B.R.; Kemberling, H.; Eyring, A.D.; Skora, A.D.; Luber, B.S.; Azad, N.S.; Laheru, D.; et al. PD-1 Blockade in Tumors with Mismatch-Repair Deficiency. *N. Engl. J. Med.* **2015**, *372*, 2509–2520. [CrossRef] [PubMed]
63. Schumacher, T.N.; Schreiber, R.D. Neoantigens in cancer immunotherapy. *Science* **2015**, *348*, 69–74. [CrossRef]
64. Chen, D.S.; Irving, B.A.; Hodi, F.S. Molecular pathways: Next-generation immunotherapy–inhibiting programmed death-ligand 1 and programmed death-1. *Clin. Cancer Res.* **2012**, *18*, 6580–6587. [CrossRef] [PubMed]

65. McGranahan, N.; Furness, A.J.S.; Rosenthal, R.; Ramskov, S.; Lyngaa, R.; Saini, S.K.; Jamal-Hanjani, M.; Wilson, G.A.; Birkbak, N.J.; Hiley, C.T.; et al. Clonal neoantigens elicit T cell immunoreactivity and sensitivity to immune checkpoint blockade. *Science* **2016**, *351*, 1463–1469. [CrossRef]
66. Subramanian, A.; Tamayo, P.; Mootha, V.K.; Mukherjee, S.; Ebert, B.L.; Gillette, M.A.; Paulovich, A.; Pomeroy, S.L.; Golub, T.R.; Lander, E.S.; et al. Gene set enrichment analysis: A knowledge-based approach for interpreting genome-wide expression profiles. *Proc. Natl. Acad. Sci. USA* **2005**, *102*, 15545–15550. [CrossRef]
67. Friedman, J.; Hastie, T.; Tibshirani, R. Regularization Paths for Generalized Linear Models via Coordinate Descent. *J. Stat. Softw.* **2010**, *33*, 1–22. [CrossRef]
68. Iasonos, A.; Schrag, D.; Raj, G.V.; Panageas, K.S. How to build and interpret a nomogram for cancer prognosis. *J. Clin. Oncol.* **2008**, *26*, 1364–1370. [CrossRef] [PubMed]
69. Lawrence, M.S.; Stojanov, P.; Polak, P.; Kryukov, G.V.; Cibulskis, K.; Sivachenko, A.; Carter, S.L.; Stewart, C.; Mermel, C.H.; Roberts, S.A.; et al. Mutational heterogeneity in cancer and the search for new cancer-associated genes. *Nature* **2013**, *499*, 214–218. [CrossRef]
70. Newman, A.M.; Liu, C.L.; Green, M.R.; Gentles, A.J.; Feng, W.; Xu, Y.; Hoang, C.D.; Diehn, M.; Alizadeh, A.A. Robust enumeration of cell subsets from tissue expression profiles. *Nat. Methods* **2015**, *12*, 453–457. [CrossRef] [PubMed]
71. Yang, W.; Soares, J.; Greninger, P.; Edelman, E.J.; Lightfoot, H.; Forbes, S.; Bindal, N.; Beare, D.; Smith, J.A.; Thompson, I.R.; et al. Genomics of Drug Sensitivity in Cancer (GDSC): A resource for therapeutic biomarker discovery in cancer cells. *Nucleic Acids Res.* **2013**, *41*, D955–D961. [CrossRef] [PubMed]
72. Wilkerson, M.D.; Hayes, D.N. ConsensusClusterPlus: A class discovery tool with confidence assessments and item tracking. *Bioinformatics* **2010**, *26*, 1572–1573. [CrossRef] [PubMed]

International Journal of *Molecular Sciences*

MDPI

Article

Identification of *NRAS* Diagnostic Biomarkers and Drug Targets for Endometrial Cancer—An Integrated in Silico Approach

Larsen Alessandro [1,2], Kat-Jun Eric Low [1], Aisha Abushelaibi [3], Swee-Hua Erin Lim [3], Wan-Hee Cheng [4], Sook-keng Chang [4], Kok-Song Lai [3], Yap Wai Sum [5,*] and Sathiya Maran [6,*]

1 Department of Biotechnology, Faculty of Applied Sciences, UCSI University, Kuala Lumpur 56000, Malaysia
2 Advanced Membrane Technology Research Centre (AMTEC), School of Chemical and Energy Engineering (FCEE), Universiti Teknologi Malaysia, Johor 81310, Malaysia
3 Health Sciences Division, Abu Dhabi Women's College, Higher Colleges of Technology, Abu Dhabi P.O. Box 25026, United Arab Emirates
4 Faculty of Health and Life Sciences, INTI International University, Persiaran Perdana BBN, Putra Nilai, Nilai 71800, Malaysia
5 He & Ni Academy, Office Tower B, Northpoint Mid Valley City, Kuala Lumpur 59200, Malaysia
6 School of Pharmacy, Monash University Malaysia, Subang Jaya 47500, Malaysia
* Correspondence: yapws@heniacademy.com (Y.W.S.); sathiya.maran@monash.edu (S.M.)

Citation: Alessandro, L.; Low, K.-J.E.; Abushelaibi, A.; Lim, S.-H.E.; Cheng, W.-H.; Chang, S.-k.; Lai, K.-S.; Sum, Y.W.; Maran, S. Identification of *NRAS* Diagnostic Biomarkers and Drug Targets for Endometrial Cancer—An Integrated in Silico Approach. *Int. J. Mol. Sci.* **2022**, *23*, 14285. https://doi.org/10.3390/ijms232214285

Academic Editor: Laura Paleari

Received: 19 August 2022
Accepted: 30 September 2022
Published: 18 November 2022

Abstract: The diagnosis of endometrial cancer involves sequential, invasive tests to assess the thickness of the endometrium by a transvaginal ultrasound scan. In 6–33% of cases, endometrial biopsy results in inadequate tissue for a conclusive pathological diagnosis and 6% of postmenopausal women with non-diagnostic specimens are later discovered to have severe endometrial lesions. Thus, identifying diagnostic biomarkers could offer a non-invasive diagnosis for community or home-based triage of symptomatic or asymptomatic women. Herein, this study identified high-risk pathogenic nsSNPs in the NRAS gene. The nsSNPs of NRAS were retrieved from the NCBI database. PROVEAN, SIFT, PolyPhen-2, SNPs&GO, PhD-SNP and PANTHER were used to predict the pathogenicity of the nsSNPs. Eleven nsSNPs were identified as "damaging", and further stability analysis using I-Mutant 2.0 and MutPred 2 indicated eight nsSNPs to cause decreased stability (DDG scores < -0.5). Post-translational modification and protein–protein interactions (PPI) analysis showed putative phosphorylation sites. The PPI network indicated a GFR-MAPK signalling pathway with higher node degrees that were further evaluated for drug targets. The P34L, G12C and Y64D showed significantly lower binding affinity towards GTP than wild-type. Furthermore, the Kaplan–Meier bioinformatics analyses indicated that the NRAS gene deregulation affected the overall survival rate of patients with endometrial cancer, leading to prognostic significance. Findings from this could be considered novel diagnostic and therapeutic markers.

Keywords: endometrial carcinoma; diagnostic markers; prognostic gene; protein–ligand interactions; *NRAS* gene

1. Introduction

Endometrial cancer (EC) is the sixth most frequent malignancy among women, and the cases were reported to be 417,000 in 2020 [1]. The incidence has increased by 132% over the past 30 years, attributed to increased risk factors, predominantly obesity and population aging [2]. Women from low- and middle-income countries (LMICs) are more likely to die from endometrial cancer than those from high-income countries (HICs) due to poor access to medical care and a higher proportion of aggressive, non-endometrioid tumors diagnosis [3]. Seventy percent of patients with atypical perimenopausal or postmenopausal vaginal bleeding are diagnosed early, allowing for timely management [4]. However, there are cases diagnosed at an advanced stage, thus decreasing proper treatment. The different stages of EC significantly influence the prognosis; in stage I, the five-year survival rate is approximately 95%, whereas, in stage IV, it is only approximately 14% [5]. Thus, the identification of diagnostic biomarkers for early diagnosis is important, especially in high-risk individuals with extreme obesity, diabetes, hypertension, and Lynch syndrome.

The RAS family is extensively studied in cancer research. RAS consists of KRAS, NRAS and HRAS, and mutations in these genes have been frequently reported in colorectal cancer, pancreatic ductal adenocarcinoma, lung adenocarcinoma, melanoma and some hematological cancers [6].

NRAS gene is a well-known driver oncogene, and mutations in NRAS are reported to cause a poor response to anti-EGFR targeted therapy. Mutations in several pathways have been reported to be responsible for its development. Studies have reported causative roles of mutations in the PI3K pathway, WNT signaling, RAS–RAF pathways, transcriptional regulation, DNA damage response, and FBXW7-related genes [7]. Liu and colleagues (2019) reported that estrogen receptor alpha (ERα) activates the MAPK signaling pathway to promote the development of EC [8]. Recent studies have demonstrated that the RAS/MAPK pathway is the most frequently mutated pathway in patients with cancer, with driver mutations in NRAS or KRAS occurring in 40 to 55% of newly diagnosed patients [9]. RAS and its isoforms are GTPases which will be activated when GTP binds to it, relay signals and activate some downstream pathways responsible for cell growth. Mutations in RAS lock them in an active state. Thus, there will be a continuous unstoppable downstream signaling process resulting in cancer.

Considering the pathological role of NRAS in the RAS/MAPK pathway, its pathogenesis towards EC remains to be elucidated. We hypothesize that understanding the structural and functional effects of NRAS could shed light on discovering the diagnosis and prognosis of EC. Therefore, this study, for the first time, examines the role of nsSNPs of NRAS using bioinformatics tools in understanding its pathogenesis towards EC. This study utilized multi-level functional and structural as described in our previous study (Lim et al., 2021) in predicting the novel biomarker and elucidating the role of nsSNPs of NRAS in the RAS/MAPK pathway.

2. Results

2.1. Prediction of High-Risk Pathogenic and Stability of nsSNPs

A total of 147 nsSNPs of NRAS were extracted from the NCBI database. High-risk pathogenic nsSNPs were predicted using PROVEAN, SIFT, PolyPhen-2, PredictSNP, SNPs&GO, PANTHER and PhD-SNP. Eleven nsSNPs, including rs121913248, rs267606920, rs1465850103, rs121913250, rs121434595, rs121434596, rs1557982817, rs869025573, rs397514553, rs1308441238 and rs752508313, were predicted as high-risk pathogenic by at least six of the tools. The high-risk damaging nsSNPs were then submitted to the MutPred server to confirm the pathogenicity (Table 1). Structural stability analysis by I-Mutant2.0 (Table 2) indicated nsSNPs, including rs267606920, rs121913250, rs121434595, rs1557982817, rs869025573, rs397514553, rs1308441238 and rs752508313, cause a decrease in stability to the resultant proteins, with DDG value < -0.5, indicating its greater impact towards the proteins.

Table 1. High-risk nsSNPs of NRAS predicted using PROVEAN, SIFT, PolyPhen-2, PredictSNP, SNPs&GO, PhD-SNP, PANTHER and MutPred2.

SNP ID	Amino Acid Change	PROVEAN	SIFT	PolyPhen-2	PredictSNP	SNPs&GO		PhD-SNP		PANTHER		MutPred2	
		Score	Score	Score	Score	RI	Prob	RI	Prob	RI	Prob	Score	Pred
rs121913248	A18P	−4.32	0.001	1	1 (C > G), 0.1257 (C > T)	9	0.972	8	0.915	2	0.603	0.93	Pathogenic
rs267606920	G60E	−7.49	0	1	1.0000	9	0.933	7	0.826	8	0.881	0.95	Pathogenic
rs1465850103	D57N	−4.52		0.996	1.0000	8	0.912	6	0.814	0	0.509	0.92	Pathogenic
rs121913250	G12C	−7.09		0.656	1.0000	9	0.937	8	0.881	3	0.644	0.92	Pathogenic

Table 1. *Cont.*

SNP ID	Amino Acid Change	PROVEAN	SIFT	PolyPhen-2	PredictSNP	SNPs&GO		PhD-SNP		PANTHER		MutPred2	
		Score	Score	Score	Score	RI	Prob	RI	Prob	RI	Prob	Score	Pred
rs121434595	G13C	−7.72		0.999	1.0000	9	0.957	9	0.926	7	0.863	0.94	Pathogenic
rs121434596	G13V	−7.65		0.996	1.0000	9	0.963	8	0.924	6	0.78	0.93	Pathogenic
rs1557982817	G60R	−7.49		1	1.0000	8	0.923	7	0.844	8	0.91	0.96	Pathogenic
rs869025573	I24N	−5.32		1	1.0000	9	0.954	4	0.699	3	0.648	0.94	Pathogenic
rs397514553	P34L	−8.56		1	1.0000	8	0.916	3	0.641	8	0.917	0.85	Pathogenic
rs1308441238	V14G	−5.86		1	1.0000	8	0.915	7	0.84	4	0.686	0.93	Pathogenic
rs752508313	Y64D	−8.84		1	0.1578	10	0.975	8	0.91	1	0.527	0.96	Pathogenic

Table 2. Stability prediction of mutated NRAS protein using I-Mutant2.0.

SNP ID	Amino Acid Change	Stability	RI
rs267606920	G60E	Decrease	1
rs121913250	G12C	Decrease	5
rs121434595	G13C	Decrease	5
rs1557982817	G60R	Decrease	7
rs869025573	I24N	Decrease	7
rs397514553	P34L	Decrease	2
rs1308441238	V14G	Decrease	10
rs752508313	Y64D	Decrease	4
rs121913248	A18P	Increase	1
rs1465850103	D57N	Increase	1
rs121434596	G13V	Increase	2

2.2. Identification of Post-Translational Modification (PTM) Sites

Post-translational modifications (PTM) are the process of proteins undergoing chemical modifications to become functional and participate in respective cellular activities. Putative PTM sites in the NRAS proteins and the eight high-risk pathogenic nsSNPs were predicted using BDM-PUB, NetPhos-3.1 and MusiteDeep. BDM-PUB predicted six ubiquitination sites on lysine residues of NRAS protein (K5, K16, K101, K128, K169, K170), while NetPhos-3.1 predicted twelve sites of phosphorylation which occurred on all three possible amino acids namely lysine, threonine and tyrosine (S106, S65, S87, T122, T127, T144, T178, T50, T58, Y157, Y64, Y71). MusiteDeep predicted only one site of each methylation (K5), lipidation (C181), hydroxylation (P185) and acetylation (K104), followed by two sites of glycosylation that happened in asparagine residues of positions 85 and 172. Figure 1 shows the putative PTM sites.

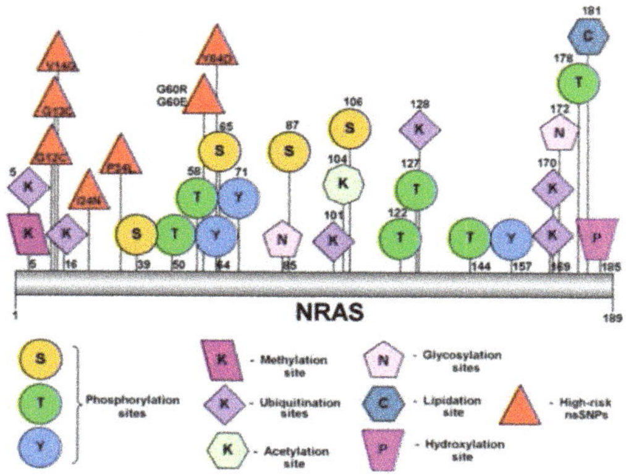

Figure 1. Putative PTM sites of high-risk nsSNPs of NRAS protein.

2.3. Protein–Protein Interaction (PPI) and Molecular Network Analysis

PPI showed that NRAS interacts with RAF1, BRAF, NF1, PTPN11, PIK3CA, HRAS, EGFR, KRAS, MAPK3, and SOS1 proteins (Figure 2a). Molecular networks showed that the NRAS gene is involved in MAPK cascade pathways involving apoptosis, cell growth, cell proliferation and cell cycle (Figure 2b).

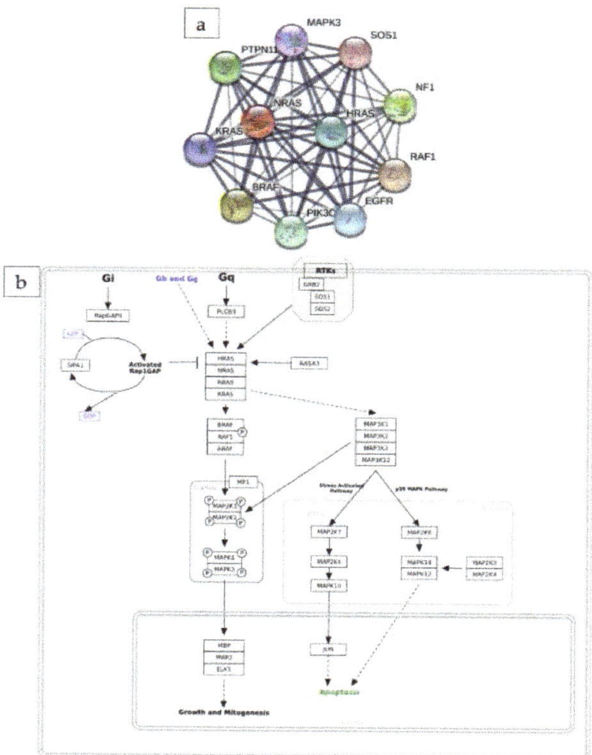

Figure 2. PPI and molecular network analysis. (**a**) The interaction of NRAs with ten partners, and (**b**) the molecular network analysis.

2.4. Prognosis of NRS in EC Malignancy

A Kaplan–Meier plotter was used to determine the prognostic value of the NRAS gene by combining gene expression and EC cancer patient survival. The analysis showed a hazard ratio (HR) = 0 and logrank p-value = 0.028 for EC cancer (Figure 3).

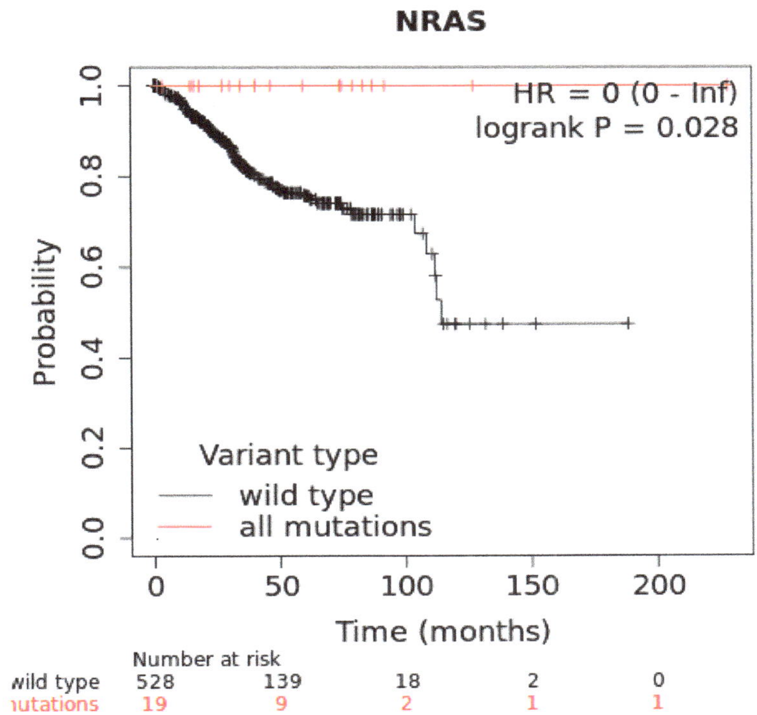

Figure 3. Kaplan–Meier plot showing the correlation between the deregulation of NRAS and overall survival rate EC.

2.5. Prediction of Structural Alteration of NRAS nsSNPs

As hydrophobicity has a significant contribution to protein function and structure, the hydrophobicity of wild-type and mutant residues were analyzed in SWISS-Model to investigate their physicochemical properties. All the predicted high-risk nsSNPs showed a hydrophobic to hydrophilic conversion except for rs397514553 and rs1308441238. Hence, mutant 3D models of nsSNPs, rs267606920, rs121913250, rs121434595, rs1557982817, rs869025573, and rs752508313, localized in the NRAS domain and predicted high-risk pathogenic with structural change from hydrophobic to hydrophilic were modeled using SWISS-Model (Figure 4).

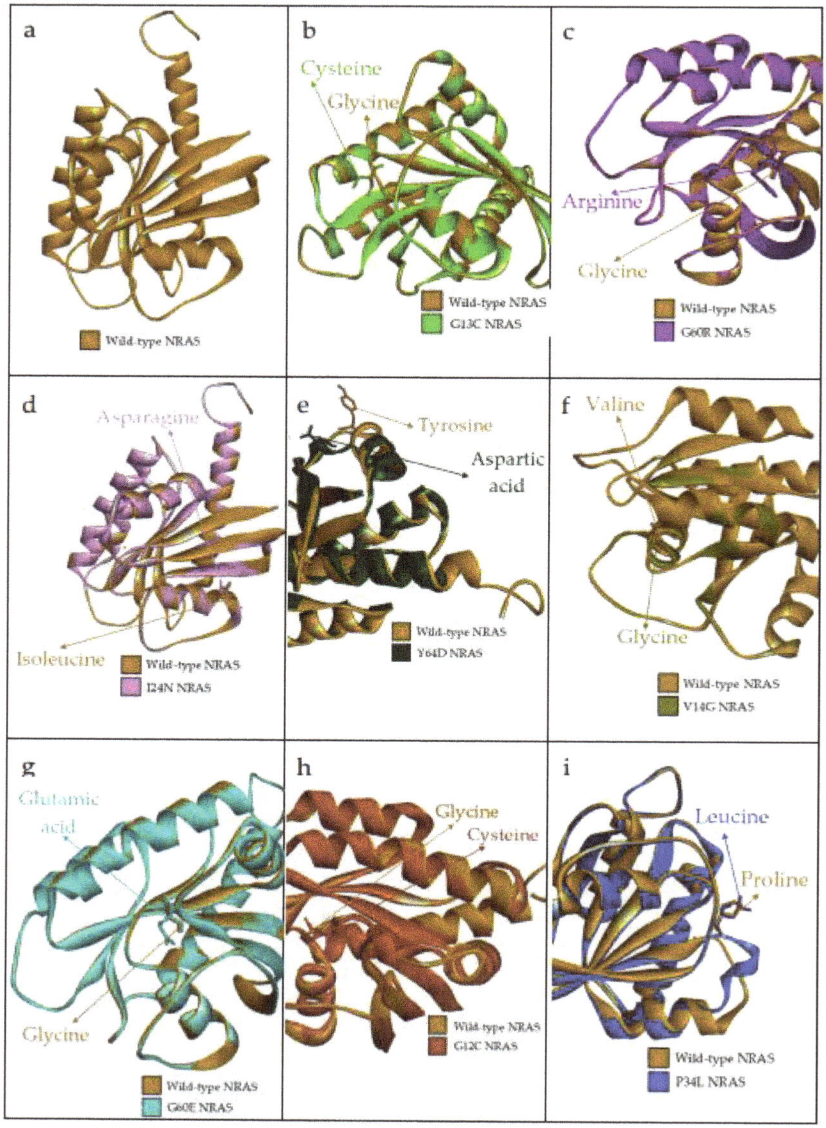

Figure 4. 3D protein structure of superimposed wild-type NRAS and its mutated protein predicted by SWISS-MODEL. (**a**) The 3D structure of wild-type NRAS protein; (**b**) superimposed structure of wild-type NRAS protein with mutated I24N NRAS protein; (**c**) superimposed structure of wild-type NRAS protein with mutated G60E NRAS protein; (**d**) superimposed structure of wild-type NRAS protein with mutated G13C NRAS protein; (**e**) superimposed structure of wild-type NRAS protein with mutated Y64D NRAS protein; (**f**) superimposed structure of wild-type NRAS protein with mutated V14G NRAS protein; (**g**) superimposed structure of wild-type NRAS protein with mutated G60R NRAS protein; (**h**) superimposed structure of wild-type NRAS protein with mutated G12C NRAS protein; (**i**) superimposed structure of wild-type NRAS protein with mutated P34L NRAS protein.

2.6. NRAS Protein Docking

NRAS proteins were individually docked with GTP using the AutoDock Vina (ver 1.1.2). Protein docking was carried out to the changes and compare the binding affinity of the GTP to wild-type NRAS protein and the mutated NRAS protein. The results were generated in kcal/mol with negative numbers indicating greater binding affinity. Binding affinity refers to the strength of binding interaction between protein to its ligand; the smaller the value, the lesser energy required by the protein and ligand to bind to each other.

Table 3 shows the binding affinity between both wild-type and mutated NRAS and GTP as ligands. The rs121913250 (G12C), rs397514553 (P34L) and rs752508313 (Y64D) showed negative values of binding affinity; however, the degree of spontaneousness is not as high as the wild-type NRAS. Hydrogen bonds have the most influence in stabilizing protein and its ligand binding and predominantly contribute to the specificity of molecular recognition. Hydrogen bonds surrounding the wild-type NRAS–GTP complex are Gly-13, Val-14, Gly-15, Lys-16, Ser-17, Thr-35, Thr-58, Gly-60, Asn-116, Asp-119, Ala-146, Lys-147. The G13C mutation made hydrogen bonds of Val-14, Gly-15, Thr-58 and Asp-119 disappear, while NRAS V14G lost hydrogen bonds of Gly-13, Thr-35 (changed into a carbon–hydrogen bond) and Thr-58 but formed new hydrogen bonds in Ala-18 (previously Pi-Alkyl bond), Val-29 and Asp-33. Furthermore, G60R mutation did not change/lose the hydrogen bond in position 60; however, it lost a few hydrogen bonds (Gly-13, Gly-15, Lys-16, Ser-17, Thr-35, Thr-58) and formed a few hydrogen bonds (Val-29, Asp-30, Asp-33). The Van der Waals forces only appeared in NRAS P34L, whereas Gln-25 and His-27 hydrogen bonds were the only bonds involved in binding three NRAS mutated proteins with the lowest binding affinity. Remarkably, these Gln-25 and His-27 were not involved in the wild-type NRAS–GTP complex. Figure 5 is the 2D diagram of the interaction between NRAS proteins and GTP.

Table 3. Binding affinity between NRAS proteins and GTP.

Protein–Ligand Complex	Amino Acid Residues Involved in NRAS–GTP Complex Stabilization	Binding Affinity (kcal/mol)
Wild-Type NRAS–GTP	Gly-13, Val-14, Gly-15, Lys-16, Ser-17, Ala-18, Phe-28, Tyr-32, Pro-34, Thr-35, Thr-58, Gly-60, Asn-116, Lys-117, Asp-119, Ala-146, Lys-147	−10.8
NRAS G13C–GTP	Cys-13, Lys-16, Ser-17, Ala-18, Phe-28, Tyr-32, Thr-35, Gly-60, Asn-116, Lys-117, Ala-146, Lys-147	−10.6
NRAS V14G–GTP	Gly-12, Gly-14, Gly-15, Lys-16, Ser-17, Ala-18, Phe-28, Val-29, Tyr-32, Asp-33, Thr-35, Gly-60, Asn-116, Lys-117, Asp-119, Ala-146, Lys-147	−10.5
NRAS I24N–GTP	Gly-13, Val-14, Gly-15, Lys-16, Ser-17, Ala-18, Phe-28, Val-29, Asp-30, Tyr-32, Thr-35, Thr-58, Gly-60, Asn-116, Lys-117, Asp-119, Ala-146, Lys-147	−10.4
NRAS G60R–GTP	Gly-13, Val-14, Lys-16, Ser-17, Ala-18, Phe-28, Val-29, Asp-30, Tyr-32, Asp-33, Arg-60, Asn-116, Lys-117, Asp-119, Ala-146, Lys-147	−10.3
NRAS G60E–GTP	Gly-13, Val-14, Lys-16, Ser-17, Ala-18, Phe-28, Asp-30, Tyr-32, Thr-35, Asp-57, Thr-58, Glu-60, Lys-117, Leu-120	−9
NRAS P34L–GTP	Gln-25, Asn-26, His-27	−2.9
NRAS G12C–GTP	Gln-25, His-27	−2.8
NRAS Y64D–GTP	Gln-25, His-27	−2.6

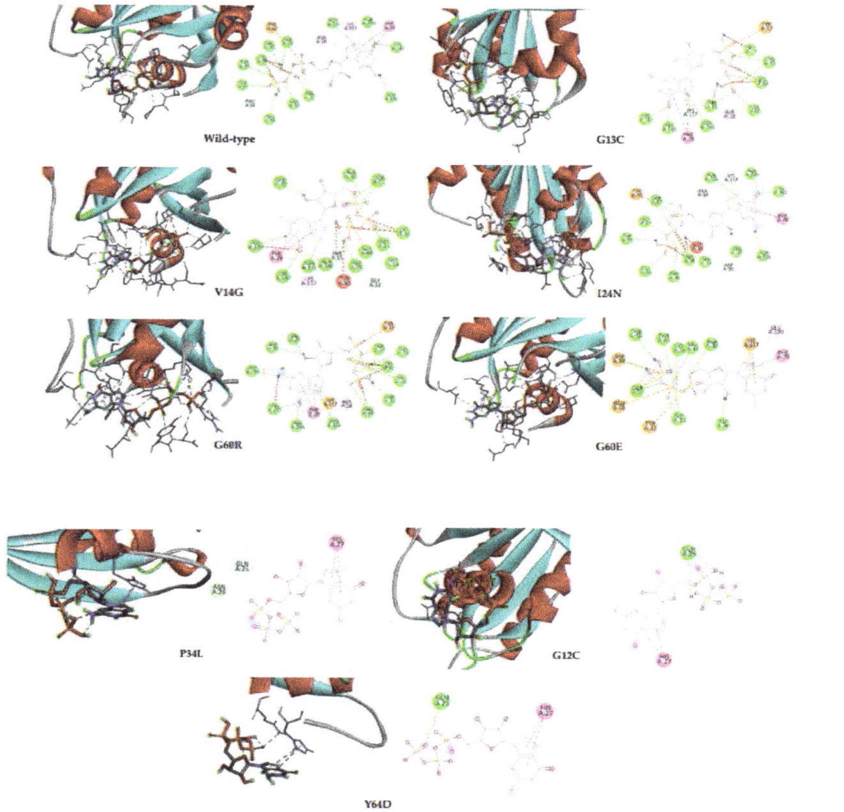

Figure 5. Docking profile of NRAS protein (wild-type and mutated) and GTP. The right panel shows the 2D representation of the interaction with ligands and the receptors in the binding pocket.

3. Discussion

Endometrial cancer (EC) is the most common malignancy affecting women in developed countries, and the incidence rate has increased since 2000 [10]. EC, which occurs mostly in developed countries, is commonly carcinoma instead of sarcoma [7]. EC originates from the uterine epithelium, and epidemiologically, EC is related to diabetes, obesity, late menopause and increasing age. Additionally, Lynch Syndrome, Polymerase Proofreading Associated Polyposis, and Cowden Syndrome increase the risk of EC genetically [7]. The two most common pathways responsible for apoptosis, cell growth, proliferation and differentiation, are highly involved in endometrial cancer: the PI3K/Akt and MAPK pathways [11]. The MAPK pathway involves three main kinases, namely MAP3K, MAP2K and MAPK. They activate the cascade and phosphorylate downstream proteins [12].

The NRAS gene and its family KRAS and HRAS have been explained to be related to different types of cancers [13]. NRAS are the prevalent oncogenes contributing 16–25% among all cancers [14]. The study also reported that mutated NRAS affects melanoma. NRAS is mostly involved in the mitogen-activated protein kinase (MAPK) pathway and the phosphoinositide 3-kinase (PI3K)/protein kinase B (AKT) cascade. These two pathways/cascades are responsible for cell proliferation, survival, differentiation and apoptosis. Therefore, mutations in the NRAS gene would disrupt these pathways and result in uncontrollable cell growth. Unlike Out of 147 nsSNPs extracted from NCBI, 11 nsSNPs; rs121913248 (A18P), rs267606920 (G60E), rs1465850103 (D57N), rs121913250

(G12C), rs121434595 (G13C), rs121434596 (G13V), rs1557982817 (G60R), rs869025573 (I24N), rs397514553 (P34L), rs1308441238 (V14G), and rs752508313 (Y64D) were predicted as high-risk deleterious. Further stability analysis predicted rs267606920, rs121913250, rs121434595, rs1557982817, rs869025573, rs397514553, rs1308441238 and rs752508313 nsSNPs to be pathogenic and to decrease stability of the proteins. Destabilized protein may cause protein degradation, improper folding/misfolding and eventually cause diseases such as neurodegenerative diseases and genetic disorders [15].

Post-translational modifications analysis was carried out to determine modification of the side chain of amino acids of proteins, which may affect protein structure and functions and disrupt biological processes [16]. BDM-PUB predicted ubiquitination sites of NRAS are in positions 5, 16, 101, 128, 169 and 170. Campbell and Philips (2021) reported that NRAS had a few PTM sites of C118 for nitrosylation, K42 for sumoylation and K5 for ubiquitination. Furthermore, the ubiquitination of NRAS has an adverse effect, reducing MAPK signaling. Twelve phosphorylation sites consisting of T50, T58, T122, T127, T144, T178, Y64, Y71, Y157, S65, S87 and S106 were predicted in NRAS, which is in agreement with a study by Yin and colleagues, who showed that NRAS was activated by phosphorylation on S89 by STK19 [17]. Methylation (K5), glycosylation (N85 and N172), lipidation (C181), hydroxylation (P185) and acetylation (K104) sites of NRAS were also predicted. The lipidation of C181 was also predicted by Ahearn et al. (2012). Palmitoylation (lipidation) is necessary for the NRAS protein trafficking from the endomembrane system to the plasma membrane and is involved in the transformation of NRAS-driven myeloid [18,19].

Protein–protein interactions showed the connections between RAF1, BRAF, HRAS, KRAS, MAPK3, SOS1, EGFR, NF1, PTPN11, and PIK3CA. RAF1 and BRAF are members of the RAF family and strongly interact with NRAS in the MAPK cascade (Figure 2b). The role of RAF proteins in NRAS-driven melanoma was discussed by Dorard and colleagues [20]. RAF/MAPK is a main downstream effector of oncogenic RAS in melanoma. Figure 2b also shows that SOS1 and MAPK3 protein were directly related to the activation of NRAS. It is reported that mutations in SOS1 and MAPK3 could lead to the over-activation of RAS pathway (including NRAS), which results in lung adenocarcinoma and melanoma [21,22]. PI3K/Akt/mTOR and MAPK pathways are constitutively activated by phosphorylation [23]. However, the data regarding biomarkers of the two pathways have not really been discovered.

The Kaplan–Meier plot was used to assess the correlation between the expression of the NRAS gene and survival in different types of tumors. The analysis showed that mutation in the NRAS gene would significantly affect the survival of patients having uterine corpus endometrial carcinoma (p-value = 0.0282). NRAS mutation, specifically G12V, is present as a mutation hotspot in endometrial cancer [24]. However, NRAS mutation generally contributes only 5% to endometrial carcinoma [25]. Although the occurrence is rare, NRAS mutation consistently happens in type I tumors whenever the RAS–MAPK pathway gets activated. Moreover, a genotyping assessment showed that mutations in NRAS exclusively happened in uterine-origin tumors [26]. NRAS mutation (e.g., Q61K) has also been identified in low-grade serous ovarian cancer [27].

Molecular recognition refers to the interaction between macromolecules with small molecules through noncovalent interactions to form a complex. There are two main characteristics: specificity and affinity [28]. Knowing protein–ligand binding affinity is useful in designing new drugs and detecting the effects of mutated protein on the binding of the ligand. Protein docking confirmed that three (P34L, G12C and Y64D) out of eight nsSNPs had significantly lower binding affinity than wild-type NRAS. While one other variant (G60E) had a slightly lower binding affinity than wild-type. The other four had a similar binding affinity. As observed in Table 3, wild-type NRAS had the greatest binding affinity with GTP. Alongside the binding affinity, Figure 5 showed that the 3 mutations with weaker binding relationships also have lesser binding sites with the GTP. As opposed to the other five NRAS proteins with more than 10 binding sites on average, the four proteins have only two binding sites each. Reduced binding affinity may affect or end the protein function [29].

The prognostic factors of EC are directly correlated with its mortality. Despite the stringent guidelines of The European Society for Medical Oncology (ESMO), the European Society of Gynecological Oncology (ESGO), the European Society for Radiotherapy and Oncology (ESTRO), and the European Society of Pathology (ESP) consortiums in managing diagnosis treatment and follow-ups, the mortality remains elevated [30]. Recent studies have highlighted the importance of prognostic indicators, which should be evaluated at the time of diagnosis [31]. This then leads to the importance of an accurate and timely diagnosis.

Considering the above, findings from this study pave the way toward identifying diagnostic biomarkers with potential clinical application. Furthermore, we consign confidence that these biomarkers can be applied as a prognostic factor towards determining the preoperative risk of recurrence and directing surgical treatment.

4. Materials and Methods

4.1. Retrieving SNPs

The nsSNPs of NRAS were obtained from NCBI dbSNP (Gene ID: 4893) database (National Center for Biological Information) (https://www.ncbi.nlm.nih.gov/snp/ (accessed on 24 May 2022)). "NRAS" was submitted as the query and was set limited to an only missense mutation. SNP IDs, wild-type nucleotides and their variations were retrieved. Duplicates were removed before proceeding to the next step. Amino acid sequence of the NRAS gene was also retrieved from UniProt (P01111) (accessed in 24 May 2022). The methodology was conducted as previously described by Lim and colleagues [32] (Figure 6).

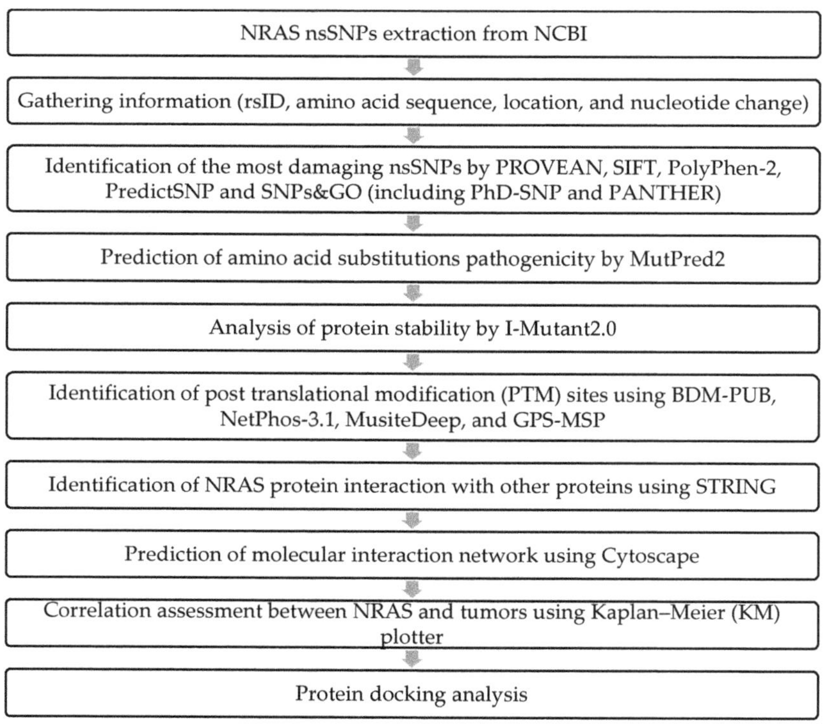

Figure 6. Diagrammatic representation of methodology.

4.2. Identification of High-Risk nsSNPs

A total of seven tools were used to predict the damaging and deleterious nsSNPs of the NRAS gene; PROVEAN (Protein Variation Effect Analyzer) [http://provean.jcvi.org/index.php (accessed on 24 May 2022)] [33], SIFT (Sorting Intolerant From Tolerant) [https://sift.bii.a-star.edu.sg/ (accessed on 24 May 2022)] [34] PolyPhen-2 (Polymorphism Phenotyping v2) [http://genetics.bwh.harvard.edu/pph2/ (accessed on 24 May 2022)] [35], PredictSNP [https://loschmidt.chemi.muni.cz/predictsnp/ (accessed on 21 June 2022)] [36], SNPs&GO [https://snps.biofold.org/snps-and-go/snps-and-go.html (accessed on 29 May 2022)] [37], PhD-SNP (Predictor of human Deleterious Single Nucleotide Polymorphisms) and PANTHER (Protein ANalysis THrough Evolutionary Relationships). The cut-off score of PROVEAN is −2.5, while SIFT is 0.05. A score above those numbers was considered benign. For PolyPhen2, the score varies between 0 and 1, where 0.45 to 0.95 is considered possibly damaging and 0.95 to 1.0 probably damaging. Mutations in PredictSNP are considered deleterious by having a score interval of (0, +1>). By having a score of more than 0.5, the nsSNPs predicted by SNPs&GO, PhD-SNP and PANTHER will be reported as "Disease" with a higher score resulting in a higher reliability index. nsSNPs predicted as "Disease/Deleterious" by at least six tools were picked for further analysis.

4.3. Prediction of Pathogenicity of Amino Acid Substitutions

MutPred2 [http://mutpred.mutdb.org/ (accessed on 30 May 2022)] [38] was used to verify the pathogenicity of the nsSNPs predicted high-risk deleterious. The FASTA format of the protein and the mutation information were submitted. A score above 0.5 was considered pathogenicity (above 0.68 for a 10% false positive rate and above 0.8 for a 5% false positive rate).

4.4. Protein Stability Analysis

The stability of protein was predicted using I-Mutant2.0 [https://folding.biofold.org/i-mutant/i-mutant2.0.html (accessed on 30 May 2022)] [39]. The protein sequence was submitted, and the default setting of 25 °C temperature was at pH 7. Amino acid mutations predicted to be "decreasing" were subjected to further analysis.

4.5. Post Translational Modification (PTM) Sites Identification

The BDM-PUB (Prediction of Ubiquitination sites with Bayesian Discriminant Method) [http://bdmpub.biocuckoo.org/ (accessed on 30 May 2022)] [40], NetPhos-3.1 (phosphorylation of serine, threonine and tyrosine) [https://services.healthtech.dtu.dk/service.php?NetPhos-3.1 (accessed on 30 May 2022)] [41] and MusiteDeep (many types of prediction model—phosphorylation, glycosylation, ubiquitination, SUMOylation, methylation, acetylation, hydroxylation, palmitoylation and cyclization) [https://www.musite.net/ (accessed on 30 May 2022)] [42,43] were used to predict the PTM sites. The FASTA format of wild-type and mutated NRAS protein sequence were submitted.

4.6. Prediction of Protein-Protein Interaction and Molecular Interaction Network

The protein–protein interaction was determined using STRING [https://string-db.org/ (accessed on 1 July 2022)] [44,45]. Cytoscape [https://cytoscape.org/ (accessed on 1 July 2022)] was used to retrieve the prediction of molecular interaction networks [46].

4.7. Prognosis Analysis

A Kaplan–Meier (KM) plot was shown to assess the correlation between NRAS and cancers. The KM plotter was used to determine the prognostic p-value [https://kmplot.com/analysis/ (accessed on 1 July 2022)] [47]. Pan-cancer DNA was chosen, and "NRAS" was submitted as the query. The survival curves and log p-values were obtained. Only cancer with a log p-value below 0.05 was considered as significantly affected by NRAS nsSNPs.

4.8. Molecular Docking Analysis

Molecular docking determined whether the mutation would reduce the binding affinity between the mutated protein and the ligand. The Amino acid sequence of wild-type NRAS and all high-risk pathogenic nsSNPs was submitted to SWISS-MODEL [https://swissmodel.expasy.org/ (accessed on 11 July 2022)] for 3D structure modelling [48]. The predicted 3D structure was modelled using homology templates from the known protein structure having similar sequences with both wild-type and mutated NRAS. Simplified Molecular-Input Line-Entry System (SMILES) sequence of Guanosine Triphosphate (GTP), as the ligand of NRAS, was retrieved from PubChem [https://pubchem.ncbi.nlm.nih.gov/ (accessed on 8 August 2022)] and converted into PDB file using OPENBABEL [http://www.cheminfo.org/Chemistry/Cheminformatics/FormatConverter/index.html (accessed on 11 July 2022)] [49].

GTP was processed in AutoDock 4.2.6 to automatically detect the root and choose rotatable bonds, respectively [50]. AutoDock 4.2.6 was also used to remove water, add hydrogen and add Kollman charges into the protein. Protein docking was carried out with coordinates $x = 18.065000$, $y = -0.343000$ and $z = 7.238000$ [51]. The docking results were viewed using BIOVIA Discovery Studio Visualizer.

5. Conclusions

This study contributed to understanding the causative roles of high-risk pathogenic nsSNPs towards endometrial cancer. NRAS protein is involved in many cell-cycle-regulating pathways, and nsSNPs in NRAS are associated with cancer and poor prognosis of endometrial cancer, thus serving as a novel diagnostic biomarker. In addition, this study also reports structural-based evidence indicating molecular changes due to nsSNPs.

Author Contributions: Conceptualization, S.M. and Y.W.S.; methodology, S.M.; software, L.A. and K.-J.E.L.; formal analysis, L.A.; data curation, L.A.; writing—original draft preparation, L.A. and S.M.; writing—review and editing, A.A., S.-H.E.L. and S.-k.C.; supervision, K.-S.L. and W.-H.C.; funding acquisition, W.-H.C. and K.-S.L. All authors have read and agreed to the published version of the manuscript.

Funding: This research was funded by Interdisciplinary Research Grant from the Higher Colleges of Technology (Interdisciplinary_213346). The APC was funded by INTI International University.

Institutional Review Board Statement: Not applicable.

Informed Consent Statement: Not applicable.

Data Availability Statement: All data are available in the manuscript.

Conflicts of Interest: The authors declare no conflict of interest.

References

1. Sung, H.; Ferlay, J.; Siegel, R.L.; Laversanne, M.; Soerjomataram, I.; Jemal, A.; Bray, F. Global Cancer Statistics 2020: GLOBOCAN Estimates of Incidence and Mortality Worldwide for 36 Cancers in 185 Countries. *CA Cancer J. Clin.* **2021**, *71*, 209–249. [CrossRef] [PubMed]
2. Gu, B.; Shang, X.; Yan, M.; Li, X.; Wang, W.; Wang, Q.; Zhang, C. Variations in incidence and mortality rates of endometrial cancer at the global, regional, and national levels, 1990–2019. *Gynecol. Oncol.* **2021**, *161*, 573–580. [CrossRef] [PubMed]
3. Crosbie, E.J.; Kitson, S.J.; McAlpine, J.N.; Mukhopadhyay, A.; Powell, M.E.; Singh, N. Endometrial cancer. *Lancet* **2022**, *399*, 1412–1428. [CrossRef]
4. Aksel, T.Z.; Çakir, A.T. Tumor Markers in Endometrial Cancer. *Curr. Obstet. Gynecol. Rep.* **2020**, *9*, 15–20. [CrossRef]
5. OFFICE FOR NATIONAL STATISTICS. CANCER SURVIVAL IN ENGLAND: Adult, Stage at Diagnosis and Childhood-Patients Followed Up ... to 2018. DANDY BOOKSELLERS Limited 2019. Available online: https://www.ons.gov.uk/ (accessed on 1 July 2022).
6. Moore, A.R.; Rosenberg, S.C.; McCormick, F.; Malek, S. RAS-targeted therapies: Is the undruggable drugged? *Nat. Rev. Drug Discov.* **2020**, *19*, 533–552. [CrossRef]
7. Urick, M.E.; Bell, D.W. Clinical Actionability of Molecular Targets in Endometrial Cancer. *Nat. Rev. Cancer* **2019**, *19*, 510–521. [CrossRef]
8. Liu, A.; Zhang, D.; Yang, X.; Song, Y. Estrogen receptor alpha activates MAPK signaling pathway to promote the development of endometrial cancer. *J. Cell. Biochem.* **2019**, *120*, 17593–17601. [CrossRef]

9. Han, E.; Kwon, A.; Kim, Y.; Han, K.; Park, S.; Min, C.; Kim, M. Prevalence of ras/mapk pathway mutation (KRAS, NRAS, and BRAF) in plasma cell myeloma at a single institute in Korea. *Cytotherapy* **2019**, *21*, S25–S26. [CrossRef]
10. Zhang, H.-H.; Wang, H.-S.; Qian, X.-W.; Fan, C.-Q.; Li, J.; Miao, H.; Zhu, X.-H.; Yu, Y.; Meng, J.-H.; Cao, P.; et al. Genetic variants and clinical significance of pediatric acute lymphoblastic leukemia. *Ann. Transl. Med.* **2019**, *7*. [CrossRef]
11. Han, J.; Zhang, L.; Guo, H.; Wysham, W.Z.; Roque, D.R.; Willson, A.K.; Sheng, X.; Zhou, C.; Bae-Jump, V.L. Glucose promotes cell proliferation, glucose uptake and invasion in endometrial cancer cells via AMPK/mTOR/S6 and MAPK signaling. *Gynecol. Oncol.* **2015**, *138*, 668–675. [CrossRef]
12. Guo, Y.J.; Pan, W.W.; Liu, S.B.; Shen, Z.F.; Xu, Y.; Hu, L.L. ERK/MAPK signalling pathway and tumorigenesis (Review). *Exp. Ther. Med.* **2020**, *19*, 1997–2007. [CrossRef] [PubMed]
13. Banys-Paluchowski, M.; Milde-Langosch, K.; Fehm, T.; Witzel, I.; Oliveira-Ferrer, L.; Schmalfeldt, B.; Müller, V. Clinical relevance of H-RAS, K-RAS, and N-RAS mRNA expression in primary breast cancer patients. *Breast Cancer Res. Treat.* **2020**, *179*, 403–414. [CrossRef] [PubMed]
14. Randic, T.; Kozar, I.; Margue, C.; Utikal, J.; Kreis, S. NRAS mutant melanoma: Towards better therapies. *Cancer Treat. Rev.* **2021**, *99*, 102238. [CrossRef] [PubMed]
15. Clausen, L.; Abildgaard, A.B.; Gersing, S.K.; Stein, A.; Lindorff-Larsen, K.; Hartmann-Petersen, R. Protein stability and degradation in health and disease. *Adv. Protein Chem. Struct. Biol.* **2019**, *114*, 61–83. [CrossRef] [PubMed]
16. Ramazi, S.; Zahiri, J. Post-translational modifications in proteins: Resources, tools and prediction methods. *Database* **2021**, *2021*, baab012. [CrossRef]
17. Yin, C.; Zhu, B.; Zhang, T.; Liu, T.; Chen, S.; Liu, Y.; Li, X.; Miao, X.; Li, S.; Mi, X.; et al. Pharmacological Targeting of STK19 Inhibits Oncogenic NRAS-Driven Melanomagenesis. *Cell* **2019**, *176*, 1113–1127.e16. [CrossRef] [PubMed]
18. Ahearn, I.M.; Haigis, K.; Bar-Sagi, D.; Philips, M.R. Regulating the Regulator: Post-Translational Modification of Ras. *Nat. Rev. Mol. Cell Biol.* **2012**, *13*, 39–51. [CrossRef] [PubMed]
19. Zambetti, N.A.; Firestone, A.J.; Remsberg, J.R.; Huang, B.J.; Wong, J.C.; Long, A.M.; Predovic, M.; Suciu, R.M.; Inguva, A.; Kogan, S.C.; et al. Genetic disruption of N-RasG12D palmitoylation perturbs hematopoiesis and prevents myeloid transformation in mice. *Blood* **2020**, *135*, 1772–1782. [CrossRef]
20. Dorard, C.; Estrada, C.; Barbotin, C.; Larcher, M.; Garancher, A.; Leloup, J.; Beermann, F.; Baccarini, M.; Pouponnot, C.; Larue, L.; et al. RAF proteins exert both specific and compensatory functions during tumour progression of NRAS-driven melanoma. *Nat. Commun.* **2017**, *8*, 15262. [CrossRef]
21. Orouji, E.; Orouji, A.; Gaiser, T.; Larribère, L.; Gebhardt, C.; Utikal, J. MAP kinase pathway gene copy alterations in NRAS/BRAF wild-type advanced melanoma. *Int. J. Cancer* **2016**, *138*, 2257–2262. [CrossRef]
22. Cai, D.; Choi, P.S.; Gelbard, M.; Meyerson, M. Identification and characterization of oncogenic SOS1 mutations in lung adenocarcinoma. *Mol. Cancer Res.* **2019**, *17*, 1002–1012. [CrossRef] [PubMed]
23. Kourea, H.P.; Nikolaou, M.; Tzelepi, V.; Adonakis, G.; Kardamakis, D.; Tsapanos, V.; Scopa, C.D.; Kalofonos, C.; Decavalas, G. Expression of Phosphorylated Akt, mTOR and MAPK in Type I Endometrial Carcinoma: Clinical Significance. *Anticancer Res.* **2015**, *35*, 2321–2331. [PubMed]
24. Gockley, A.A.; Kolin, D.L.; Awtrey, C.S.; Lindeman, N.I.; Matulonis, U.A.; Konstantinopoulos, P.A. Durable response in a woman with recurrent low-grade endometrioid endometrial cancer and a germline BRCA2 mutation treated with a PARP inhibitor. *Gynecol. Oncol.* **2018**, *150*, 219–226. [CrossRef] [PubMed]
25. van der Putten, L.J.M.; van Hoof, R.; Tops, B.B.J.; Snijders, M.P.L.M.; van den Berg-van Erp, S.H.; van der Wurff, A.A.M.; Bulten, J.; Pijnenborg, J.M.A.; Massuger, L.F.A.G. Molecular profiles of benign and (pre)malignant endometrial lesions. *Carcinogenesis* **2017**, *38*, 329–335. [CrossRef] [PubMed]
26. Growdon, W.B.; Roussel, B.N.; Scialabba, V.L.; Foster, R.; Dias-Santagata, D.; Iafrate, A.J.; Ellisen, L.W.; Tambouret, R.H.; Rueda, B.R.; Borger, D.R. Tissue-specific signatures of activating PIK3CA and RAS mutations in carcinosarcomas of gynecologic origin. *Gynecol. Oncol.* **2011**, *121*, 212–217. [CrossRef]
27. Champer, M.; Miller, D.; Kuo, D.Y.S. Response to trametinib in recurrent low-grade serous ovarian cancer with NRAS mutation: A case report. *Gynecol. Oncol. Rep.* **2019**, *28*, 26–28. [CrossRef]
28. Du, X.; Li, Y.; Xia, Y.L.; Ai, S.M.; Liang, J.; Sang, P.; Ji, X.L.; Liu, S.Q. Insights into Protein–Ligand Interactions: Mechanisms, Models, and Methods. *Int. J. Mol. Sci.* **2016**, *17*, 144. [CrossRef]
29. Li, M.; Petukh, M.; Alexov, E.; Panchenko, A.R. Predicting the impact of missense mutations on protein-protein binding affinity. *J. Chem. Theory Comput.* **2014**, *10*, 1770–1780. [CrossRef]
30. Colombo, E.-E. ESTRO consensus conference on endometrial cancer: Diagnosis, treatment and follow-up. *Ann. Oncol.* **2016**, *27*, 16. [CrossRef]
31. Coll-de la Rubia, E.; Martinez-Garcia, E.; Dittmar, G.; Nazarov, P.V.; Bebia, V.; Cabrera, S.; Gil-Moreno, A.; Colás, E. In silico Approach for Validating and Unveiling New Applications for Prognostic Biomarkers of Endometrial Cancer. *Cancers* **2021**, *13*, 5052. [CrossRef]
32. Lim, E.C.; Lim, S.W.; Tan, K.J.; Sathiya, M.; Cheng, W.H.; Lai, K.S.; Loh, J.Y.; Yap, W.S. In-Silico Analysis of Deleterious SNPs of FGF4 Gene and Their Impacts on Protein Structure, Function and Bladder Cancer Prognosis. *Life* **2022**, *12*, 1018. [CrossRef] [PubMed]

33. Choi, Y.; Chan, A.P. PROVEAN web server: A tool to predict the functional effect of amino acid substitutions and indels. *Bioinformatics* **2015**, *31*, 2745–2747. [CrossRef] [PubMed]
34. Sim, N.-L.; Kumar, P.; Hu, J.; Henikoff, S.; Schneider, G.; Ng, P.C. SIFT web server: Predicting effects of amino acid substitutions on proteins. *Nucleic Acids Res.* **2012**, *40*, W452–W457. [CrossRef] [PubMed]
35. Adzhubei, I.; Jordan, D.M.; Sunyaev, S.R. Predicting functional effect of human missense mutations using PolyPhen-2. *Curr. Protoc. Hum. Genet.* **2013**, *76*, 7–20. [CrossRef] [PubMed]
36. Bendl, J.; Stourac, J.; Salanda, O.; Pavelka, A.; Wieben, E.D.; Zendulka, J.; Brezovsky, J.; Damborsky, J. PredictSNP: Robust and accurate consensus classifier for prediction of disease-related mutations. *PLoS Comput. Biol.* **2014**, *10*, e1003440. [CrossRef] [PubMed]
37. Capriotti, E.; Calabrese, R.; Fariselli, P.; Martelli, P.L.; Altman, R.B.; Casadio, R. WS-SNPs&GO: A web server for predicting the deleterious effect of human protein variants using functional annotation. *BMC Genom.* **2013**, *14*, S6.
38. Pejaver, V.; Urresti, J.; Lugo-Martinez, J.; Pagel, K.A.; Lin, G.N.; Nam, H.-J.; Mort, M.; Cooper, D.N.; Sebat, J.; Iakoucheva, L.M. Inferring the molecular and phenotypic impact of amino acid variants with MutPred2. *Nat. Commun.* **2020**, *11*, 5918. [CrossRef]
39. Capriotti, E.; Fariselli, P.; Casadio, R. I-Mutant2.0: Predicting stability changes upon mutation from the protein sequence or structure. *Nucleic Acids Res.* **2005**, *33*, W306–W310. [CrossRef]
40. Li, H.; Handsaker, B.; Wysoker, A.; Fennell, T.; Ruan, J.; Homer, N.; Marth, G.; Abecasis, G.; Durbin, R. The sequence alignment/map format and SAMtools. *Bioinformatics* **2009**, *25*, 2078–2079. [CrossRef]
41. Blom, N.; Gammeltoft, S.; Brunak, S. Sequence and structure-based prediction of eukaryotic protein phosphorylation sites. *J. Mol. Biol.* **1999**, *294*, 1351–1362. [CrossRef]
42. Yang, Y.; He, Y.; Wang, X.; Liang, Z.; He, G.; Zhang, P.; Zhu, H.; Xu, N.; Liang, S. Protein SUMOylation modification and its associations with disease. *Open Biol.* **2017**, *7*, 170167. [CrossRef] [PubMed]
43. Sheng, Z.; Wang, X.; Ma, Y.; Zhang, D.; Yang, Y.; Zhang, P.; Zhu, H.; Xu, N.; Liang, S. MS-based strategies for identification of protein SUMOylation modification. *Electrophoresis* **2019**, *40*, 2877–2887. [CrossRef] [PubMed]
44. Szklarczyk, D.; Gable, A.L.; Nastou, K.C.; Lyon, D.; Kirsch, R.; Pyysalo, S.; Doncheva, N.T.; Legeay, M.; Fang, T.; Bork, P. The STRING database in 2021: Customizable protein–protein networks, and functional characterization of user-uploaded gene/measurement sets. *Nucleic Acids Res.* **2021**, *49*, D605–D612. [CrossRef] [PubMed]
45. Szklarczyk, D.; Gable, A.L.; Lyon, D.; Junge, A.; Wyder, S.; Huerta-Cepas, J.; Simonovic, M.; Doncheva, N.T.; Morris, J.H.; Bork, P. STRING v11: Protein–protein association networks with increased coverage, supporting functional discovery in genome-wide experimental datasets. *Nucleic Acids Res.* **2019**, *47*, D607–D613. [CrossRef]
46. Shannon, P.; Markiel, A.; Ozier, O.; Baliga, N.S.; Wang, J.T.; Ramage, D.; Amin, N.; Schwikowski, B.; Ideker, T. Cytoscape: A software environment for integrated models of biomolecular interaction networks. *Genome Res.* **2003**, *13*, 2498–2504. [CrossRef]
47. Lánczky, A.; Győrffy, B. Web-based survival analysis tool tailored for medical research (KMplot): Development and implementation. *J. Med. Internet Res.* **2021**, *23*, e27633. [CrossRef]
48. Waterhouse, A.; Bertoni, M.; Bienert, S.; Studer, G.; Tauriello, G.; Gumienny, R.; Heer, F.T.; de Beer, T.A.P.; Rempfer, C.; Bordoli, L. SWISS-MODEL: Homology modelling of protein structures and complexes. *Nucleic Acids Res.* **2018**, *46*, W296–W303. [CrossRef]
49. O'Boyle, N.M.; Banck, M.; James, C.A.; Morley, C.; Vandermeersch, T.; Hutchison, G.R. Open Babel: An open chemical toolbox. *J. Cheminformatics* **2011**, *3*, 33. [CrossRef]
50. Morris, G.M.; Huey, R.; Lindstrom, W.; Sanner, M.F.; Belew, R.K.; Goodsell, D.S.; Olson, A.J. AutoDock4 and AutoDockTools4: Automated docking with selective receptor flexibility. *J. Comput. Chem.* **2009**, *30*, 2785–2791. [CrossRef]
51. Trott, O.; Olson, A.J. AutoDock Vina: Improving the speed and accuracy of docking with a new scoring function, efficient optimization, and multithreading. *J. Comput. Chem.* **2010**, *31*, 455–461. [CrossRef]

International Journal of *Molecular Sciences*

MDPI

Review

TCGA Molecular Prognostic Groups of Endometrial Carcinoma: Current Knowledge and Future Perspectives

Damiano Arciuolo [1,2], Antonio Travaglino [1,3], Antonio Raffone [4], Diego Raimondo [4], Angela Santoro [1], Daniela Russo [3], Silvia Varricchio [3], Paolo Casadio [4], Frediano Inzani [1], Renato Seracchioli [4], Antonio Mollo [5], Massimo Mascolo [3,*] and Gian Franco Zannoni [1,2]

1 Pathology Unit, Department of Woman and Child's Health and Public Health Sciences, Fondazione Policlinico Universitario Agostino Gemelli IRCCS, 00168 Rome, Italy
2 Pathology Institute, Catholic University of Sacred Heart, 00168 Rome, Italy
3 Pathology Unit, Department of Advanced Biomedical Sciences, University of Naples "Federico II", 80138 Naples, Italy
4 Division of Gynaecology and Human Reproduction Physiopathology, Department of Medical and Surgical Sciences (DIMEC), IRCCS Azienda Ospedaliera Univeristaria di Bologna. S. Orsola Hospital, University of Bologna, 40126 Bologna, Italy
5 Gynecology and Obstetrics Unit, Department of Medicine, Surgery and Dentistry "Schola Medica Salernitana", University of Salerno, 84081 Baronissi, Italy
* Correspondence: massimo.mascolo@unina.it

Abstract: The four TCGA-based molecular prognostic groups of endometrial carcinoma (EC), i.e., POLE-mutant, mismatch repair (MMR)-deficient, p53-abnormal, and "no specific molecular profile" (NSMP), have recently been integrated into ESGO-ESTRO-ESP guidelines. The POLE-mutant and MMR-deficient groups are associated with high mutational load, morphological heterogeneity, and inflammatory infiltration. These groups are frequent in high-grade endometrioid, undifferentiated/dedifferentiated, and mixed histotypes. POLE-mutant ECs show good prognosis and do not require adjuvant treatment, although the management of cases at stage >II is still undefined. MMR-deficient ECs show intermediate prognosis and are currently substratified based on clinicopathological variables, some of which might not have prognostic value. These groups may benefit from immunotherapy. P53-mutant ECs are typically high-grade and often morphologically ambiguous, accounting for virtually all serous ECs, most carcinosarcomas and mixed ECs, and half of clear-cell ECs. They show poor prognosis and are treated with chemoradiotherapy; a subset may benefit from HER2 inhibitors or PARP inhibitors. The NSMP group is the most frequent TCGA group; its prognosis is highly variable and affected by clinicopathological/molecular factors, most of which are still under evaluation. In conclusion, the TCGA classification has improved diagnosis, risk stratification, and management of EC. Further studies are needed to resolve the points of uncertainty that still exist.

Keywords: TCGA; endometrial carcinoma; mismatch-repair; p53; molecular; prognosis; treatment; histotype; POLE; microsatellite

Citation: Arciuolo, D.; Travaglino, A.; Raffone, A.; Raimondo, D.; Santoro, A.; Russo, D.; Varricchio, S.; Casadio, P.; Inzani, F.; Seracchioli, R.; et al. TCGA Molecular Prognostic Groups of Endometrial Carcinoma: Current Knowledge and Future Perspectives. *Int. J. Mol. Sci.* **2022**, *23*, 11684. https://doi.org/10.3390/ijms231911684

Academic Editor: Laura Paleari

Received: 19 September 2022
Accepted: 30 September 2022
Published: 2 October 2022

1. Introduction

The Cancer Genome Atlas (TCGA) Research Network has revolutionized our approach to endometrial carcinoma (EC). For decades, the risk stratification of EC has been based on histopathological features, such as tumor grade and histotype, depth of myometrial invasion, and cervical and adnexal involvement. In 2013, an integrated molecular characterization of EC by TCGA showed that EC could be subdivided into four prognostically relevant groups based on mutational burden and somatic copy-number variations [1–4]. Subsequent studies have found that cheaper immunohistochemical and molecular tests can serve as surrogates of the complex and expensive analyses by TCGA [4–8]. The four molecular prognostic groups identified by surrogate tests are POLE-mutated, mismatch repair (MMR)-deficient, p53-abnormal, and "no specific molecular profile" (NSMP) (Table 1). These groups have now been integrated into the European (ESGO-ESTRO-ESP) guidelines for management of EC [9]. However, several issues must still be resolved, such as the significance of the four groups across different histotypes and tumor stages, as well as the value of novel immunohistochemical and molecular prognostic markers [4].

Table 1. Definition of the 4 molecular prognostic groups of endometrial carcinoma.

Molecular Prognostic Group	Original Name	Identified by	Surrogate Marker
POLE-mutated	POLE/ultramutated	High mutational load (232×10^{-6} mutations per megabase)	POLE exonuclease domain mutation
MMR-deficient	MSI/hypermutated	High mutational load (18×10^{-6} mutations per megabase)	Loss of MMR proteins expression
p53-abnormal	Copy number—high/serous	Low mutational load; high copy-number variations	Abnormal p53 expression
NSMP	Copy number—low/endometrioid	Low mutational load; low copy-number variations	Absence of the other markers

In this review, we deal with each molecular prognostic group, discussing their clinicopathological and molecular features, their significance across different histotypes, their possible integration with additional prognostic markers, and their possible predictive value for novel treatments.

2. POLE-Mutated

The POLE-mutated group was the least common molecular group in the TCGA series (7.3% of all ECs) and was labeled "ultramutated group" based on its exceptionally high mutational burden, (232×10^{-6} mutations per megabase). The authors noted that all and only ultramutated ECs showed pathogenetic mutations in the exonuclease domain of Polymerase-ε (POLE), which thus served as a surrogate of the ultramutated status [1] (Table 1). POLE-mutant tumors are characterized by an overwhelmingly favorable prognosis (progression-free survival of 92% to 100% [1,5–8,10]). Compared with other TCGA groups, POLE-mutant ECs show younger age (mean 58.6 years), lower BMI (mean 27.2), and an earlier FIGO stage (stage I in 93.7% of cases) [11] (Table 2).

Table 2. Clinico-pathological features of the 4 molecular prognostic groups of endometrial carcinoma.

Molecular Prognostic Group	Age	BMI	Stage >I	High Grade	Non-Endometrioid Histotype	LVSI	Deep Myometrial Invasion	Lymph Node Involvement
POLE-mutated	58.5 ± 2.7	27.2 ± 0.9	6.3%	39.6%	13.9%	32.7%	27.3%	0%
MMR-deficient	66.5 ± 0.6	30.6 ± 1.2	27.4%	47.4%	14.2%	41.3%	44.5%	9.9%
p53-abnormal	71.1 ± 0.5	29.1 ± 0.5	49.2%	90%	73%	13.8%	48.9%	23.7%
NSMP	64.2 ± 1.9	32.3 ± 1.4	19.5%	15.6%	3.3%	48.8%	27.4%	4.3%

More than 80% of pathogenetic POLE mutations fall into one of five hotspots (P286R, V411L, S297F, A456P, S459F); furthermore, mutations in other sites within the exonuclease domain are rarer and are pathogenetic in about 39% of cases. Mutations outside the exonuclease domain are pathogenetic in only 4% of cases [10]. Unlike the MMR-deficient and p53-abnormal groups, there are no immunohistochemical surrogate markers to identify the POLE-mutated group [4–8].

The TCGA series only included endometrioid and serous ECs, and all POLE-mutant ECs were of the endometroid type. Interestingly, in apparent contrast with their good prognosis, about half of POLE-mutant ECs were high-grade [1]. POLE mutations are indeed significantly more frequent in high-grade endometrioid carcinoma (12.1%) than in low-grade endometrioid carcinoma (6.2%) [12]. Morphological heterogeneity and marked atypia are common features in POLE-mutant ECs, which may also show giant anaplastic cells [13–15]. This has suggested that most POLE-mutant ECs are currently overtreated based on their histological appearance [16].

Subsequent studies have shown that POLE-mutations may be found in several other EC histotypes [13,17–30]. A relatively high frequency of POLE-mutations was found in undifferentiated/dedifferentiated carcinoma (12.4%) [31], while a low frequency was found in clear-cell carcinoma (3.8%) [32] and carcinosarcoma (5.3%) [33] (Table 3).

Table 3. Prevalence of the 4 TCGA molecular prognostic groups across different histotypes of endometrial carcinoma.

Molecular Prognostic Group	LG-EEC	HG-EEC	SC	CCC	Mixed	UDC/ DDC	CS	NEC ***	MLC
POLE-mutated	6.2%	12.1%	0% *	3.8%	5.6%	12.4%	5.3%	7.1%	0%
MMR-deficient	24.7%	39.7%	0% *	9.8%	33.3%	44%	7.3%	42.9%	0%
p53-abnormal	4.7%	21.3%	100% **	42.5%	61.1%	18.6%	73.9%	35.7%	0%
NSMP	63.5%	28%	0% *	40.9%	0%	25%	13.5%	14.3%	100%

LG-EEC: low-grade endometrioid carcinoma; HG-EEC: high-grade endometrioid carcinoma; SC: serous carcinoma; CCC: clear-cell carcinoma; Mixed: mixed carcinoma; UDC/DDC: undifferentiated/dedifferentiated carcinoma; CS: carcinosarcoma; NEC: neuroendocrine carcinoma; MLC: mesonephric-like carcinoma. * Endometrial carcinomas with a serous morphology and POLE mutation or MMR deficiency are diagnosed as serous-like high-grade endometrioid carcinoma. ** Serous carcinomas with normal p53 expression in the presence of TP53 mutation, or with no TP53 mutation but with high copy-number variation, may rarely occur. *** The only published series of endometrial neuroendocrine carcinoma assessed with the TCGA classifier was constituted of 4 pure neuroendocrine carcinomas and 10 mixed carcinomas with a neuroendocrine component [30].

Remarkably, POLE mutation was found in a significant proportion of mixed endometrioid-serous carcinomas arising in young women (16%); furthermore, these tumors are thought to arise as endometrioid ECs that secondarily develop a serous morphology, with or without p53 mutations [34]. Gynecological pathologists agree that POLE-mutant ECs with a serous morphology and immunophenotype should be deemed as endometrioid carcinomas [4,15,34,35]. In these tumors, both the high-grade features and the p53 mutation are consequence of the high mutational load and have no clinical significance. Similarly, the presence of MMR deficiency in POLE-mutant ECs seems to have no prognostic value. Therefore, POLE mutations prognostically supersede both MMR deficiency and p53 mutations [4,10,36]. Even histotype seems to have no value in the case of POLE mutation, since POLE-mutant non-endometrioid ECs still show a good prognosis [9,18,37–39]. The mutational load itself causes a strong immune response due to the exposition of many neoantigens, which is reflected in a lymphocytic infiltration that accompanies most (~79%) POLE-mutant ECs and could be responsible for their good prognosis [13,40]. However, Talhouk et al. found that the immune response was not independently associated with prognosis, suggesting that other factors drive prognosis in POLE-mutant ECs [41].

The POLE-mutated group was found to be the least prognostically affected by clinicopathological factors [42]. This is the reason why the ESGO-ESTRO-ESP guidelines consider POLE-mutant EC up to FIGO stage II as low-risk tumors that do not need adjuvant treatment [9]. This approach has been criticized by some authors due to the lack of prospective studies supporting it. The authors also highlighted that >10% of POLE-mutant ECs present at a FIGO stage >II, and it is unclear how these cases should be treated [43]. Given the prominent lymphocytic infiltrate found in most POLE-mutant ECs, it is reasonable to hypothesize that these tumors may benefit from immunotherapy [41].

3. MMR-Deficient

The MMR-deficient group was first defined by TCGA as "hypermutated group", since it showed a high mutational load (18×10^{-6} mutations per megabase), which was, however, lower than that of the ultramutated/POLE-mutated group. The hypermutated group was the second most common group after the NSMP group, accounting for 28% of EC cases [1]. All hypermutated ECs showed high microsatellite instability, a condition associated with the rapid accumulation of genomic mutations [1,44]. Since microsatellite instability is typically caused by a deficiency in the MMR system, immunohistochemistry for MMR proteins has been used as a surrogate test to identify the hypermutated group (Table 1). The four MMR proteins assessed are MLH1, MSH2, MSH6, and PMS2. These proteins form two heterodimers, which are MLH1-PMS2 and MSH2-MSH6. When MLH1 or MSH2 are lost, there is a consequent loss of PMS2 or MSH6, respectively. On the other hand, a loss of MSH6 or PMS2 expression can occur as an isolated event. For this reason, the immunohistochemical assessment of only MSH6 and PMS2 has been suggested to have the same accuracy as the full MMR panel in identifying MMRd cases [45]. The assessment of MMR immunohistochemical expression may be difficult and affected by fixation issues. A positive internal control in endometrial stroma and lymphocytes should be clearly evaluable. A normal, retained expression of MMR proteins consists of positivity in tumor cell nuclei, which should be stronger than stromal positivity [46].

Histologically, the MMR-deficient group showed several similarities with the POLE-mutated group. In fact, the MMR-deficient group are mostly endometrioid (85.8%), with a high frequency of high-grade cases (47.4%) [16] (Table 2). MMR deficiency is significantly more frequent in high-grade than in low-grade endometrioid carcinoma (39.7% vs. 24.7%) [21] and is particularly frequent in undifferentiated/dedifferentiated carcinoma (44%) [31] and in mixed ECs with an endometrioid component (16–66%) [30,34,47]. On the other hand, the frequency is lower in clear-cell carcinoma (9.8%) [32] and carcinosarcoma (7.3%) [33] (Table 3). As discussed for POLE-mutant ECs, MMR-deficient ECs with a serous morphology and/or immunophenotype are considered to be endometrioid EC [4,15,34,35]. MMRd ECs often show a prominent lymphocytic infiltration and striking morphological heterogeneity [15].

The overall prognosis of MMR-deficient ECs is intermediate. MMR deficiency prognostically supersedes p53 abnormalities but is superseded by POLE mutations [4,10,36]. Compared to POLE-mutant ECs, MMR-deficient ECs seem to be more affected by clinicopathological variables, although not as much as NSMP ECs [42]. The ESGO-ESTRO-ESP guidelines substratify MMR-deficient ECs into different risk groups based on FIGO grade, histotype, depth of myometrial invasion, and LVSI [9]. However, it is unclear whether these factors are all prognostically significant in MMR-deficient ECs [4]. For instance, while deep myometrial invasion and LVSI significantly worsen the prognosis of MMR-deficient ECs, tumor grade seems not to have independent prognostic value [48]. It has been suggested that even histotype has no prognostic value in MMR-deficient ECs [4]. In fact, MMR deficiency seems to be consistently associated with an intermediate prognosis across different histotypes, leading to worsened outcomes in early-stage, low-grade ECs [49], and improved outcomes in non-endometrioid ECs [8,38,50]. According to this view, differences in grade and histotype might be part of the morphological heterogeneity of MMR-deficient ECs, with no impact on prognosis [4]. Undifferentiated/dedifferentiated carcinoma is an exception. About 2/3 of undifferentiated/dedifferentiated carcinoma show the loss of one of three crucial proteins of the SWI/SNF complex, i.e., ARID1B, SMARCA4/BRG1, and SMARCB1/INI1. SWI/SNF-deficient carcinomas have shown an exceedingly bad prognosis, even in the presence of an MMR-deficient signature [37]. Interestingly, it has been suggested that MMR-deficient ECs associated with MLH1 promoter methylation have a poorer prognosis than MMR-deficient EC associated with mutations in the MMR genes [37]. However, it is unclear whether this difference might require a different treatment.

With regard to therapy, MMR-deficient ECs have shown higher susceptibility to radiotherapy than MMR-proficient ECs [51]. Based on their high mutational load and immune infiltrate, MMR-deficient ECs are a candidate for immunotherapy [52].

4. p53-Abnormal

After the exclusion of ultramutated and hypermutated tumors, TCGA subdivided ECs with a low mutational load into two groups based on somatic copy-number variation: "copy number—high" and "copy number—low". The copy number—high group was characterized by a high frequency of TP53 mutation (85%) and serous morphology (73.3%) and was therefore defined as the "serous group" [1]. Such a group represents the prototypical "type II" EC as it is associated with older age, non-endometrioid morphology, an advanced stage, and poor prognosis [1,11,16,42] (Table 2).

Since p53 immunohistochemistry has been used as a cheaper surrogate of TP53 molecular testing, the copy number—high/serous group has subsequently been termed as the p53-abnormal group [4–9] (Table 1). The systematic assessment of p53 immunohistochemical expression in TP53-mutant tumors has shown three possible aberrant patterns of p53: overexpression (strong expression in >70–80% of tumor cell, accounting for 85.6% of cases); complete loss (11.5% of cases); and cytoplasmic expression (1.9% of cases) [53]. An optimized immunohistochemical protocol is crucial to correctly identify these patterns. For example, tumors with low proliferation may show a very focal and faint p53 positivity, which can be misinterpreted as a "complete loss" pattern, especially when a positive internal control is not clearly assessable [54,55]. The presence of a subclonal p53-abnormal pattern is often associated with MMR deficiency or POLE mutation [53]; in the presence of these signatures, p53 abnormalities have no prognostic value [36]. A small subset of TP53-mutant tumors (~5%) do not show abnormalities in p53 expression and thus cannot be identified by immunohistochemistry [54]. In addition, there is a subset of copy number—high ECs that do not show TP53 mutations and can only be classified by a molecular analysis of copy-number variations [1]. Despite not being a perfect surrogate of copy-number analysis, p53 immunohistochemistry has shown sufficient accuracy to be used in the common practice [4–8].

P53-abnormal tumors are typically high-grade and show striking nuclear atypia [15]. In fact, a p53-abnormal signature is by far more common in high-grade than in low-grade endometrioid ECs (21.3% vs. 4.7%) [12]. A recent study suggested that p53-abnormal low-grade endometrioid ECs can be observed in elderly patients [56]. A p53-abnormal signature is virtually present in all serous ECs [4]; furthermore, morphologically serous ECs with POLE-mutant or MMR-deficient signatures are considered as serous-like, high-grade endometrioid ECs, as discussed above. The p53-abnormal group accounts for the vast majority of carcinosarcomas (73.9%), which commonly arise from serous ECs [33], and almost half of clear-cell ECs (42.5%) [32] (Table 3). The p53-abnormal signature is often associated with ambiguous morphology and is frequent in mixed ECs with a serous component and in ambiguous ECs [15,34,35,57].

The biological behavior of p53-abnormal ECs is consistently aggressive across different histotypes [4–8,38,39,42]. ESGO-ESTRO-ESP guidelines include all p53-abnormal ECs in the high-risk group (except for non-myoinvasive cases) [9]. It has been suggested that prognostic differences do exist among p53-abnormal tumors. For instance, serous carcinoma may be more aggressive than p53-abnormal endometroid carcinoma but less aggressive than carcinosarcoma [39,58]. However, these differences have not consistently been reported and it is unclear if they are such as to need different treatments. Moreover, differentiating between serous carcinoma and p53-abnormal endometrioid carcinoma can be difficult, especially considering the morphological ambiguity that often accompanies p53-abnormal ECs. Therefore, it appears appropriate to lump all p53-abnormal ECs together in the same risk group [4,15,59].

Regarding treatment, p53-abnormal ECs always need adjuvant treatment. In all myoinvasive cases, chemoradiotherapy is indicated [9]. HER2 amplification has been identified as a therapeutic target in a subset of p53-abnormal carcinomas, regardless of the histotype [60–63]. Frequent high DNA damage and high PARP-1 expression have also been observed p53-abnormal ECs, suggesting the possibility to use PARP inhibitors [64].

5. NSMP

The remaining TCGA group showed neither high mutational load nor significant copy-number variations. Such group was deemed "copy number—low/endometrioid group" and was considered to represent the prototypical type I EC [1]. Since this group is identified by the absence of the molecular signatures of the other groups, it has been deemed "NSMP" [4,9] (Table 1). The NSMP group is the most frequent TCGA group (~40% of cases) at intermediate prognosis, and is similar to the MMRd group [1,4–8].

While the vast majority (84.4%) of NSMP ECs are low-grade endometrioid tumors [16] (Table 2), the NSMP group can be found in almost any EC histotype, accounting for all mesonephric-like ECs (100%) [65,66], almost half of clear-cell ECs (40.9%) [31], and sizable quantities of neuroendocrine ECs (36%) [30], high-grade endometrioid ECs (28%) [12], undifferentiated/dedifferentiated ECs (25%) [31], and carcinosarcomas (13.5%) [33]. The NSMP group is virtually never found in serous EC; however, serous ECs with a "copy number—high" signature may lack abnormal immunohistochemical expression of p53 and even TP53 mutations, resulting in a NSMP classification. These cases are considered biologically and prognostically analogous to p53-abnormal serous ECs [1,53,54]. Mixed ECs typically are not NSMP, although data in this regard are based on small series [67] (Table 3).

The ESGO-ESTRO-ESP guidelines recommend substratifying the NSMP group based on the same criteria as the MMR-deficient group [9]. However, there is evidence suggesting that the NSMP is more prognostically heterogeneous and more heavily affected by other clinicopathological factors than the MMR-deficient group [4]. In fact, NSMP non-endometrioid ECs showed a bad prognosis similar to that of p53-abnormal ECs [8,38,39,65,66], while NSMP endometrioid ECs showed a highly heterogeneous prognosis, ranging from as good as POLE-mutant ECs to as bad as p53-abnormal ECs [1,8,14,58,68]. Several authors proposed a possible substratification of the NSMP group based on histological, immunohistochemical, and molecular markers, some of which may constitute therapeutic targets [8,64,69,70]. The Leiden group found that NSMP endometrioid ECs could be subdivided into three groups: high-risk, intermediate-risk, and low-risk. High-risk cases showed LVSI and/or overexpression of L1CAM (positivity in $\geq$10% of tumor cells). Intermediate-risk cases lacked LVSI and L1CAM overexpression but showed CTNNB1 exon 3 mutations. Low-risk cases were CTNNB1 wild-type with no LVSI or L1CAM overexpression [8]. Such subclassification of NSMP cases is currently under evaluation in the PORTEC-4a study [69]. Other proposed substratifications are based on the presence of DNA damage biomarkers [64] and the status of several genes such as PTEN, AKT, PI3KCA, PI3KR1, and KRAS [70]. All these findings highlight how the NSMP group is a heterogenous admixture of clinically and molecularly different entities, which require different management.

6. Conclusions

The TCGA classification has offered the possibility to improve the risk stratification and management of EC. Grouping ECs based on molecular signatures may help reduce inter- and intraobserver variability in the assignment of grades and histotypes, especially for morphologically heterogeneous and ambiguous ECs. Moreover, specific therapeutic possibilities can be found in each molecular group. Further research is needed to resolve the issues that still exist, such as the substratification of the NSMP group, the prognostic value of clinicopathological variables in MMR-deficient ECs, and how to treat advanced POLE-mutant ECs.

Author Contributions: Conceptualization, D.A., A.T., A.S. and A.R.; methodology, D.A., A.T., A.S. and A.R., software, D.R. (Diego Raimondo), S.V., D.R. (Daniela Russo) and P.C.; validation, A.M. and R.S.; formal analysis, D.A., A.T., A.S. and A.R.; investigation, A.T., D.A. and A.R.; resources, D.R. (Diego Raimondo) and A.S.; data curation, S.V. and D.R. (Daniela Russo); writing—original draft preparation, A.T., D.A., A.R., A.S., D.R. (Diego Raimondo), D.R. (Daniela Russo) and S.V.; writing—review and editing, P.C., R.S., A.M., F.I., M.M. and G.F.Z.; visualization, F.I. and P.C.; supervision, A.M., R.S., F.I., M.M. and G.F.Z.; project administration, M.M. and G.F.Z. All authors have read and agreed to the published version of the manuscript.

Funding: This research received no external funding.

Institutional Review Board Statement: Not applicable.

Informed Consent Statement: Not applicable.

Data Availability Statement: Not applicable.

Conflicts of Interest: The authors declare no conflict of interest.

References

1. Levine, D.A.; Cancer Genome Atlas Research Network. Integrated genomic characterization of endometrial carcinoma. *Nature* **2013**, *497*, 67–73. [CrossRef] [PubMed]
2. WHO Classification of Tumours Editorial Board. *Female Genital Tumours*, 5th ed.; WHO Classification of Tumours Series; International Agency for Research on Cancer: Lyon, France, 2020; Volume 4.
3. McAlpine, J.; Leon-Castillo, A.; Bosse, T. The rise of a novel classification system for endometrial carcinoma; integration of molecular subclasses. *J. Pathol.* **2018**, *244*, 538–549. [CrossRef] [PubMed]
4. Santoro, A.; Angelico, G.; Travaglino, A.; Inzani, F.; Arciuolo, D.; Valente, M.; D'Alessandris, N.; Scaglione, G.; Fiorentino, V.; Raffone, A.; et al. New Pathological and Clinical Insights in Endometrial Cancer in View of the Updated ESGO/ESTRO/ESP Guidelines. *Cancers* **2021**, *13*, 2623. [CrossRef]
5. Talhouk, A.; McConechy, M.K.; Leung, S.; Li-Chang, H.H.; Kwon, J.S.; Melnyk, N.; Yang, W.; Senz, J.; Boyd, N.; Karnezis, A.N.; et al. A clinically applicable molecular-based classification for endometrial cancers. *Br. J. Cancer* **2015**, *113*, 299–310. [CrossRef] [PubMed]
6. Talhouk, A.; McConechy, M.K.; Leung, S.; Yang, W.; Lum, A.; Senz, J.; Boyd, N.; Pike, J.; Anglesio, M.; Kwon, J.S.; et al. Confirmation of ProMisE: A simple, genomics-based clinical classifier for endometrial cancer. *Cancer* **2017**, *123*, 802–813. [CrossRef]
7. Kommoss, S.; McConechy, M.K.; Kommoss, F.; Leung, S.; Bunz, A.; Magrill, J.; Britton, H.; Kommoss, F.; Grevenkamp, F.; Karnezis, A.; et al. Final validation of the ProMisE molecular classifier for endometrial carcinoma in a large population-based case series. *Ann. Oncol.* **2018**, *29*, 1180–1188. [CrossRef]
8. Stelloo, E.; Nout, R.A.; Osse, E.M.; Jürgenliemk-Schulz, I.J.; Jobsen, J.J.; Lutgens, L.C.; van der Steen-Banasik, E.M.; Nijman, H.W.; Putter, H.; Bosse, T.; et al. Improved Risk Assessment by Integrating Molecular and Clinicopathological Factors in Early-stage Endometrial Cancer-Combined Analysis of the PORTEC Cohorts. *Clin. Cancer Res.* **2016**, *22*, 4215–4224. [CrossRef]
9. Concin, N.; Matias-Guiu, X.; Vergote, I.; Cibula, D.; Mirza, M.R.; Marnitz, S.; Ledermann, J.; Bosse, T.; Chargari, C.; Fagotti, A.; et al. ESGO/ESTRO/ESP Guidelines for the management of patients with endometrial carcinoma. *Virchows Arch.* **2021**, *478*, 153–190. [CrossRef]
10. León-Castillo, A.; Britton, H.; McConechy, M.K.; McAlpine, J.N.; Nout, R.; Kommoss, S.; Brucker, S.Y.; Carlson, J.W.; Epstein, E.; Rau, T.T.; et al. Interpretation of somatic *POLE* mutations in endometrial carcinoma. *J. Pathol.* **2019**, *250*, 323–335. [CrossRef]
11. Raffone, A.; Travaglino, A.; Gabrielli, O.; Micheli, M.; Zuccalà, V.; Bitonti, G.; Camastra, C.; Gargiulo, V.; Insabato, L.; Zullo, F. Clinical features of ProMisE groups identify different phenotypes of patients with endometrial cancer. *Arch. Gynecol. Obstet.* **2021**, *303*, 1393–1400. [CrossRef]
12. Travaglino, A.; Raffone, A.; Mollo, A.; Borrelli, G.; Alfano, P.; Zannoni, G.F.; Insabato, L.; Zullo, F. TCGA molecular subgroups and FIGO grade in endometrial endometrioid carcinoma. *Arch. Gynecol. Obstet.* **2020**, *301*, 1117–1125. [CrossRef] [PubMed]
13. Bakhsh, S.; Kinloch, M.; Hoang, L.N.; Soslow, R.A.; Köbel, M.; Lee, C.H.; McAlpine, J.N.; McConechy, M.K.; Gilks, C.B. Histopathological features of endometrial carcinomas associated with POLE mutations: Implications for decisions about adjuvant therapy. *Histopathology* **2016**, *68*, 916–924. [CrossRef]
14. Joehlin-Price, A.; Van Ziffle, J.; Hills, N.K.; Ladwig, N.; Rabban, J.T.; Garg, K. Molecularly Classified Uterine FIGO Grade 3 Endometrioid Carcinomas Show Distinctive Clinical Outcomes But Overlapping Morphologic Features. *Am. J. Surg. Pathol.* **2021**, *45*, 421–429. [CrossRef] [PubMed]
15. Soslow, R.A.; Tornos, C.; Park, K.J.; Malpica, A.; Matias-Guiu, X.; Oliva, E.; Parkash, V.; Carlson, J.; McCluggage, W.G.; Gilks, C.B. Endometrial Carcinoma Diagnosis: Use of FIGO Grading and Genomic Subcategories in Clinical Practice: Recommendations of the International Society of Gynecological Pathologists. *Int. J. Gynecol. Pathol.* **2019**, *38* (Suppl. 1), S64–S74. [CrossRef]

16. Raffone, A.; Travaglino, A.; Mascolo, M.; Carotenuto, C.; Guida, M.; Mollo, A.; Insabato, L.; Zullo, F. Histopathological characterization of ProMisE molecular groups of endometrial cancer. *Gynecol. Oncol.* **2020**, *157*, 252–259. [CrossRef] [PubMed]
17. Hoang, L.N.; McConechy, M.K.; Meng, B.; McIntyre, J.B.; Ewanowich, C.; Gilks, C.B.; Huntsman, D.G.; Köbel, M.; Lee, C.-H. Targeted mutation analysis of endometrial clear cell carcinoma. *Histopathology* **2014**, *66*, 664–674. [CrossRef]
18. Stelloo, E.; Bosse, T.; Nout, R.A.; Mackay, H.J.; Church, D.N.; Nijman, H.W.; Leary, A.; Edmondson, R.J.; Powell, M.E.; Crosbie, E.J.; et al. Refining prognosis and identifying targetable pathways for high-risk endometrial cancer; a TransPORTEC initiative. *Mod. Pathol.* **2015**, *28*, 836–844. [CrossRef]
19. Le Gallo, M.; Rudd, M.L.; Urick, M.E.; Hansen, N.F.; Zhang, S.; NISC Comparative Sequencing Program; Lozy, F.; Sgroi, D.C.; Vidal Bel, A.; Matias-Guiu, X.; et al. Somatic mutation profiles of clear cell endometrial tumors revealed by whole exome and targeted gene sequencing. *Cancer* **2017**, *123*, 3261–3268. [CrossRef]
20. Delair, D.F.; A Burke, K.; Selenica, P.; Lim, R.S.; Scott, S.N.; Middha, S.; Mohanty, A.S.; Cheng, D.T.; Berger, M.F.; A Soslow, R.; et al. The genetic landscape of endometrial clear cell carcinomas. *J. Pathol.* **2017**, *243*, 230–241. [CrossRef]
21. Baniak, N.; Fadare, O.; Köbel, M.; DeCoteau, J.; Parkash, V.; Hecht, J.L.; Hanley, K.Z.; Gwin, K.; Zheng, W.; Quick, C.M.; et al. Targeted molecular and immunohistochemical analyses of endometrial clear cell carcinoma show that POLE mutations and DNA mismatch repair protein deficiencies are uncommon. *Am. J. Surg. Pathol.* **2019**, *43*, 531–537. [CrossRef]
22. Rosa-Rosa, J.M.; Leskelä, S.; Cristóbal-Lana, E.; Santón, A.; López-García, M.; Muñoz, G.; Pérez-Mies, B.; Biscuola, M.; Prat, J.; Esther, O.E.; et al. Molecular genetic heterogeneity in undifferentiated endometrial carcinomas. *Mod. Pathol.* **2016**, *29*, 1390–1398. [CrossRef] [PubMed]
23. Espinosa, I.; Lee, C.H.; D'Angelo, E.; Palacios, J.; Prat, J. Undifferentiated and dedifferentiated endometrial carcinomas with POLE exonuclease domain mutations have a favorable prognosis. *Am. J. Surg. Pathol.* **2017**, *41*, 1121–1128. [CrossRef] [PubMed]
24. Köbel, M.; Hoang, L.N.; Tessier-Cloutier, B.; Meng, B.; Soslow, R.A.; Stewart, C.J.; Lee, C.-H. Undifferentiated Endometrial Carcinomas Show Frequent Loss of Core Switch/Sucrose Nonfermentable Complex Proteins. *Am. J. Surg. Pathol.* **2018**, *42*, 76–83. [CrossRef] [PubMed]
25. Tessier-Cloutier, B.; Coatham, M.; Carey, M.; Nelson, G.S.; Hamilton, S.; Lum, A.; A Soslow, R.; Stewart, C.J.; Postovit, L.M.; Köbel, M.; et al. SWI/SNF—Deficiency defines highly aggressive undifferentiated endometrial carcinoma. *J. Pathol. Clin. Res.* **2020**, *7*, 144–153. [CrossRef]
26. McConechy, M.K.; Hoang, L.N.; Chui, M.H.; Senz, J.; Yang, W.; Rozenberg, N.; Mackenzie, R.; McAlpine, J.N.; Huntsman, D.G.; Clarke, B.A.; et al. In-depth molecular profiling of the biphasic components of uterine carcinosarcomas. *J. Pathol. Clin. Res.* **2015**, *1*, 173–185. [CrossRef]
27. Cherniack, A.D.; Shen, H.; Walter, V.; Stewart, C.; Murray, B.A.; Bowlby, R.; Hu, X.; Ling, S.; Soslow, R.A.; Broaddus, R.R.; et al. Integrated Molecular Characterization of Uterine Carcinosarcoma. *Cancer Cell* **2017**, *31*, 411–423. [CrossRef]
28. Le Gallo, M.; Rudd, M.L.; Urick, M.E.; Hansen, N.F.; National Institutes of Health Intramural Sequencing Center Comparative Sequencing Program; Merino, M.J.; Mutch, D.G.; Goodfellow, P.J.; Mullikin, J.C.; Bell, D.W. The FOXA2 transcription factor is frequently somatically mutated in uterine carcinosarcomas and carcinomas. *Cancer* **2018**, *124*, 65–73. [CrossRef]
29. Gotoh, O.; Sugiyama, Y.; Takazawa, Y.; Kato, K.; Tanaka, N.; Omatsu, K.; Takeshima, N.; Nomura, H.; Hasegawa, K.; Fujiwara, K.; et al. Clinically relevant molecular subtypes and genomic alteration-independent differentiation in gynecologic carcinosarcoma. *Nat. Commun.* **2019**, *10*, 4965. [CrossRef]
30. Howitt, B.E.; Dong, F.; Vivero, M.; Shah, V.; Lindeman, N.; Schoolmeester, J.K.; Baltay, M.; MacConaill, L.; Sholl, L.M.; Nucci, M.R.; et al. Molecular Characterization of Neuroendocrine Carcinomas of the Endometrium: Representation in All 4 TCGA Groups. *Am. J. Surg. Pathol.* **2020**, *44*, 1541–1548. [CrossRef]
31. Travaglino, A.; Raffone, A.; Mascolo, M.; Guida, M.; Insabato, L.; Zannoni, G.F.; Zullo, F. TCGA Molecular Subgroups in Endometrial Undifferentiated/Dedifferentiated Carcinoma. *Pathol. Oncol. Res.* **2019**, *26*, 1411–1416. [CrossRef]
32. Travaglino, A.; Raffone, A.; Mascolo, M.; Guida, M.; Insabato, L.; Zannoni, G.F.; Zullo, F. Clear cell endometrial carcinoma and the TCGA classification. *Histopathology* **2019**, *76*, 336–338. [CrossRef] [PubMed]
33. Travaglino, A.; Raffone, A.; Gencarelli, A.; Mollo, A.; Guida, M.; Insabato, L.; Santoro, A.; Zannoni, G.F.; Zullo, F. TCGA Classification of Endometrial Cancer: The Place of Carcinosarcoma. *Pathol. Oncol. Res.* **2020**, *26*, 2067–2073. [CrossRef] [PubMed]
34. Conlon, N.; Paula, A.D.C.; Ashley, C.W.; Segura, S.; De Brot, L.; da Silva, E.M.; Soslow, R.A.; Weigelt, B.; DeLair, D.F. Endometrial Carcinomas with a "Serous" Component in Young Women Are Enriched for DNA Mismatch Repair Deficiency, Lynch Syndrome, and POLE Exonuclease Domain Mutations. *Am. J. Surg. Pathol.* **2020**, *44*, 641–648. [CrossRef] [PubMed]
35. Murali, R.; Davidson, B.; Fadare, O.; Carlson, J.A.; Crum, C.P.; Gilks, C.B.; Irving, J.A.; Malpica, A.; Matias-Guiu, X.; McCluggage, W.G.; et al. High-grade Endometrial Carcinomas: Morphologic and Immunohistochemical Features, Diagnostic Challenges and Recommendations. *Int. J. Gynecol. Pathol.* **2019**, *38* (Suppl. 1), S40–S63. [CrossRef] [PubMed]
36. León-Castillo, A.; Gilvazquez, E.; Nout, R.; Smit, V.T.; McAlpine, J.N.; McConechy, M.; Kommoss, S.; Brucker, S.Y.; Carlson, J.W.; Epstein, E.; et al. Clinicopathological and molecular characterisation of 'multiple-classifier' endometrial carcinomas. *J. Pathol.* **2020**, *250*, 312–322. [CrossRef] [PubMed]
37. Santoro, A.; Angelico, G.; Travaglino, A.; Raffone, A.; Arciuolo, D.; D'Alessandris, N.; Inzani, F.; Zannoni, G.F. Clinico-pathological significance of TCGA classification and SWI/SNF proteins expression in undifferentiated/dedifferentiated endometrial carcinoma: A possible prognostic risk stratification. *Gynecol. Oncol.* **2021**, *161*, 629–635. [CrossRef] [PubMed]

38. Travaglino, A.; Raffone, A.; Santoro, A.; Raimondo, D.; Angelico, G.; Valente, M.; Arciuolo, D.; Scaglione, G.; D'Alessandris, N.; Casadio, P.; et al. Clear cell endometrial carcinomas with mismatch repair deficiency have a favorable prognosis: A systematic review and meta-analysis. *Gynecol. Oncol.* **2021**, *162*, 804–808. [CrossRef] [PubMed]
39. Travaglino, A.; Raffone, A.; Raimondo, D.; Arciuolo, D.; Angelico, G.; Valente, M.; Scaglione, G.; D'Alessandris, N.; Casadio, P.; Inzani, F.; et al. Prognostic value of the TCGA molecular classification in uterine carcinosarcoma. *Int. J. Gynecol. Obstet.* **2021**, *158*, 13–20. [CrossRef] [PubMed]
40. van Gool, I.C.; Eggink, F.A.; Freeman-Mills, L.; Stelloo, E.; Marchi, E.; de Bruyn, M.; Palles, C.; Nout, R.A.; de Kroon, C.D.; Osse, E.M.; et al. POLE Proofreading Mutations Elicit an Antitumor Immune Response in Endometrial Cancer. *Clin. Cancer Res.* **2015**, *21*, 3347–3355. [CrossRef]
41. Talhouk, A.; Derocher, H.; Schmidt, P.; Leung, S.; Milne, K.; Gilks, C.B.; Anglesio, M.S.; Nelson, B.H.; McAlpine, J.N. Molecular Subtype Not Immune Response Drives Outcomes in Endometrial Carcinoma. *Clin. Cancer Res.* **2019**, *25*, 2537–2548. [CrossRef]
42. Raffone, A.; Travaglino, A.; Mascolo, M.; Carbone, L.; Guida, M.; Insabato, L.; Zullo, F. TCGA molecular groups of endometrial cancer: Pooled data about prognosis. *Gynecol. Oncol.* **2019**, *155*, 374–383. [CrossRef] [PubMed]
43. Amant, F.; McCluggage, W.G.; Werner, H.M.J.; Kasius, J.C.; Krakstad, C.; Pijnenborg, J. Incorporating molecular profiling into endometrial cancer management requires prospective studies. *Int. J. Gynecol. Cancer* **2021**, *31*, 944–945. [CrossRef] [PubMed]
44. Catasus, L.; Matias-Guiu, X.; Machin, P.; Zannoni, G.F.; Scambia, G.; Benedetti-Panici, P.; Prat, J. Frameshift mutations at coding mononucleotide repeat microsatellites in endometrial carcinoma with microsatellite instability. *Cancer* **2000**, *88*, 2290–2297. [CrossRef]
45. Raffone, A.; Travaglino, A.; Cerbone, M.; Gencarelli, A.; Mollo, A.; Insabato, L.; Zullo, F. Diagnostic Accuracy of Immunohistochemistry for Mismatch Repair Proteins as Surrogate of Microsatellite Instability Molecular Testing in Endometrial Cancer. *Pathol. Oncol. Res.* **2020**, *26*, 1417–1427. [CrossRef] [PubMed]
46. Singh, N.; Wong, R.; Tchrakian, N.; Allen, S.G.; Clarke, B.; Gilks, B. Interpretation and Reporting Terminology for Mismatch Repair Protein Immunohistochemistry in Endometrial Cancer. In *BAGP Guidance Document: MMR Immunohistochemistry Interpretation and Terminology*; Version 1.1; 2020. Available online: https://www.thebagp.org/wp-content/uploads/download-manager-files/1593411202wpdm_BAGP%20MMR%20IHC%20Interpretation%20June%202020.pdf (accessed on 18 September 2022).
47. Köbel, M.; Tessier-Cloutier, B.; Leo, J.; Hoang, L.N.; Gilks, C.B.; Soslow, R.A.; Delair, D.; Stewart, C.J.; Lee, C.-H. Frequent Mismatch Repair Protein Deficiency in Mixed Endometrioid and Clear Cell Carcinoma of the Endometrium. *Int. J. Gynecol. Pathol.* **2017**, *36*, 555–561. [CrossRef] [PubMed]
48. Pasanen, A.; Loukovaara, M.; Bützow, R. Clinicopathological significance of deficient DNA mismatch repair and MLH1 promoter methylation in endometrioid endometrial carcinoma. *Mod. Pathol.* **2020**, *33*, 1443–1452. [CrossRef]
49. Moroney, M.R.; Davies, K.D.; Wilberger, A.C.; Sheeder, J.; Post, M.D.; Berning, A.A.; Fisher, C.; Lefkowits, C.; Guntupalli, S.R.; Behbakht, K.; et al. Molecular markers in recurrent stage I, grade 1 endometrioid endometrial cancers. *Gynecol. Oncol.* **2019**, *153*, 517–520. [CrossRef]
50. Segura, S.E.; Nobre, S.P.; Hussein, Y.R.; Abu-Rustum, N.R.; Weigelt, B.; Soslow, R.A.; DeLair, D.F. DNA Mismatch Repair-deficient Endometrial Carcinosarcomas Portend Distinct Clinical, Morphologic, and Molecular Features Compared With Traditional Carcinosarcomas. *Am. J. Surg. Pathol.* **2020**, *44*, 1573–1579. [CrossRef]
51. Reijnen, C.; Küsters-Vandevelde, H.V.; Prinsen, C.F.; Massuger, L.F.; Snijders, M.P.; Kommoss, S.; Brucker, S.Y.; Kwon, J.S.; McAlpine, J.N.; Pijnenborg, J.M. Mismatch repair deficiency as a predictive marker for response to adjuvant radiotherapy in endometrial cancer. *Gynecol. Oncol.* **2019**, *154*, 124–130. [CrossRef]
52. Swift, B.E.; Gien, L.T. Incorporating Molecular Diagnostics into Treatment Paradigms for Endometrial Cancer. *Curr. Treat. Options Oncol.* **2022**, *23*, 1121–1134. [CrossRef]
53. Vermij, L.; Léon-Castillo, A.; Singh, N.; Powell, M.E.; Edmondson, R.J.; Genestie, C.; Khaw, P.; Pyman, J.; McLachlin, C.M.; Ghatage, P.; et al. p53 immunohistochemistry in endometrial cancer: Clinical and molecular correlates in the PORTEC-3 trial. *Mod. Pathol.* **2022**, *35*, 1475–1483. [CrossRef] [PubMed]
54. Köbel, M.; Ronnett, B.M.; Singh, N.; Soslow, R.A.; Gilks, C.B.; McCluggage, W.G. Interpretation of P53 Immunohistochemistry in Endometrial Carcinomas: Toward Increased Reproducibility. *Int. J. Gynecol. Pathol.* **2019**, *38* (Suppl. 1), S123–S131. [CrossRef] [PubMed]
55. Raffone, A.; Travaglino, A.; Cerbone, M.; De Luca, C.; Russo, D.; Di Maio, A.; De Marco, M.; Turco, M.C.; Insabato, L.; Zullo, F. Diagnostic accuracy of p53 immunohistochemistry as surrogate of TP53 sequencing in endometrial cancer. *Pathol.—Res. Pract.* **2020**, *216*, 153025. [CrossRef] [PubMed]
56. Hachisuga, K.; Ohishi, Y.; Tomonobe, H.; Yahata, H.; Kato, K.; Oda, Y. Endometrial endometrioid carcinoma, grade 1, is more aggressive in the elderly than in the young. *Histopathology* **2021**, *79*, 708–719. [CrossRef] [PubMed]
57. Espinosa, I.; D'Angelo, E.; Palacios, J.; Prat, J. Mixed and Ambiguous Endometrial Carcinomas: A Heterogenous Group of Tumors With Different Clinicopathologic and Molecular Genetic Features. *Am. J. Surg. Pathol.* **2016**, *40*, 972–981. [CrossRef]
58. Bosse, T.; Nout, R.A.; McAlpine, J.N.; McConechy, M.K.; Britton, H.; Hussein, Y.R.; Gonzalez, C.; Ganesan, R.; Steele, J.C.; Harrison, B.T.; et al. Molecular Classification of Grade 3 Endometrioid Endometrial Cancers Identifies Distinct Prognostic Subgroups. *Am. J. Surg. Pathol.* **2018**, *42*, 561–568. [CrossRef]

59. Brett, M.A.; Atenafu, E.G.; Singh, N.; Ghatage, P.; Clarke, B.A.; Nelson, G.S.; Bernardini, M.Q.; Köbel, M. Equivalent Survival of p53 Mutated Endometrial Endometrioid Carcinoma Grade 3 and Endometrial Serous Carcinoma. *Int. J. Gynecol. Pathol.* **2020**, *40*, 116–123. [CrossRef]
60. Ross, D.S.; Devereaux, K.A.; Jin, C.; Lin, D.Y.; Zhang, Y.; Marra, A.; Makker, V.; Weigelt, B.; Ellenson, L.H.; Chui, M.H. Histopathologic features and molecular genetic landscape of HER2-amplified endometrial carcinomas. *Mod. Pathol.* **2022**, *35*, 962–971. [CrossRef]
61. Buza, N.; Hui, P. Characteristics of HER2 Gene Amplification by Fluorescence In Situ Hybridization in Endometrial Serous Carcinoma. *Arch. Pathol. Lab. Med.* **2022**, *146*. [CrossRef]
62. Jenkins, T.M.; Cantrell, L.A.; Stoler, M.H.; Mills, A.M. HER2 Overexpression and Amplification in Uterine Carcinosarcomas With Serous Morphology. *Am. J. Surg. Pathol.* **2022**, *46*, 435–442. [CrossRef]
63. Cagaanan, A.; Stelter, B.; Vu, N.; Rhode, E.N.; Stewart, T.; Hui, P.; Buza, N.; Al-Niaimi, A.; Flynn, C.; Weisman, P.S.; et al. HER2 Expression in Endometrial Cancers Diagnosed as Clear Cell Carcinoma. *Int. J. Gynecol. Pathol.* **2021**, *41*, 132–141. [CrossRef] [PubMed]
64. Auguste, A.; Genestie, C.; De Bruyn, M.; Adam, J.; Le Formal, A.; Drusch, F.; Pautier, P.; Crosbie, E.J.; Mackay, H.; Kitchener, H.C.; et al. Refinement of high-risk endometrial cancer classification using DNA damage response biomarkers: A TransPORTEC initiative. *Mod. Pathol.* **2018**, *31*, 1851–1861. [CrossRef] [PubMed]
65. Ma, T.; Chai, M.; Shou, H.; Ru, G.; Zhao, M. Mesonephric-Like Adenocarcinoma of Uterine Corpus: A Clinicopathological and Targeted Genomic Profiling Study in a Single Institution. *Front. Oncol.* **2022**, *12*, 911695. [CrossRef]
66. Horn, L.-C.; Höhn, A.K.; Krücken, I.; Stiller, M.; Obeck, U.; Brambs, C.E. Mesonephric-like adenocarcinomas of the uterine corpus: Report of a case series and review of the literature indicating poor prognosis for this subtype of endometrial adenocarcinoma. *J. Cancer Res. Clin. Oncol.* **2020**, *146*, 971–983. [CrossRef]
67. Köbel, M.; Meng, B.; Hoang, L.N.; Almadani, N.; Li, X.; Soslow, R.; Gilks, C.B.; Lee, C.-H. Molecular Analysis of Mixed Endometrial Carcinomas Shows Clonality in Most Cases. *Am. J. Surg. Pathol.* **2016**, *40*, 166–180. [CrossRef] [PubMed]
68. Weisman, P.; Park, K.J.; Xu, J. FIGO Grade 3 Endometrioid Adenocarcinomas With Diffusely Aberrant β-Catenin Expression: An Aggressive Subset Resembling Cutaneous Pilomatrix Carcinomas. *Int. J. Gynecol. Pathol.* **2021**, *41*, 126–131. [CrossRef] [PubMed]
69. Van Den Heerik, A.S.V.; Horeweg, N.; Nout, R.A.; Lutgens, L.C.; Van Der Steen-Banasik, E.M.; Westerveld, G.H.; Van Den Berg, H.A.; Slot, A.; Koppe, F.L.; Kommoss, S.; et al. PORTEC-4a: International randomized trial of molecular profile-based adjuvant treatment for women with high-intermediate risk endometrial cancer. *Int. J. Gynecol. Cancer* **2020**, *30*, 2002–2007. [CrossRef]
70. Momeni-Boroujeni, A.; Nguyen, B.; Vanderbilt, C.M.; Ladanyi, M.; Abu-Rustum, N.R.; Aghajanian, C.; Ellenson, L.H.; Weigelt, B.; Soslow, R.A. Genomic landscape of endometrial carcinomas of no specific molecular profile. *Mod. Pathol.* **2022**, *Epub ahead of print*, 1–10. [CrossRef]

International Journal of
Molecular Sciences

Article

Identification of Candidate Biomarker and Drug Targets for Improving Endometrial Cancer Racial Disparities

Pouya Javadian [1,*], Chao Xu [2], Virginie Sjoelund [2], Lindsay E. Borden [1], Justin Garland [3] and Doris Mangiaracina Benbrook [1,3,*]

1 Section of Gynecologic Oncology, Department of Obstetrics and Gynecology, Stephenson Cancer Center, University of Oklahoma Health Sciences Center, Oklahoma City, OK 73104, USA; lindsay-borden@ouhsc.edu
2 Department of Biostatistics and Epidemiology, Hudson College of Public Health, University of Oklahoma Health Sciences Center, Oklahoma City, OK 73104, USA; chao-xu@ouhsc.edu (C.X.); virginie-sjoelund@ouhsc.edu (V.S.)
3 Department of Pathology, University of Oklahoma Health Sciences Center, Oklahoma City, OK 73104, USA; justin-garland@ouhsc.edu
* Correspondence: pouyajavadian@yahoo.com (P.J.); doris-benbrook@ouhsc.edu (D.M.B.)

Abstract: Racial disparities in incidence and survival exist for many human cancers. Racial disparities are undoubtedly multifactorial and due in part to differences in socioeconomic factors, access to care, and comorbidities. Within the U.S., fundamental causes of health inequalities, including socioeconomic factors, insurance status, access to healthcare and screening and treatment biases, are issues that contribute to cancer disparities. Yet even these epidemiologic differences do not fully account for survival disparities, as for nearly every stage, grade and histologic subtype, survival among Black women is significantly lower than their White counterparts. To address this, we sought to investigate the proteomic profiling molecular features of endometrial cancer in order to detect modifiable and targetable elements of endometrial cancer in different racial groups, which could be essential for treatment planning. The majority of proteins identified to be significantly altered among the racial groups and that can be regulated by existing drugs or investigational agents are enzymes that regulate metabolism and protein synthesis. These drugs have the potential to improve the worse outcomes of endometrial cancer patients based on race.

Keywords: endometrial cancer; racial disparity; molecular profiling; proteomic

Citation: Javadian, P.; Xu, C.; Sjoelund, V.; Borden, L.E.; Garland, J.; Benbrook, D.M. Identification of Candidate Biomarker and Drug Targets for Improving Endometrial Cancer Racial Disparities. *Int. J. Mol. Sci.* **2022**, *23*, 7779. https://doi.org/10.3390/ijms23147779

Academic Editor: Laura Paleari

Received: 31 May 2022
Accepted: 12 July 2022
Published: 14 July 2022

1. Introduction

Endometrial cancer continues to rise in both incidence and mortality, in contrast to the decline in both statistics for most other cancers. Racial disparity is the major factor affecting endometrial cancer patient survival in the United States. Black women with endometrial cancer experience 2-fold higher mortality compared to White patients [1]. This represents one of the largest racial disparities in mortality among cancers [2]. Socioeconomic status, higher incidence of aggressive histology and comorbid conditions are known factors contributing to endometrial cancer racial disparity [3]. As these factors do not account for the entire racial disparity, it is likely that specific molecular defects and their associated pathways in the endometrial cancers also contribute [4].

A variety of cancer types, such as breast, prostate and uterine, vary clinically and pathologically based on the race of the patient. Research efforts are beginning to focus on understanding the molecular mechanisms of these differences [5]. The Cancer Genome Atlas (TCGA) analysis identified molecular subtypes of endometrial cancer associated with significantly different progression-free survival (PFS) rates [6]. These results were generated with data of endometrial cancer specimens primarily from White patients; there were insufficient Black patients to power statistical comparisons between the races. Published global profiling studies that identified mRNA and proteins that could contribute to endometrial cancer racial disparities are limited to two studies comparing endometrial cancer specimens from Black and White patients. One reported on whole transcriptome sequencing and mass spectrometry analysis of endometrial cancer samples from 17 Black and 13 White patients, which identified 89 genes that had consistent mRNA and protein differences between the two racial groups, and also concordant alterations in TCGA data of endometrial cancer patients from 49 Black and 216 White patients [7]. According to the multivariate analysis, PFS was associated with 10 of the mRNA transcripts for White patients, 9 transcripts for Black patients, and 2 transcripts for both White and Black patients. The other study reported higher transcript levels for baculoviral IAP repeat containing 7 (BIRC7), polo-like kinase 1 (PLK1) and multiple cell cycle regulatory proteins in endometrial cancer specimens from Black compared to White patients [8]. The differential molecules identified in both of these studies are involved in pathways that regulate cell cycle and cell death. In these studies, there was the limitation that the patients in the different racial groups were not matched for characteristics that may have affected the aggressiveness of their cancers or their treatment outcomes, such as age, tumor stage, weight and body mass index (BMI), although they did focus on one histology. Other studies that matched endometrial cancer tissue specimens from Black and White patients found no significant differences in gene expression patterns evaluated by microarrays. One of these studies matched the specimens from the two racial groups by stage, grade and histological subtype at various ratios [9]. The other study matched specimens from 25 Black and 25 White women for histology (endometrioid and serous), grade (1A through IVB) and stage (1–3) [10]. Each of these studies described above was limited because it compared only two races, Black and White.

In this retrospective study, we conducted a proteomic analysis comparing Black, White, American Indian and Asian groups of 10–12 endometrial cancers, each matched for age, BMI and histology. This matching was based on known involvement of these factors in endometrial cancer risk and patient outcomes. Age and obesity are well-established risk factors for increased risk and mortality of endometrial cancer [10]. Histology is known to affect patient prognosis: endometrioid is associated with more favorable outcomes, as opposed to serous, clear cell and other non-endometrioid histologies [9]. Multiple proteins were identified to be present at significantly different levels in the tumors of different races and were evaluated with bioinformatics for their potential use as biomarkers and drug targets in strategies to improve endometrial cancer patient outcomes.

2. Results

A total of 46 patients were included in this study (12 African Americans, 12 Whites, 12 Native Americans and 10 Asians). Table 1 summarizes the patient demographics and tumor characteristics. All samples were from patients with grade 1 endometrioid endometrial histology and stage 1. The distribution of these groups did not differ according to age; BMI; smoking status; alcohol or NSAID use; or cardiovascular or autoimmune disease. There were significantly different levels of hypertension and diabetes. The rate of diabetes was 67% in Black patients, which was significantly higher than in Whites (25%) and Asians (16%) (p = 0.011). The hypertension rate was significantly higher among White patients (75%) in contrast to (33%) the Asian group (p = 0.012).

Table 1. Comparison of patient characteristics between groups.

	Black (*n*:12)	White (*n*:12)	American Indian (*n*:12)	Asian (*n*:10)	*p*
Age (yrs)	63.6 ± 13.5	59.2 ± 6.2	58.1 ± 10.5	64.1 ± 12.5	0.45
BMI	39.2 ± 10.3	35.4 ± 7.1	42.2 ± 8.7	38.6 ± 8.4	0.27
Smoking	2	7	2	1	0.29
Hypertension	5	9	7	4	0.012
Alcohol use	1	2	1	0	0.84
NSAID use	1	0	1	0	0.15
Cardiovascular disease	3	3	5	2	0.21
Autoimmune disease	1	0	0	0	0.44
Diabetes	8	3	7	2	0.011

2.1. Proteomic Discovery

Protein extracts from all (n = 46) cases were analyzed by tandem mass tag (TMT) liquid chromatography-tandem mass spectrometry (LC-MS/MS) for protein identification. As Table S1 shows, a total of 1611 proteins were identified by TMT LC-MS/MS in all endometrial samples from all groups. The gene ID listed in Table S2 is used instead of the full name for each gene in all tables of this article. The normalized data after log 2 transformation are shown in a heat map; they were analyzed using ANOVA with the adjustment of the first principal component (Table S3). Figure 1 shows a heat map of the normalized data for the 58 proteins found to be present at significantly different levels among the races.

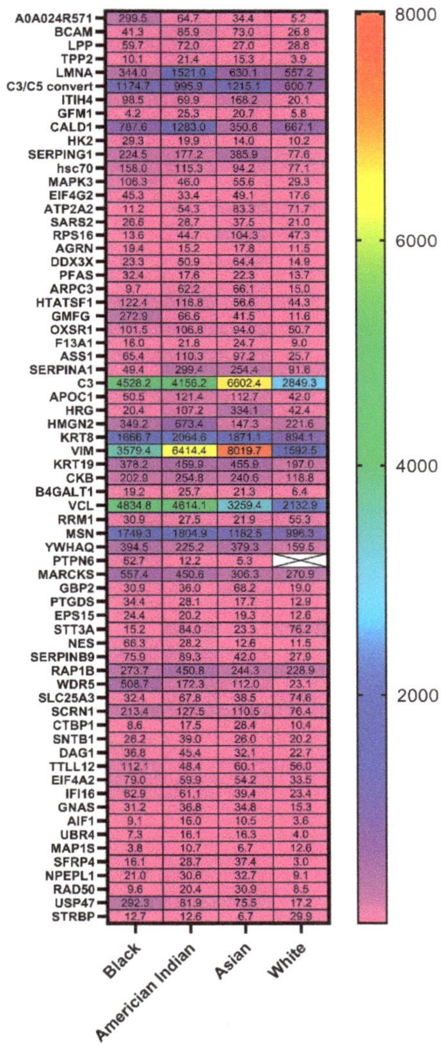

Figure 1. Heat map of log2 normalized levels for each protein found to be present at significantly different levels across the four races.

Some of these 58 proteins had significant differences among all four races, and a few were uniquely differentially present in one specific race compared to the other three (Figure 2 and Tables S3 and S4).

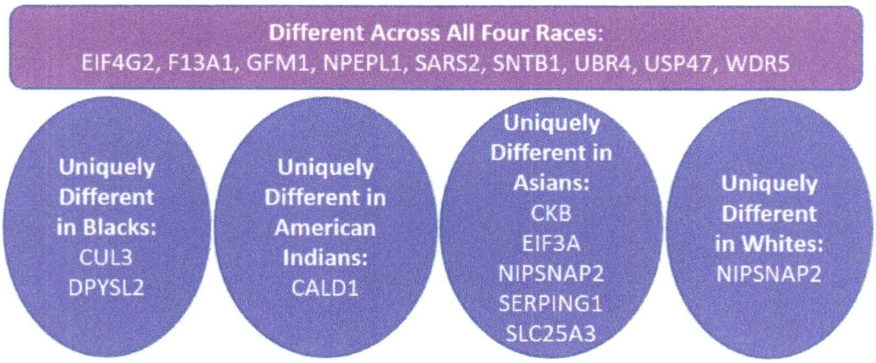

Figure 2. Comparison of protein differences across all racial groups.

Significant differences in individual protein levels in the three other racial groups compared to the White racial group were identified for 44 proteins in the Black racial group, 24 proteins in the American Indian racial group, and 28 proteins in the Asian racial group (Table 2). Some of these protein levels were significantly different from those of the White racial group in all three races, whereas others were unique to individual Black, American Indian, or Asian racial groups.

Table 2. List of gene IDs for proteins significantly different in endometrial cancer specimens from Black, American Indian and Asian racial groups compared to the White racial group.

Race	Gene IDs	
	Higher Concentration	**Lower Concentration**
Black	AIF1, AGRN, ASS1, CUL3, DAG1, DPYSL2, EHD1, EIF4A2, EIF4G2, EPS15, F13A1, GMFG, GFM1, HK2, HTATSF1 IFI16, MAPK3, NES, NPEPL1, OXSR1, PTPN6, PTGDS, PFAS, RAB5B, RAD50, SCRN1, SNX1, SNTB1, SERPINB9, SARS2, TPP2, UBR4, USP47, WDR5, YWHAQ,	ATP2A2, APOC1, MAP1S, RPL4, RPL23, RPS16, SERPINA1, STT3A, ZMYND8
American Indian	AGRN, ASS1, AIF1, CLINT1, CALD1, DAG1, EIF4G2, EPS15, F13A1, GFM1, HK2, HMGN2, KRT19, NPEPL1, SNX1, SNTB1, SARS2, UBR4, USP47, TPP2, WDR5	DX3X, MAP1S, PFAS
Asian	APOC1, CKB, EIF4G2, F13A1, GBP2, GFM1, HMGN2, KRT19, NPEPL1, RAB5B, SNTB1, SERPING1, SARS2, SERPINA1, UBR4, USP47, VIM, WDR5	CLINT1, DDX3X, EIF3E, GBAS, IFI16, PTPN6, OXSR1, RPL5, SLC25A3, STRBP

2.2. Bioinformatic Analysis

The 44 proteins listed in Table 2 were evaluated by Ingenuity Pathway Analysis (IPA) to identify pathways that could be targeted for improving endometrial cancer treatments for Black, American Indian and Asian races. The top three canonical pathways identified by IPA to be the most different in endometrial cancers across racial groups (EIF2 signaling, regulation of eIF4 and p70S6K signaling and mTOR signaling) are all involved in the regulation of protein synthesis (Figure 3). These pathways were most associated with endometrial cancers from White patients; the least association was with endometrial cancers from American Indian patients. Endometrial cancers from the Asian racial group exhibited the most disparate profile of pathways compared to the other groups, due to less association with multiple pathways primarily involved in metabolism. Multiple pathways were upregulated across the races compared to the White racial group, including Coagulation System, Clathrin-mediated Endocytosis, Regulation of EIF4 and p70S, TOR Signaling and IL-12 Signaling and Production. Other pathways were upregulated in the Blacks and the Native Americans, but not in the Asians, compared to the Whites, including nNOS Signaling in Skeletal Muscle, Agrin Interactions at Neuromotor, 5 aminoimidazole Ribonulease, Citrulline-Nitric Oxide Cycle, Arginine Biosynthesis IV and Urea Cycle.

Figure 3. IPA-identified canonical pathways associated with 44 proteins present at significantly different levels in endometrial cancer specimens from Black, African American or Asian compared to the White race. B: Black, AI: American Indian, A: Asian, W: White.

IPA analysis of the fold differences between these proteins in each non-White group compared to the White group documented their associations with endometrial cancer (Figure 4), which provides a certain level of validation to the results of this study.

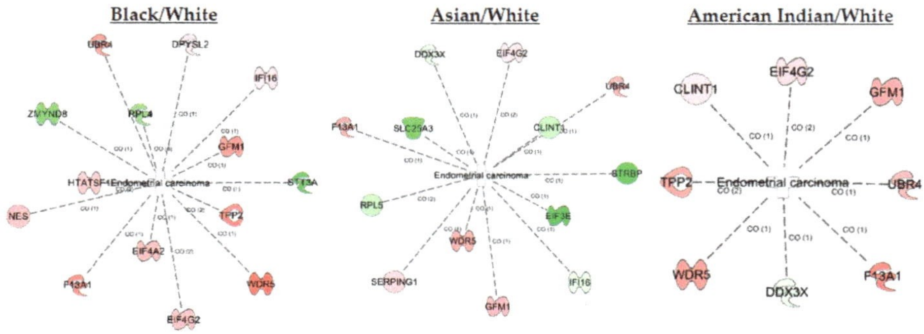

Figure 4. Association of differential proteins with endometrial cancer. These pathways provide a visualization of fold change increase (red) or decrease (green) in the specific racial group compared to the White racial group.

To explore the potential of the 58 identified differential proteins to be used as biomarkers and drug targets in endometrial cancer, we evaluated the differences in their mRNA and protein levels in endometrial cancers compared to healthy tissue, and their relevance to endometrial cancer patient survival using the UALCAN website (http://ualcan.path.uab.edu, accessed on 29 June 2022) [11] to probe TCGA and Clinical Proteomic Tumor Analysis Consortium CPTAC databases [11]. Table 3 shows the list of nine proteins that were found to either have an impact in endometrial cancer survival based on the TCGA database or are already targeted in cancer treatment based on literature reviews. All these proteins are expressed at significantly different levels in endometrial cancers from Black, American Indian or Asian groups compared to the White racial group. We used CPTAC data on the ULACAN website as our baseline racial data in endometrial cancer vs. normal tissue for these nine target proteins. This UALCAN website also provided a breakdown of the patient samples by race; however, the numbers of minority race patients were too low for statistical significance.

Table 3. Targetable differential proteins that have reported survival and prognostic value.

Gene Symbol	Expression in Endometrial Cancer vs. Normal Tissue ¥ [11]	Biomarker-Driven Therapy	Disease or Use	Clinical Trial Phase	Impact on Endometrial Cancer Patient Survival (TCGA) † [11]	Expression in Our Study Cohort
ASS1	Lower ($p < 0.0001$)	Rapamycin/ mTOR inhibtors, or fluorouracil (5-FU) [12]	Multiple cancers	Rapamycin Approved/Phase 2	Worse outcome with higher expression	AI (High) W (Low)
CKB	Lower ($p < 0.0001$)	RGX-202 [13,14]	Gastrointestinal cancer	Phase 1	No	A (High) W (low)
EIF4A2	No significant difference	Zotatifin [15]	Solid tumors	Phase 1–2	Worse outcome with higher expression	B (High) W (Low)
HK2	Higher ($p < 0.0001$)	2-DG and analogs [16]	Prostate cancer, PET imaging	Phase 2	No	B and AI (High) W (Low)
MAPK3	Lower ($p < 0.0001$)	Ulixertinib [17]	Solid tumors	Phase 2	No	B (High) W (Low)
OXSR1	No significant difference	Lutein [18]	Oral cancer	Preclinical	No	B (High) A (Low)
PFAS	Higher ($p = 0.0002$)	Acivicin [19–22]	Liver cancer	Preclinical	No	B (High) AI (Low)
PTPN6	Higher ($p < 0.0001$)	Sodium stibogluconate [23]	Melanoma	Phase 1	Worse outcome with higher expression	B (High) A (Low)
SERPINA1	Lower ($p < 0.0001$)	Trastuzumab [24].	Breast cancer	Phase 2 (Exploratory Biomarker)	No	A and AI (High) B (low)

B: Black, AI: American Indian, A: Asian, W: White—with the highest expression indicated by (High) and the lowest expression indicated by (Low); PET: positiron emission tomography, ¥ based on proteomic expression in the CPTAC dataset [11,25], † survival based on the TCGA dataset [11,25].

3. Discussion

This is the largest study to date examining the molecular profiles of stage I endometrial cancer with matched endometrioid histology, age and BMI across more than two races. Several of the proteins identified in this study have known associations with patient prognosis and survival, which supports their candidacy as biomarkers and drug targets, as described below and illustrated in Figure 5.

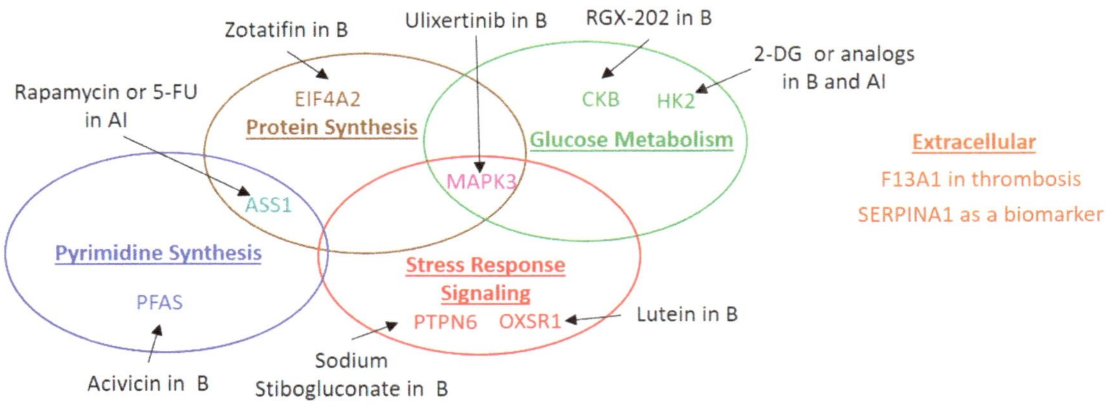

Figure 5. Illustration of functions and available drugs or drug candidates for identified targetable proteins.

Two of the proteins identified in this study, ASS1 (argininosuccinate synthase 1) and PFAS (phosphoribosylformylglycinamidine synthase), regulate pyrimidine biosynthesis, which is essential for cellular growth and homeostasis maintenance. Pyrimidine biosynthesis is elevated when ASS1 is reduced or PFAS increased.

In this study, ASS1 was found to be present at significantly higher levels in endometrial cancers from patients of American Indian race, compared to White ones (mean $\pm$ SD 142.31 $\pm$ 126.13 vs. 53.22 $\pm$ 76.60 relative units (RU), $p = 0.005$). Although there is no difference in ASS1 mRNA expression between endometrial cancer samples and normal tissue in TCGA data, CPTAC data show significantly lower ASS1 expression in the primary tumor compared to normal samples, and another study found decreased ASS1 expression at the invading fronts of endometrial cancer and increased migration and invasion in endometrial cancer cells upon ASS1 knockout [26]. ASS1 is the rate-limiting enzyme in arginine biosynthesis that is deficient in multiple tumor types, resulting in decreased arginine levels [27]. Attempts to kill the tumor cells by further depleting arginine with catabolic enzymes proved unsuccessful due to development of resistance caused by upregulation of ASS1 [28]. Alternatively, an arginine-independent ASS1 effect can be taken advantage of in developing cancer therapeutics [12]. A consequence of ASS1 deficiency is accumulation aspartate, which serves a substrate and positive inducer of CAD (carbamoyl-phosphate synthase 2, aspartate transcarbamylase and dihydroorotase complex). The CAD upregulation as a consequence of ASS1 deficiency increases pyrimidine production. An alternate approach to target ASS1 deficiency that has proven effective in preclinical studies was to inhibit the upregulation of mTOR enzyme that occurs as a consequence of arginine depletion, or directly inhibiting pyrimidine synthesis using 5-fluorouracil (5-FU) [12]. Everolimus, an mTOR inhibitor, combined with letrozole, an aromatase inhibitor, is currently used for treatment of recurrent endometrial cancer based upon a positive phase 2 study [29]. 5-FU is not currently used for endometrial cancer but is widely used for other solid tumors, including first-line therapy for colorectal cancer (https://pubmed.ncbi.nlm.nih.gov/15051767/ accessed on 28 June 2022). 5-FU has been previously studied in combination with other chemotherapies for endometrial cancer but is not part of current practice (NCT00612495) [30]. Based on their significantly lower levels of ASS1, American Indian endometrial cancer patients would be expected to benefit from these approaches to deplete arginine.

PFAS, an enzyme that is directly involved in catalyzing purine salvage and de novo pyrimidine biosynthesis, was elevated in the Black (mean ± SD 152.30 ± 377.24 vs. 42.66 ± 62.94 RU, (p = 0.036)) and White endometrial cancer patients. CPTAC data show significantly higher PFAS protein expression in endometrial cancer compared to normal tissues. In preclinical studies, acivicin, an analog of glutamine, was shown to inhibit PFAS and other enzymes involved in purine salvage [19], resulting in growth inhibition of various cancer cell lines [20,22]. A large-scale screening study of 2000 compounds identified acivicin as a potent inhibitor of *Drosophila* tumor formation; however, RNAi knockdown studies demonstrated that this anti-tumor activity was due to inhibition of CTP synthase, another pyrimidine biosynthesis enzyme [21]. Thus, a significant amount of validation of PFAS as a drug target, and drug discovery and development research, would be needed to target PFAS in endometrial cancer. Before targeting PFAS or purine savage to improve Black endometrial cancer patient outcomes, the upregulation of ASS1 and the ultimate level of pyrimidine biosynthesis in endometrial cancer specimens would need to be determined.

Another enzyme involved in metabolism, CKB (creatine kinase B CKB), was found in this study to have higher expression in endometrial cancers in Asian patients compared to White patients (mean ± SD 755 ± 1015.9 vs. 391.39 ± 906.7 RU, p = 0.034). CKB had lower expression in endometrial cancer samples vs. normal tissue based on a CPTAC database of all races. CKB supports metabolism by using creatine to produce ATP. Preclinical studies demonstrated that inhibiting phosphocreatine import with the small molecule drug, RGX-202, inhibited primary, metastatic and patient derived xenograft (PDX) colorectal tumors in association with reducing phosphocreatine levels [14]. A phase 1 trial of RGX-202 in advanced gastrointestinal cancers documented increased serum and urine creatine levels consistent with the decreased creatine uptake in tumors of preclinical models treated with RGX-202 [14]. This phase 1 trial showed no dose-limiting toxicity in the first 17 patients enrolled, and a durable partial response in the highest dose cohort [13]. Combinations of various metabolic inhibitors also have potential for improving endometrial cancer outcomes based on preclinical studies. For example, RGX-202 demonstrated anti-cancer synergy with F-FU and leflunomide, an inhibitor of dihydroorotate dehydrogenase-induced nucleotide biosynthesis levels [14].

Our results showed HK2 (hexokinase 2) was elevated in endometrial cancer specimens from the Black (mean ± SD 55.31 ± 61.38 vs. 12.29 ± 9.19 RU, p = 0.008) and American Indian (mean ± SD 48.78 ± 64.32 vs. 12.29 ± 9.19 RU, p = 0.002) racial groups; the lowest expression was in the White racial group. Moreover, HK2 has higher expression in endometrial cancer vs. normal tissue based on the CPTAC database of all races. HK2 is the enzyme that catalyzes the conversion of hexoses, such as D-glucose and D-fructose, into hexose 6-phosphates (D-glucose 6-phosphate and D-fructose 6-phosphate, respectively) [31–33]. HK2 phosphorylation of D-glucose to D-glucose 6-phosphate is step the first of glycolysis [33]. Additionally, by preventing the release of apoptogenic molecules from the mitochondrial intermembrane space and subsequent intrinsic apoptosis, HK2 is necessary for maintaining the integrity of the outer mitochondrial membrane [34]. Various preclinical studies validate HK2 as a potential target. The long non-coding RNA (lncRNA) SNHG16 promoted the proliferation of endometrial carcinoma cells and glycolysis by competitively sponging miR-490-3p and upregulating HK2, which is miR-490-3p's target gene [35].

Inhibition of HK2 activity with 2-deoxyglucose (2-DG) blocks the glycolysis pathway [36]. 2-DG has multiple attributes as a candidate therapeutic agent. There is substantial clinical information about 2-DG and its 2-[18F]-fluoro-2-deoxy-D-glucose (FDG) analog, as they have been extensively utilized to enhance diagnostic imaging [37]. Glucose, 2-DG and FDG have increased uptake in cancer tissues due to elevation of glucose transporters in cancer cells, which allows for the imaging of malignancies and metabolically active tissues in contrast to nonmalignant tissues [38,39]. Additionally, 2-DG is reasonably well-tolerated in cancer patients [40]. In a phase I study of oral 2-DG in patients with solid tumors, 45 mg/kg 2-DG given on a daily schedule of the first two weeks of a three-week cycle was defined as the recommended dose. This was based on cardiac QTc prolongation occurring in two of four patients at the 60 mg/kg dose level, and in none of the five patients treated at the 45 mg/kg dose level.

2-DG also has considerable potential for use in combination therapies. In preclinical studies, 2-DG sensitized cancer cells to multiple types of chemotherapeutic agent [41–44], and paclitaxel enhanced the uptake of 2-DG [45]. 2-DG sensitized pancreatic cancer cells and tumors to inhibition of MEK [46], the kinase immediately upstream of MAPK3 in the RAS-induced kinase activation cascade. Thus, combining 2-DG with conventional chemotherapies or other agents that induce glycolysis is a reasonable strategy for improving the cancer therapeutics' efficacies without significantly increasing toxicity [47]. In a phase I trial in advanced solid tumors, daily oral 2-DG at 63 mg/kg in combination with docetaxel was chosen as the recommended phase 2 dose [48]. QTc prolongation was observed in 22% of patients, and the authors speculated that this was related to reversible hyperglycemia that occurred in 100% of the patients at the 63 to 88 mg dose levels. There was no pharmacokinetic interaction between 2-DG and docetaxel. Efficacy was not evaluated in the single-agent 2-DG trial; however, of the 34 patients in the combination trial, eleven (32%) had stable disease, one (3%) had a partial response and twenty-two (66%) had progressive disease.

Two kinases, MAPK3 and OXSR1, and a phosphatase, PTPN6, were also identified in this study to be significantly present at different levels in the races. MAPK3 (mitogen-activated protein kinase 3, also called ERK1) was present at higher levels in endometrial cancer specimens from Black compared to White patients in our study (mean $\pm$ SD 113.23 $\pm$ 45.03 vs. 27 $\pm$ 17.88 RU, $p = 0.001$); however, CPTAC data of all races demonstrated lower MAPK3 expression in endometrial cancer compared to control tissue. MAP kinases act in a signaling cascade that regulates various cellular processes, such as proliferation, differentiation, and cell cycle progression in response to a variety of extracellular signals [49,50]. BVD-523 (Ulixertinib) is a MAPK3/1 kinase inhibitor that inhibited xenograft growth in multiple cancer types and synergized with BRAF inhibitors in BRAF mutant cell lines [51]. A phase 1 study of BVD-523 in advanced solid tumors had treatment responses in subsets of patients with acceptable toxicities and pharmacokinetics [17].

OXSR1 (oxidative stress responsive 1) (mean $\pm$ SD 130.91 $\pm$ 86.07 vs. 95 $\pm$ 47.43 RU, $p = 0.047$) was highly expressed in endometrial cancer samples from Black patients; the lowest expression was in specimens from the Asian racial group. Based on CPTAC data, OXSR1 is not overexpressed in endometrial cancer compared to control tissue. OXSR1 is a serine/threonine kinase that regulates downstream kinases in response to environmental stress, and may play a role in regulating the actin cytoskeleton. The natural carotenoid lutein has been shown to reduce OXSR1 expression in oral cancer squamous cells [18].

PTPN6 (protein tyrosine phosphatase non-receptor type 6, also called Src homology region 2 domain-containing phosphatase-1 or SHP-1) was present at the highest levels in endometrial cancer specimens from Black patients and the lowest in Asian patients in this study (mean 62.72 $\pm$ 2.01 vs. 5.34 $\pm$ 0.21 RU, $p = 0.029$). In CPTAC data, PTPN6 was expressed at significantly higher levels in endometrial cancer compared to normal tissue. In endometrial cancer, PTPN6 was expressed at significantly higher levels in endometrioid in comparison to serous histology, and in association with worse prognosis in patients with endometrioid histology, but there was no association with survival of patients with serous histology [52]. Preclinical studies demonstrated that inhibition of protein tyrosine phosphatases with sodium stibogluconate inhibits hematopoietic cell line response to cytokines [53]. A phase 1 trial of sodium stibogluconate in solid tumors demonstrated safe inhibition of PTPN6 substrate dephosphorylation in peripheral blood cells [54].

Another protein identified in this study that has a related drug in development is involved in protein synthesis: EIF4A2 (eukaryotic translation initiation factor 4A2 (EIF4A2). It was elevated in endometrial cancers of the Black racial group compared to the White racial group in our study (mean 156.35 ± 163.26 vs. 41.09 ± 24.32 RU, $p = 0.04$). In CPTAC data comparing all races, there is no difference of EIF4A2 protein expression in endometrial cancer specimens compared to normal tissue. EIF4A2 is an ATP-dependent RNA helicase involved in the binding of mRNA to ribosomes [55]. Preclinical studies show that eFT266, a first in its class EIF4A inhibitor, is effective at inhibiting B-cell tumor growth in association with mTOR signaling [56]; however, its specificity for EIF4A1 versus EIF4A2 has not been robustly determined [57]. Currently, eFT266 (zotatifin) is in a phase 1 trial in combination with other drugs in patients with solid tumors (clinicaltrial.gov: NCT04092673).

The serine protease inhibitor SERPINA1 was highly expressed in endometrial cancer specimens from Asian (mean 483.88 ± 609.04 vs. 57.05 ± 52.81 RU, $p = 0.011$) and American Indian races (mean 306.92 ± 219.87 vs. 57.05 ± 52.81 RU, $p = 0.009$). The patients with the lowest expression in endometrial cancer were Black (73.36 RU) in this study. In the CPTAC dataset, SERPINA1 was significantly less expressed in endometrial cancer specimens compared to control tissues ($p < 0.001$), but had no association with patient survival. SERPINA1 is being used as an exploratory biomarker of breast cancer patient survival in a clinical trial of trastuzumab [24].

Additional proteins identified in this study warrant further investigation to determine their validity as endometrial cancer drug targets. Validation of our results with a different technology and with an independent set of specimens is needed; furthermore, the association of the biomarker with drug response should be validated before designing a clinical trial that is race-specific or race-enriched. Another limitation of our study is power limitation. With the current sample size, we had 80% power to identify a significant difference of protein level among races after multiple testing correction. There was a 0.9 SD difference among mean protein level in race groups. Furthermore, this study did not include normal tissue for comparison. The main objective was to identify differences in proteins between races that could provide biological clues to the racial disparity. Followup research using public databases was then used to determine what is already known about the identified proteins in endometrial cancer so that the proteins could be prioritized for further research as candidate biomarkers and drug targets. While Table 3 lists CPTAC data of expression of the identified proteins in endometrial cancer compared to normal tissue, it does not account for the potential that the alteration of the protein's expression in cancer compared to normal tissue might be race-specific.

4. Materials and Methods

Clinically-annotated snap-frozen endometrial cancer specimens were obtained from the OUHSC Stephenson Cancer Center (SCC) Biospecimen and Tissue Pathology Shared Resource under OUHSC IRB approved protocol. Eligibility criteria for all subjects were: (a) stage I endometrial cancer; (b) matched endometrioid type histology; (c) documented age and BMI.

4.1. Isolation of Proteins from Endometrial Cancer Specimens

Protein was isolated from 48 matching endometrial cancer specimens from women of different races (12 Black, 12 White, 12 Native American and 12 Asian). Briefly, proteins were isolated from ~10 mg frozen endometrial cancer specimens using T-PER Tissue Protein Extraction Reagent (ThermoFisher Scientific, Waltham, MA USA). Total protein concentration was determined using bicinchoninic acid assay (BCA) reagent (Abcam; Cambridge, UK) as per manufacturer's instruction, as previously described.

4.2. Mass Spectrometry Analysis

Ten micrograms of protein from each sample was subjected to the FASP protocol (Wisniewski, Methods Mol Biol (2018)) and digested with 0.2 µg Promega Sequencing Grade Modified Trypsin (Promega, Madison, WI, USA, V5111, using manufacturer's protocols) for overnight incubation at 37 °C in 40 mM NH4HCO3. One sample containing a mix of equal amount of each sample was prepared to a total amount of 10 μg and was subjected to the same procedure as the individual samples.

After digestion, the samples were desalted, dried and resuspended in 100 mM triethylamine ammonium bicarbonate buffer. The samples were then labeled with TMT-11plex (Thermo Fischer Scientific, Waltham, MA, USA) according to manufacturer's instructions. The TMT-11 channel was used for the "mix" sample to serve as a reference between the different TMT runs.

The TMT labeled tryptic peptides were then desalted and concentrated using Pierce™ C 18 spin columns (Thermo Fischer Scientific, Waltham, MA, USA). A total of 1 ug of tryptic peptides were loaded onto a C18 sequencing column (Acclaim™ PepMap™ 100 C18, ThermoFisher) and then eluted using a 120 min acetonitrile gradient for quantification. Eluted peptides were analyzed by liquid chromatography–tandem mass spectrometry (LC-MS/MS) analysis using a Thermo Lumos Fusion tribrid Orbitrap mass spectrometer coupled to an Ultimate 3000 RSLC nano ultra-high-performance liquid chromatography (UHPLC). Protein identification was performed by Proteome Discoverer version 2.4 (ThermoFisher Scientific, Waltham, MA USA) utilizing SEQUEST as the search engine and the human Uniprot reference proteome database version 20201123 (42,412 reviewed proteins). Protein identification required detection of at least two peptides per protein. Database search parameters were restricted to three missed tryptic cleavage sites, a precursor ion mass tolerance of 10 ppm, a fragment ion mass tolerance of 0.05 Da and a false discovery rate of $\leq$1%. Fixed protein modification was Cys carboxymethylation (+58 Da). Variable protein modifications included Met oxidation (+16 Da) and N-terminal acetylation (+42). Reporter ion intensities were bias corrected for the overlapping isotope contributions from the TMT tags according to the manufacturer's certificate. Reporting of proteins followed the general recommended guidelines. In brief, 3 biological replicates; minimum peptide length was $\geq$7 amino acids; $\geq$2 peptide matches; protein FDR cutoff $\geq$ 1%. Proteins with a "Protein false discovery rate (FDR) Confidence Combined" as "High" were used for analysis; all other parameters were set as default. We corrected the total signals of each channel by computing normalization factors in order to equalize the amounts of protein labeled by each TMT reagent. In order to conduct the subsequent analysis, the data were exported to Microsoft Excel and Perseus (version 1.6.15.0, https://maxquant.net/perseus/ accessed on 28 June 2022).

4.3. Statistical Analysis

The proteomics data were compared between racial groups. The original data for test specimens were normalized by several steps. First, the blank values were subtracted from each sample value for each protein, and negative values were forced to zero. Then, all resulting values were multiplied by a normalization factor derived for each pool using the protein level values in the mixed control sample. Lastly, the trimmed mean of M values (TMM) was applied to remove the compositional bias [58]. The normalized data after log 2 transformation were analyzed using ANOVA with the adjustment of the first principal component, which contained the unobserved confounding effect. Further, two post-hoc multiple-comparison corrected tests for each protein having p-values < 0.05 in the overall ANOVA test were performed. One was Tukey's test for pairwise comparison. The other was Dunnett's test for the comparison between each of the minority groups and the White group. Missing data were excluded from the analysis for each protein. The significance level was 0.05. All analyses were implemented in R version 4.0 (R Core Team, 2014, Vienna, Austria) using package edgeR [59] and other base functions.

4.4. Bioinformatics

The proteins found to be present in endometrioid endometrial cancer specimens from women of each race at significantly different levels in comparison to at least one other set of specimens from a different race were subjected to bioinformatic analysis. The analysis goals were to validate the protein's roles in cancer and racial disparity, and identify pathways and actionable targets in endometrial cancer specimens from patients of each race. Public databases were used, including The Cancer Genome Atlas (TCGA) and Clinical Proteomic Tumor Analysis Consortium (CPTAC), for the individual proteins. The UALCAN website (http://ualcan.path.uab.edu/index.html, accessed on 29 June 2022) was used to determine if the specific protein or its mRNA are significantly: altered in endometrial cancer compared to healthy tissue; altered in Black compared to other races; and/or are associated with survival and expression of other relevant proteins. UALCAN provides statistical analysis for the data, although in many instances the numbers of people an individual race are too low to provide sufficient power for significance. This information was carefully collated and studied for each identified gene.

Ingenuity Pathway Analysis (IPA, Qiagen) was used to gather additional information, such as the proteins' mechanisms of action; interactions with other proteins; associations with various diseases and cellular processes; and identification of drugs that target the specific proteins. IPA utilizes published literature and existing databases to provide extensive information on individual proteins/genes and how they interact with other factors. The list of proteins found to be present at differential levels in endometrial cancer specimens from women of different races were uploaded into IPA web-based software. IPA was also used to identify how the proteins are involved in canonical pathways, cellular and disease processes and upstream/downstream molecules that can be targeted with FDA-approved or investigational agents in clinical trials.

5. Conclusions

In our study, specific proteins have been identified as potential biomarkers of patient prognosis and as drug targets in patients with endometrial cancer, for several races. Some of these proteins are targeted by specific drugs that are currently being, or have already been, tested in clinical trials and may be adaptable for endometrial cancer. Drug discovery efforts for regulators of the other proteins that do not already have specific drugs would be warranted upon their validation of the proteins being differentially expressed across races and functionally involved in endometrial cancer development or aggressiveness. The goal of these efforts would not be to develop a race-specific treatment, but more importantly, to improve the worse outcomes of races that experience endometrial cancer disparities.

Supplementary Materials: The following supporting information can be downloaded at: https://www.mdpi.com/article/10.3390/ijms23147779/s1.

Author Contributions: Conceptualization, P.J. and D.M.B.; methodology, P.J., D.M.B. and C.X.; software, P.J., J.G. and D.M.B.; validation, P.J., V.S., L.E.B. and D.M.B.; formal analysis, P.J., C.X. and V.S.; investigation, P.J., L.E.B. and D.M.B.; resources D.M.B.; data curation, P.J., C.X. and D.M.B.; writing—original draft preparation, P.J.; writing—review and editing, P.J., D.M.B., J.G., V.S., C.X. and L.E.B.; visualization P.J. and J.G.; supervision D.M.B.; project administration, P.J.; funding acquisition, D.M.B. All authors have read and agreed to the published version of the manuscript.

Funding: This research received no external funding.

Institutional Review Board Statement: Not applicable.

Informed Consent Statement: Not applicable.

Data Availability Statement: Data results for the mass spectrometry experiments are provided in the supplemental materials.

Conflicts of Interest: The authors declare no conflict of interest.

References

1. Sud, S.; Holmes, J.; Eblan, M.; Chen, R.; Jones, E. Clinical characteristics associated with racial disparities in endometrial cancer outcomes: A surveillance, epidemiology and end results analysis. *Gynecol. Oncol.* **2018**, *148*, 349–356. [CrossRef]
2. Jemal, A.; Ward, E.M.; Johnson, C.J.; Cronin, K.A.; Ma, J.; Ryerson, B.; Mariotto, A.; Lake, A.J.; Wilson, R.; Sherman, R.L.; et al. Annual Report to the Nation on the Status of Cancer, 1975–2014, Featuring Survival. *J. Natl. Cancer Inst.* **2017**, *109*, djx030. [CrossRef]

3. Rodriguez, V.E.; LeBron, A.M.W.; Chang, J.; Bristow, R.E. Racial-Ethnic and Socioeconomic Disparities in Guideline-Adherent Treatment for Endometrial Cancer. *Obstet. Gynecol.* **2021**, *138*, 21–31. [CrossRef]
4. Fucinari, J.; Elshaikh, M.A.; Ruterbusch, J.J.; Khalil, R.; Dyson, G.; Shultz, D.; Ali-Fehmi, R.; Cote, M.L. The impact of race, comorbid conditions and obesity on survival endpoints in women with high grade endometrial carcinoma. *Gynecol. Oncol.* **2021**, *162*, 134–141. [CrossRef]
5. Javadian, P.; Washington, C.; Mukasa, S.; Benbrook, D.M. Histopathologic, Genetic and Molecular Characterization of Endometrial Cancer Racial Disparity. *Cancers* **2021**, *13*, 1900. [CrossRef]
6. Cancer Genome Atlas Research, N.; Kandoth, C.; Schultz, N.; Cherniack, A.D.; Akbani, R.; Liu, Y.; Shen, H.; Robertson, A.G.; Pashtan, I.; Shen, R.; et al. Integrated genomic characterization of endometrial carcinoma. *Nature* **2013**, *497*, 67–73. [CrossRef]
7. Bateman, N.W.; Dubil, E.A.; Wang, G.; Hood, B.L.; Oliver, J.M.; Litzi, T.A.; Gist, G.D.; Mitchell, D.A.; Blanton, B.; Phippen, N.T.; et al. Race-specific molecular alterations correlate with differential outcomes for black and white endometrioid endometrial cancer patients. *Cancer* **2017**, *123*, 4004–4012. [CrossRef]
8. Dubil, E.A.; Tian, C.; Wang, G.; Tarney, C.M.; Bateman, N.W.; Levine, D.A.; Conrads, T.P.; Hamilton, C.A.; Maxwell, G.L.; Darcy, K.M. Racial disparities in molecular subtypes of endometrial cancer. *Gynecol. Oncol.* **2018**, *149*, 106–116. [CrossRef]
9. Ferguson, S.E.; Olshen, A.B.; Levine, D.A.; Viale, A.; Barakat, R.R.; Boyd, J. Molecular profiling of endometrial cancers from African–American and Caucasian women. *Gynecol. Oncol.* **2006**, *101*, 209–213. [CrossRef]
10. Maxwell, G.L.; Allard, J.; Gadisetti, C.V.R.; Litzi, T.; Casablanca, Y.; Chandran, U.; Darcy, K.M.; Levine, D.A.; Berchuck, A.; Hamilton, C.A.; et al. Transcript expression in endometrial cancers from Black and White patients. *Gynecol. Oncol.* **2013**, *130*, 169–173. [CrossRef]
11. Available online: http://ualcan.path.uab.edu (accessed on 29 June 2022).
12. Rabinovich, S.; Adler, L.; Yizhak, K.; Sarver, A.; Silberman, A.; Agron, S.; Stettner, N.; Sun, Q.; Brandis, A.; Helbling, D.; et al. Diversion of aspartate in ASS1-deficient tumours fosters de novo pyrimidine synthesis. *Nature* **2015**, *527*, 379–383. [CrossRef] [PubMed]
13. Bendell, J.C.; Strauss, J.F.; Fakih, M.; McRee, A.J.; Hendifar, A.E.; Rosen, L.S.; Cercek, A.; Rowinsky, E.K.; Szarek, M.; Gonsalves, F.; et al. Phase I monotherapy dose escalation of RGX-202, a first-in-class oral inhibitor of the SLC6a8/CKB pathway, in patients with advanced gastrointestinal (GI) solid tumors. *J. Clin. Oncol.* **2020**, *38*, 3504. [CrossRef]
14. Kurth, I.; Yamaguchi, N.; Andreu-Agullo, C.; Tian, H.S.; Sridhar, S.; Takeda, S.; Gonsalves, F.C.; Loo, J.M.; Barlas, A.; Manova-Todorova, K.; et al. Therapeutic targeting of SLC6A8 creatine transporter suppresses colon cancer progression and modulates human creatine levels. *Sci. Adv.* **2021**, *7*, eabi7511. [CrossRef]
15. Study of eFT226 in Subjects With Selected Advanced Solid Tumor Malignancies (Zotatifin). Available online: https://www.clinicaltrials.gov/ct2/show/NCT04092673 (accessed on 29 June 2022).
16. Available online: https://clinicaltrials.gov/ct2/show/NCT04898634?term=HK2&draw=2&rank=7 (accessed on 29 June 2022).
17. Sullivan, R.J.; Infante, J.R.; Janku, F.; Wong, D.J.L.; Sosman, J.A.; Keedy, V.; Patel, M.R.; Shapiro, G.I.; Mier, J.W.; Tolcher, A.W.; et al. First-in-Class ERK1/2 Inhibitor Ulixertinib (BVD-523) in Patients with MAPK Mutant Advanced Solid Tumors: Results of a Phase I Dose-Escalation and Expansion Study. *Cancer Discov.* **2018**, *8*, 184–195. [CrossRef]
18. Enasescu, D.A.; Moisescu, M.G.; Imre, M.; Greabu, M.; Ripszky Totan, A.; Stanescu-Spinu, I.; Burcea, M.; Albu, C.; Miricescu, D. Lutein Treatment Effects on the Redox Status and Metalloproteinase-9 (MMP-9) in Oral Cancer Squamous Cells-Are There Therapeutical Hopes? *Materials* **2021**, *14*, 2968. [CrossRef]
19. Natsumeda, Y.; Ikegami, T.; Olah, E.; Weber, G. Significance of purine salvage in circumventing the action of antimetabolites in rat hepatoma cells. *Cancer Res.* **1989**, *49*, 88–92.
20. Roy, S.; Ghosh, S.; Mallick, P.; Maity, P. Acivicin with glutaminase regulates proliferation and invasion of human MCF-7 and OAW-42 cells–an in vitro study. *Indian J. Exp. Biol.* **2008**, *46*, 22–26.
21. Willoughby, L.F.; Schlosser, T.; Manning, S.A.; Parisot, J.P.; Street, I.P.; Richardson, H.E.; Humbert, P.O.; Brumby, A.M. An in vivo large-scale chemical screening platform using Drosophila for anti-cancer drug discovery. *Dis. Models Mech.* **2013**, *6*, 521–529. [CrossRef]
22. Peters, G.J. Antipyrimidine effects of five different pyrimidine de novo synthesis inhibitors in three head and neck cancer cell lines. *Nucleosides Nucleotides Nucleic Acids* **2018**, *37*, 329–339. [CrossRef]
23. Sodium Stibogluconate and IFNa-2b Followed by CDDP, VLB and DTIC Treating Pts. With Advanced Melanoma or Other Cancers. Available online: https://clinicaltrials.gov/ct2/show/NCT00498979?term=PTPN6&draw=2&rank=1 (accessed on 29 June 2022).
24. Copper (Cu) 64-DOTA-Trastuzumab PET Imaging in Predicting Response to Treatment with Trastuzumab and Pertuzumab Before Surgery in Patients with HER2 Positive Breast Cancer. Available online: https://clinicaltrials.gov/ct2/show/NCT02827877?term=serpina&cond=Cancer&draw=2&rank=1 (accessed on 29 June 2022).
25. Chandrashekar, D.S.; Bashel, B.; Balasubramanya, S.A.H.; Creighton, C.J.; Rodriguez, I.P.; Chakravarthi, B.V.S.K.; Varambally, S. UALCAN: A portal for facilitating tumor subgroup gene expression and survival analyses. *Neoplasia* **2017**, *19*, 649–658. [CrossRef]
26. Ohshima, K.; Nojima, S.; Tahara, S.; Kurashige, M.; Hori, Y.; Hagiwara, K.; Okuzaki, D.; Oki, S.; Wada, N.; Ikeda, J.I.; et al. Argininosuccinate Synthase 1-Deficiency Enhances the Cell Sensitivity to Arginine through Decreased DEPTOR Expression in Endometrial Cancer. *Sci. Rep.* **2017**, *7*, 45504. [CrossRef] [PubMed]

27. Dillon, B.J.; Prieto, V.G.; Curley, S.A.; Ensor, C.M.; Holtsberg, F.W.; Bomalaski, J.S.; Clark, M.A. Incidence and distribution of argininosuccinate synthetase deficiency in human cancers. *Cancer* **2004**, *100*, 826–833. [CrossRef] [PubMed]
28. Long, Y.; Tsai, W.-B.; Wangpaichitr, M.; Tsukamoto, T.; Savaraj, N.; Feun, L.G.; Kuo, M.T. Arginine Deiminase Resistance in Melanoma Cells Is Associated with Metabolic Reprogramming, Glucose Dependence, and Glutamine Addiction. *Molecular Cancer Ther.* **2013**, *12*, 2581–2590. [CrossRef] [PubMed]
29. Slomovitz, B.M.; Jiang, Y.; Yates, M.S.; Soliman, P.T.; Johnston, T.; Nowakowski, M.; Levenback, C.; Zhang, Q.; Ring, K.; Munsell, M.F.; et al. Phase II study of everolimus and letrozole in patients with recurrent endometrial carcinoma. *J. Clin. Oncol.* **2015**, *33*, 930–936. [CrossRef] [PubMed]
30. Vale, C.L.; Tierney, J.; Bull, S.J.; Symonds, P.R. Chemotherapy for advanced, recurrent or metastatic endometrial carcinoma. *Cochrane Database Syst. Rev.* **2012**, *2012*, CD003915. [CrossRef]
31. Cheung, E.C.; Ludwig, R.L.; Vousden, K.H. Mitochondrial localization of TIGAR under hypoxia stimulates HK2 and lowers ROS and cell death. *Proc. Natl. Acad. Sci. USA* **2012**, *109*, 20491–20496. [CrossRef]
32. Lin, H.; Zeng, J.; Xie, R.; Schulz, M.J.; Tedesco, R.; Qu, J.; Erhard, K.F.; Mack, J.F.; Raha, K.; Rendina, A.R.; et al. Discovery of a Novel 2,6-Disubstituted Glucosamine Series of Potent and Selective Hexokinase 2 Inhibitors. *ACS Med. Chem. Lett.* **2016**, *7*, 217–222. [CrossRef]
33. Nawaz, M.H.; Ferreira, J.C.; Nedyalkova, L.; Zhu, H.; Carrasco-Lopez, C.; Kirmizialtin, S.; Rabeh, W.M. The catalytic inactivation of the N-half of human hexokinase 2 and structural and biochemical characterization of its mitochondrial conformation. *Biosci. Rep.* **2018**, *38*, BSR20171666. [CrossRef]
34. Chiara, F.; Castellaro, D.; Marin, O.; Petronilli, V.; Brusilow, W.S.; Juhaszova, M.; Sollott, S.J.; Forte, M.; Bernardi, P.; Rasola, A. Hexokinase II detachment from mitochondria triggers apoptosis through the permeability transition pore independent of voltage-dependent anion channels. *PLoS ONE* **2008**, *3*, e1852. [CrossRef]
35. Zhang, G.; Ma, A.; Jin, Y.; Pan, G.; Wang, C. LncRNA SNHG16 induced by TFAP2A modulates glycolysis and proliferation of endometrial carcinoma through miR-490-3p/HK2 axis. *Am. J. Transl. Res.* **2019**, *11*, 7137–7145.
36. Cerella, C.; Dicato, M.; Diederich, M. Modulatory roles of glycolytic enzymes in cell death. *Biochem. Pharmacol.* **2014**, *92*, 22–30. [CrossRef] [PubMed]
37. Pajak, B.; Siwiak, E.; Sołtyka, M.; Priebe, A.; Zieliński, R.; Fokt, I.; Ziemniak, M.; Jaśkiewicz, A.; Borowski, R.; Domoradzki, T.; et al. 2-Deoxy-d-Glucose and Its Analogs: From Diagnostic to Therapeutic Agents. *Int. J. Mol. Sci.* **2019**, *21*, 234. [CrossRef] [PubMed]
38. Zhang, D.; Li, J.; Wang, F.; Hu, J.; Wang, S.; Sun, Y. 2-Deoxy-D-glucose targeting of glucose metabolism in cancer cells as a potential therapy. *Cancer Lett.* **2014**, *355*, 176–183. [CrossRef] [PubMed]
39. Akin, O. (Ed.) *Atlas of Gynecologic Oncology Imaging*; Springer: New York, NY, USA, 2014.
40. Xi, H.; Kurtoglu, M.; Lampidis, T.J. The wonders of 2-deoxy-D-glucose. *IUBMB Life* **2014**, *66*, 110–121. [CrossRef]
41. Hernlund, E.; Ihrlund, L.S.; Khan, O.; Ates, Y.O.; Linder, S.; Panaretakis, T.; Shoshan, M.C. Potentiation of chemotherapeutic drugs by energy metabolism inhibitors 2-deoxyglucose and etomoxir. *Int. J. Cancer* **2008**, *123*, 476–483. [CrossRef]
42. Chatterjee, S.; Hirota, H.; Belfi, C.A.; Berger, S.J.; Berger, N.A. Hypersensitivity to DNA cross-linking agents associated with up-regulation of glucose-regulated stress protein GRP78. *Cancer Res.* **1997**, *57*, 5112–5116.
43. Yamada, M.; Tomida, A.; Yun, J.; Cai, B.; Yoshikawa, H.; Taketani, Y.; Tsuruo, T. Cellular sensitization to cisplatin and carboplatin with decreased removal of platinum-DNA adduct by glucose-regulated stress. *Cancer Chemother Pharm.* **1999**, *44*, 59–64. [CrossRef]
44. Mese, H.; Sasaki, A.; Nakayama, S.; Yokoyama, S.; Sawada, S.; Ishikawa, T.; Matsumura, T. Analysis of cellular sensitization with cisplatin-induced apoptosis by glucose-starved stress in cisplatin-sensitive and -resistant A431 cell line. *Anticancer Res.* **2001**, *21*, 1029–1033.
45. Singh, S.P.; Gao, Y.; Singh, L.D.; Kunapuli, S.P.; Ravindra, R. Role of microtubules in glucose uptake by C6 glioma cells. *Pharm. Toxicol.* **1998**, *83*, 83–89. [CrossRef]
46. Yan, L.; Tu, B.; Yao, J.; Gong, J.; Carugo, A.; Bristow, C.A.; Wang, Q.; Zhu, C.; Dai, B.; Kang, Y.; et al. Targeting Glucose Metabolism Sensitizes Pancreatic Cancer to MEK Inhibition. *Cancer Res.* **2021**, *81*, 4054–4065. [CrossRef]
47. Stein, M.; Lin, H.; Jeyamohan, C.; Dvorzhinski, D.; Gounder, M.; Bray, K.; Eddy, S.; Goodin, S.; White, E.; Dipaola, R.S. Targeting tumor metabolism with 2-deoxyglucose in patients with castrate-resistant prostate cancer and advanced malignancies. *Prostate* **2010**, *70*, 1388–1394. [CrossRef]
48. Raez, L.E.; Papadopoulos, K.; Ricart, A.D.; Chiorean, E.G.; Dipaola, R.S.; Stein, M.N.; Rocha Lima, C.M.; Schlesselman, J.J.; Tolba, K.; Langmuir, V.K.; et al. A phase I dose-escalation trial of 2-deoxy-D-glucose alone or combined with docetaxel in patients with advanced solid tumors. *Cancer Chemother Pharm.* **2013**, *71*, 523–530. [CrossRef] [PubMed]
49. Kashima, H.; Shiozawa, T.; Miyamoto, T.; Suzuki, A.; Uchikawa, J.; Kurai, M.; Konishi, I. Autocrine stimulation of IGF1 in estrogen-induced growth of endometrial carcinoma cells: Involvement of the mitogen-activated protein kinase pathway followed by up-regulation of cyclin D1 and cyclin E. *Endocr Relat Cancer* **2009**, *16*, 113–122. [CrossRef] [PubMed]
50. Ninomiya, Y.; Kato, K.; Takahashi, A.; Ueoka, Y.; Kamikihara, T.; Arima, T.; Matsuda, T.; Kato, H.; Nishida, J.; Wake, N. K-Ras and H-Ras activation promote distinct consequences on endometrial cell survival. *Cancer Res.* **2004**, *64*, 2759–2765. [CrossRef] [PubMed]
51. Germann, U.A.; Furey, B.F.; Markland, W.; Hoover, R.R.; Aronov, A.M.; Roix, J.J.; Hale, M.; Boucher, D.M.; Sorrell, D.A.; Martinez-Botella, G.; et al. Targeting the MAPK Signaling Pathway in Cancer: Promising Preclinical Activity with the Novel Selective ERK1/2 Inhibitor BVD-523 (Ulixertinib). *Mol. Cancer Ther.* **2017**, *16*, 2351–2363. [CrossRef] [PubMed]

52. Giordano, G.; Campanini, N.; Goldoni, M.; Rodolfi, A.M.; Brigati, F.; Merisio, C.; Berretta, R. Immunohistochemical Detection of Hematopoietic Cell-specific Protein-Tyrosine Phosphatase (Tyrosine Phosphatase SHP-1) in a Series of Endometrioid and Serous Endometrial Carcinoma. *Appl. Immunohistochem. Mol. Morphol.* **2018**, *26*, 468–477. [CrossRef] [PubMed]
53. Pathak, M.K.; Yi, T. Sodium stibogluconate is a potent inhibitor of protein tyrosine phosphatases and augments cytokine responses in hemopoietic cell lines. *J. Immunol.* **2001**, *167*, 3391–3397. [CrossRef]
54. Yi, T.; Elson, P.; Mitsuhashi, M.; Jacobs, B.; Hollovary, E.; Budd, T.G.; Spiro, T.; Triozzi, P.; Borden, E.C. Phosphatase inhibitor, sodium stibogluconate, in combination with interferon (IFN) alpha 2b: Phase I trials to identify pharmacodynamic and clinical effects. *Oncotarget* **2011**, *2*, 1155–1164. [CrossRef]
55. Pelletier, J.; Sonenberg, N. The Organizing Principles of Eukaryotic Ribosome Recruitment. *Annu. Rev. Biochem.* **2019**, *88*, 307–335. [CrossRef]
56. Thompson, P.A.; Eam, B.; Young, N.P.; Fish, S.; Chen, J.; Barrera, M.; Howard, H.; Sung, E.; Parra, A.; Staunton, J.; et al. Targeting Oncogene mRNA Translation in B-Cell Malignancies with eFT226, a Potent and Selective Inhibitor of eIF4A. *Mol. Cancer Ther.* **2021**, *20*, 26–36. [CrossRef]
57. Ernst, J.T.; Thompson, P.A.; Nilewski, C.; Sprengeler, P.A.; Sperry, S.; Packard, G.; Michels, T.; Xiang, A.; Tran, C.; Wegerski, C.J.; et al. Design of Development Candidate eFT226, a First in Class Inhibitor of Eukaryotic Initiation Factor 4A RNA Helicase. *J. Med. Chem.* **2020**, *63*, 5879–5955. [CrossRef] [PubMed]
58. Robinson, M.D.; Oshlack, A. A scaling normalization method for differential expression analysis of RNA-seq data. *Genome Biol.* **2010**, *11*, R25. [CrossRef] [PubMed]
59. Robinson, M.D.; McCarthy, D.J.; Smyth, G.K. edgeR: A Bioconductor package for differential expression analysis of digital gene expression data. *Bioinformatics* **2010**, *26*, 139–140. [CrossRef] [PubMed]

International Journal of
Molecular Sciences

MDPI

Review

Vibrational Biospectroscopy: An Alternative Approach to Endometrial Cancer Diagnosis and Screening

Roberta Schiemer [1,*], David Furniss [2], Sendy Phang [2], Angela B. Seddon [2], William Atiomo [3] and Ketankumar B. Gajjar [1]

1 Division of Child Health, Obstetrics and Gynaecology, University of Nottingham, Nottingham NG5 1PB, UK; ketankumar.gajjar@nuh.nhs.uk
2 Mid-Infrared Photonics Group, George Green Institute for Electromagnetics Research, Faculty of Engineering, University of Nottingham, Nottingham NG7 2RD, UK; david.furniss@nottingham.ac.uk (D.F.); sendy.phang@nottingham.ac.uk (S.P.); angela.seddon@nottingham.ac.uk (A.B.S.)
3 College of Medicine, Mohammed Bin Rashid University of Medicine and Health Sciences (MBRU), Dubai P.O. Box 505055, United Arab Emirates; william.atiomo@mbru.ac.ae
* Correspondence: roberta.schiemer@nottingham.ac.uk

Abstract: Endometrial cancer (EC) is the sixth most common cancer and the fourth leading cause of death among women worldwide. Early detection and treatment are associated with a favourable prognosis and reduction in mortality. Unlike other common cancers, however, screening strategies lack the required sensitivity, specificity and accuracy to be successfully implemented in clinical practice and current diagnostic approaches are invasive, costly and time consuming. Such limitations highlight the unmet need to develop diagnostic and screening alternatives for EC, which should be accurate, rapid, minimally invasive and cost-effective. Vibrational spectroscopic techniques, Mid-Infrared Absorption Spectroscopy and Raman, exploit the atomic vibrational absorption induced by interaction of light and a biological sample, to generate a unique spectral response: a "biochemical fingerprint". These are non-destructive techniques and, combined with multivariate statistical analysis, have been shown over the last decade to provide discrimination between cancerous and healthy samples, demonstrating a promising role in both cancer screening and diagnosis. The aim of this review is to collate available evidence, in order to provide insight into the present status of the application of vibrational biospectroscopy in endometrial cancer diagnosis and screening, and to assess future prospects.

Keywords: endometrial cancer; uterine neoplasm; cancer of the endometrium; spectroscopy; Raman spectroscopy; Fourier transform infrared spectroscopy; diagnosis; screening

Citation: Schiemer, R.; Furniss, D.; Phang, S.; Seddon, A.B.; Atiomo, W.; Gajjar, K.B. Vibrational Biospectroscopy: An Alternative Approach to Endometrial Cancer Diagnosis and Screening. *Int. J. Mol. Sci.* **2022**, *23*, 4859. https://doi.org/10.3390/ijms23094859

Academic Editor: Laura Paleari

Received: 1 March 2022
Accepted: 26 April 2022
Published: 27 April 2022

1. Introduction

Endometrial cancer (EC) is the sixth most common cancer in women worldwide, with rising incidence affecting women's survival and health care resources [1] and global projections of progressively increasing disease burden [2]. EC remains under-researched compared with other malignancies, such as ovarian cancer and breast cancer [3]; however, the last two decades have seen a steady increase in EC research activity [4], alongside the involvement of patients and the public in the identification and planning of research priorities [5].

Among patient-supported priorities for womb cancer, of particular interest are the need for: patient risk stratification, individualised diagnostic pathways in case of abnormal uterine bleeding and the development of minimally invasive approaches to monitor treatment and detect disease recurrence [6]. These highlight the inadequacies and limitations of current diagnostic, screening and monitoring approaches and the unmet need for minimally invasive, objective, rapid and accurate alternatives.

Increasingly more data are emerging in support of the role of vibrational biospectroscopy techniques as cancer diagnostic and screening tools [7,8], as well as the potential role in the assessment of endometrial pathology [9,10]. It is relevant, therefore, to highlight the status quo of the progress made to date in applying vibrational biospectrscopy techniques to EC.

A large number of reviews has been published examining the oncological applications of biospectroscopy [7,11–20], but to the authors' knowledge, this is the first to focus on EC.

This review will first summarise current methods of EC diagnosis and screening, with their advantages and limitations. It will then examine current advances of vibrational spectroscopic techniques in EC, assessing how biospectroscopy methods could shape diagnosis, screening and disease monitoring, their relevance to some of the EC top research priorities, and will discuss current limitations and obstacles to clinical application.

2. Endometrial Cancer

EC is the most common cancer in women in the developed world, and the fourth leading cause of death due to gynaecological cancer among women worldwide, accounting for an estimated 382,069 new cases and 89,929 deaths in 2018 [1]. Early diagnosis is usually associated with favourable prognosis, but there are significant disparities between high- and low-income countries, indeed the highest annual increase in mortality is recorded in the southern sub-Saharan African region [2,21]. In England, the 5-year survival rate ranges between over 90% for women diagnosed in stage 1 and just 15% for those in stage 4 between 2013–2017 [22,23]. Most cases of EC occur in post-menopausal women, yet the last few decades have seen an incidence increase across all age groups [24]. This trend, particularly in developed countries, is directly linked to the rising prevalence of obesity, one of the most important risk factors for EC [22,25].

EC is a heterogeneous disease [26]. Historically, it was classified by Bokhman [27] into two main types, that differ in aetiology, as well as in clinico–pathological and epidemiological characteristics: Type I, or endometrioid adenocarcinoma, and Type II, which includes non-endometrioid subtypes such as: serous carcinoma, clear cell carcinoma and carcinosarcoma/malignant mixed Müllerian tumours [26–28]. Type I cancers are the most common, particularly in Caucasian populations, accounting for approximately three-quarters of all EC. This form is associated with unopposed oestrogen exposure and often arises from a precursor lesion, known as endometrial hyperplasia.

Type II cancers, by contrast, more prevalent for instance in the African–American population, appear to be independent of hormonal risk and are usually not preceded by endometrial hyperplasia. Type II cancers are considered high grade, and display a more aggressive behaviour, with a high risk of extra-uterine disease at first presentation. The prognosis is poorer compared with Type I cancers, with a tendency to recur even in early-stage disease [29–31].

This dichotomous classification, while useful, is limited by the molecular and histological heterogeneity within each group, and the now recognised overlap between some Type I and II cancer behaviour. Advances in endometrial cancer genomic and proteomic characterisation have led to a deeper understanding of the biological, pathological and genomic characteristics of cancer subsets and have now provided the basis for a substantial integration of pathological and molecular classifications [32,33]. Indeed, The Cancer Genome Atlas (TCGA) project reported, in 2013, a comprehensive genetic analysis of the most common histological types of EC, identifying molecular mutations that correlate with their clinical behaviour [34]. This pioneering research provided important prognostic information, particularly for a sub-group of high-grade and high-risk endometrial carcinomas, the *POLE* mutated tumours, that have been found to have an excellent prognosis [35]. Due to the potential impact of molecular classification on cancer treatment, the European Society of Gynaecological Oncology (ESGO), the European Society for Radiotherapy and Oncology (ESTRO), and the European Society of Pathology (ESP) now jointly encourage molecular classification of all EC, particularly the high-grade tumours [36].

The practical integration between pathological and molecular classifications in diagnostic pathways remains, however, challenging: from gene sequencing selection and the risk of identifying unwanted or uninterpretable information, to the additional time required to process results, and costs of equipment and specialised laboratory expertise, which may limit the availability of such an integrated approach, particularly in low-income settings.

An alternative approach to a model based on sequencing a panel of selected genes [33] could be to explore all of the range of biochemical features in a sample simultaneously, with the use of vibrational biospectroscopy techniques.

2.1. Current Endometrial Cancer Diagnosis and Screening

Timely investigation of women presenting with symptoms, such as post-menopausal bleeding and persistent menstrual irregularities, allows most cases of EC to be identified in early stage.

Ultrasound imaging, hysteroscopy and endometrial biopsy, together with the histopathological tissue analysis, are the current mainstay of EC diagnosis. In addition, magnetic resonance imaging techniques (MRI) are useful in the assessment of depth of myometrium invasion, cervical stromal involvement and lymph node metastasis [37–40].

However, unnecessary procedure should be avoided, which may expose patients to complications, generate needless anxiety and take up financial resources. Indeed, hormonal imbalance, coagulopathies, benign endometrial lesions and the use of medications including hormone replacement therapy (HRT) are some of the factors associated with irregular and recurrent vaginal bleeding, which may occur in the absence of EC [41]. Consequently, the main challenge in early cancer diagnosis is the appropriate selection of those patients that require investigations and invasive procedures.

2.1.1. Ultrasound Imaging

Ultrasound imaging is a technique, which uses high frequency sound waves to provide information about tissue and organ characteristics. The procedure can be performed by the transabdominal and transvaginal access routes, does not require bowel preparation, is safe and is, overall, well tolerated by patients [42].

Ultrasonography is, however, highly operator dependent. Furthermore, excess adipose tissue interferes with sound wave signals, affecting image quality [42], thus women with high body habitus are at increased risk of suffering diagnostic delays [43].

Measurements of the endometrial thickness using ultrasound imaging are used as a surrogate marker to check for the presence of intrauterine abnormalities [44]. However, ultrasound imaging alone cannot discriminate whether an increased endometrial thickness is secondary to a benign lesion or to malignant disease [44].

Women with post-menopausal bleeding have a 8–11% risk of EC, which justifies the need for endometrial assessment in these patients [45]. The use of endometrial thickness cut-offs of 4 mm and 5 mm leads to the correct identification of 94.8% and 90.3% of EC cases, respectively [46]. However, the test specificity is poor, leading to a high risk of false positive results and, consequently, many unnecessary invasive investigations and biopsies [46,47].

In pre-menopausal women with abnormal uterine bleeding, the diagnostic role of endometrial thickness is controversial, as there can be overlap between physiological thickening caused by sex hormones and that caused by endometrial disease [48]. While it has been suggested that a thickness of <8 mm should be considered as non-hyperplastic [49], and only 1% of endometrial cancers occurs in women < 40 years of age [47], there is still no consensus on the ideal endometrial thickness cut-off in this group of patients [50], thus an alternative or complementary non-invasive triaging tool would facilitate the clinician's decision-making on when to refer for further invasive diagnostic procedures.

2.1.2. Hysteroscopy

The direct endoscopic visualisation of the endometrial cavity by hysteroscopy, using visible light at 4 to 5× magnification [51], is an invasive procedure that can be performed in order to evaluate the endometrial cavity, to remove lesions such as polyps or small fibroids and to obtain endometrial biopsies.

Hysteroscopy can be carried out both in the outpatient setting and in theatre, under regional or general anaesthetic [52]. Although, overall, the procedure has been shown to be well tolerated, safe, accurate and acceptable, regardless of the setting in which is performed [53–55], some patients do experience significant discomfort during outpatient hysteroscopy [55]. Unfortunately, it is difficult to identify this group of patients preoperatively, and for these women a routine hysteroscopic procedure may turn into a painful and traumatic experience. Furthermore, although uncommon, complications may arise, including bleeding, infection and uterine damage [56], and the failure rate of hysteroscopy, where the instrument cannot be successfully introduced into the uterine cavity, has been estimated at 4.2% [55,57].

Well-conducted systematic reviews and meta-analyses found that hysteroscopy is highly accurate for the diagnosis of EC in women with abnormal uterine bleeding [55,58,59] and it is useful at excluding endometrial disease [58,59], although the diagnostic accuracy for endometrial hyperplasia appears to be more modest [55].

Indeed, in its updated 2018 guidance, the UK National Institute for Health and Care Excellence (NICE) now recommends that hysteroscopy can be offered as a first-line investigation for heavy menstrual bleeding, in preference to pelvic ultrasound, if the woman's history suggests sub-mucosal fibroids, polyps or endometrial pathology [60].

The inevitable consequence of such a diagnostic strategy, however, is the need for a structural re-organisation of healthcare services, in order to absorb the estimated 10,000 extra procedures that would be performed in England each year [61] and their added financial costs. The availability of adequately sized and equipped facilities and investments in recruitment and training of skilled staff are some of the challenges to overcome, in order to implement the new guidance into clinical practice.

2.1.3. Endometrial Biopsy and Histological Analysis

Histological examination of an endometrial biopsy specimen is the current so-called: "Gold Standard" of EC diagnosis. The sample preparation and analyses, required to allow the visualisation of the internal architecture of cells and tissues and identify cancerous features [62] are, however, time-consuming, and can be subject to human error [63].

The condition or quality of an endometrial biopsy must be "adequate" in order to provide the histological diagnosis, but the lack of standard agreement on quality and quantity assessment criteria [64] leaves the decision regarding sample suitability to individual pathologists. This allows for a high inter-observer variability [65] and a risk of diagnostic delay and potentially detrimental consequences for patients. The reported rate of insufficient quality or quantity of endometrial tissue samples for histological diagnosis in postmenopausal women is 31% (range 7–76%) [66], while in pre-menopausal women it is lower, ranging between 2% and 10% [67]. The reasons for insufficient sampling appear unclear: the experience of the operator has not been confirmed to be a determining factor [68,69] and there is wide variance between insufficient sample rates reported in single versus multicentre studies, suggesting that study design may influence the results [55,68–71].

Importantly, obtaining an endometrial biopsy is not always a straightforward process. Indeed, endometrial sampling fails in approximately 11% of cases (range 1–53%), mostly as a consequence of cervical stenosis [66]. Factors such as the post-menopause, advanced age and nulliparity also appear to be associated with higher failure rates, likely as a consequence of variation in endometrial thickness and anatomical changes that occur in these patient groups [69].

Outpatient endometrial biopsy has mostly replaced traditional dilatation and curettage under general anaesthetic worldwide [72–74], as it has shown comparable performance, while being less invasive and more cost effective [75,76]. The diagnostic accuracy of these biopsies were investigated extensively and a number of meta-analyses were published, reporting on sensitivity and specificity in relation to endometrial cancer, endometrial hyperplasia (with and without atypia) and benign endometrial disease [66,67,74,75,77]. Overall pipelle biopsy, with conventional histopathology, appears to be an effective tool to identify endometrial cancer when adequate samples are obtained; however, the test is not as reliable in the case of endometrial hyperplasia, where a negative result only decreases the hyperplasia risk by 2-fold [77].

The potential failure to diagnose or exclude disease after invasive procedures, such as hysteroscopy and endometrial biopsy, is concerning. Coupled with the highly subjective nature of histopathological assessments, it highlights the need for alternative approaches to complement current practice and provide pathologists with additional support in achieving a more objective tissue evaluation. Furthermore, in the context of EC research priorities [6], current diagnostic modalities, despite their established advantages, are insufficient to fully address the need for patient risk stratification and the demand for minimally invasive, individualised, screening, diagnostic and treatment monitoring pathways.

2.1.4. Screening for Endometrial Cancer

Screening is defined by the World Health Organisation (WHO) as "the presumptive identification of unrecognized disease in an apparently healthy, asymptomatic population by means of tests, examinations or other procedures that can be applied rapidly and easily to the target population" [78]. We suggest that an ideal screening test should be accurate, well-tolerated, associated with minimal morbidity and cost-effective.

Women with known Lynch Syndrome already undergo a multimodal surveillance of the endometrium until hysterectomy is performed, due to their high lifetime risk of developing EC [79].

Unfortunately, there is no EC screening test which is accurate and reliable enough to be implemented for the general asymptomatic population [46–48,80,81]. Ultrasound imaging, despite its accessibility, safety and low cost, unfortunately lacks the required sensitivity and specificity, as demonstrated by the nested case-control study [80] within the 2016 United Kingdom Collaborative Trial of Ovarian Cancer Screening (UKCTOCS) [82]. The study showed that if the general UK population were screened using an endometrial thickness cut-off of 5 mm, in order to diagnose 80.5% of cancers, for each endometrial cancer or atypical endometrial hyperplasia (AEH) case detected, 58 healthy women would have to undergo additional unnecessary investigation [80]. It is apparent that the modest test accuracy, potential patient risks and added costs, do not justify the implementation of such a screening strategy. There is, therefore, an unmet need to innovate current diagnostic and screening methods, to tackle the increasing endometrial cancer disease burden and allow early disease detection and timely treatment.

3. Aim of this Review

Since the turn of the 21st Century, vibrational biospectroscopy technologies have been widely researched [7,8,12,13,83–85] and now show significant promise for clinical application as a new type of stain-free pathology on ex vivo tissue, and also promise for in vivo imaging. In the quest for clinical advance, it is timely to examine the potential added value of vibrational biospectroscopy applications with regards to EC.

Here we will review Mid-Infrared Absorption Spectroscopy and Raman Spectroscopy. We aim to highlight progress to date, with particular emphasis on implications for clinical practice and the relevance to EC diagnosis and screening.

4. Biospectroscopy

The interaction between electromagnetic radiation and any particular matter results in the measurable linear and nonlinear physical phenomena of absorption, emission, reflection and scattering of the radiation by the matter; measurement of the radiation after its interaction with matter yields information on the makeup and arrangement of the matter and is known as spectroscopy. The measurement is displayed as a spectrum, which is a graphical representation of energy absorption, emission, reflection or scattering by the material as a function of the incoming radiation photon energy (plotted usually as frequency or wavelength). The application of spectroscopic techniques to biological materials is called biospectroscopy and the name was only coined in the 1960s [86].

Mid-infrared Absorption and Raman scattering spectroscopy, are sister-vibrational absorption techniques, being complementary as they are based on different quantum mechanical rules. They are label-free, non-destructive optical methods with the ability to investigate the vibration and rotation of atoms and molecules in biological materials, induced by irradiation by light.

The vibrational spectra that are generated depend on the specific biochemical structure of the sample tested; they provide information on the whole range of molecules within the sample simultaneously, which can, therefore, be interpreted as a unique "signature" or a "fingerprint" of that sample [12].

The alteration of molecular signatures in a cell or tissue, which has undergone disease transformation, can be objectively detected, gaining vibrational spectroscopic techniques a potential role in cancer diagnosis and screening [14,87,88].

4.1. Mid-Infrared Absorption Spectroscopy

Mid-infrared (MIR) light is a radiation region of the electromagnetic spectrum of 3–50 microns wavelength, as defined by ISO 20473:2007 [89]. When biological tissues are exposed to MIR light, part of the photon energy can be resonantly absorbed, inducing vibrations; the quantum mechanical selection rules include that there must be a change in dipole moment during the vibration, hence heteropolar chemical bonds are vibrationally stimulated. The intensity and wavelength of each vibration depend on the nature of the chemical bonds and their specific molecular environment, that is its molecular structure [90]. The fraction of energy absorbed by the sample at different frequencies can be quantitatively measured by means of dispersive infrared (IR) spectroscopes [87].

The technology has been further refined and made faster since the 1970s by the introduction of Fourier transform (Ft) IR spectrometers, in which all broadband spectral information is collected simultaneously, and then many times, in order to average and then maximise the signal-to-noise ratio. The raw data obtained, called an interferogram, is then converted using the Fourier transform mathematical algorithm into wavelength intensity, from which the energy absorbed by the sample can be derived [90].

The majority of work reported to date used the technique of Fourier transform Infrared (FtIR) with Attenuated Total Reflectance (ATR) on excised tissue, or extracted body fluids, which overcomes the need for complex sample preparation [90]. Other image acquisition modes include transmission and transflection; these require the use of suitable substrates (e.g., calcium or barium fluoride slides) and longer machine and sample preparation compared with ATR [17]. Transflection was shown to introduce spectral artefacts and so has lost favour [91,92].

4.2. Raman Spectroscopy

Raman spectroscopy relies on the principle of inelastic scattering of photons, also known as Raman scattering, and was first discovered by Raman in 1928 [93]. When a monochromatic light source, such as a visible or near-infrared laser, interacts with a sample, most of the light which scatters off is unchanged in energy. However, a very small number of photons will exchange part of their energy with the molecules of the sample: the chemical bonds of the sample become temporarily excited to a virtual state, then relax to a different vibrational state, while the emitted photons shift to a lower (Stokes) or higher (Anti-Stokes) frequency [94]. The shift in frequency, measured by the Raman spectrometer, is indicative of specific vibrational modes of the sample molecules and, therefore, a unique "fingerprint" spectrum can be inferred [88]. The quantum mechanical selection rules of Raman include that the molecular bond should not undergo a change in dipole during vibration, thus favouring homopolar chemical bonds. Hence, Raman spectroscopy is unaffected by water, and is non-destructive and label-free [95]. These characteristics offer technology a high degree of flexibility, with potential applications to the study of fresh, fixed and live tissues and cells [96]. Spontaneous Raman scattering is a rare phenomenon, with a very low probability of occurrence (~1 in 10^8) [13]. In order to enhance the Raman-scattering signal level, several variations of Raman spectroscopy have been developed, including resonant Raman (RR), coherent anti-Stokes Raman scattering (CARS) and surface-enhanced Raman scattering (SERS) [14]; these are, however, expensive technologies, with a large footprint. Ultimately, the choice of instrument, desired wavelength and spatial resolution will vary depending on the required application.

4.3. Biospectroscopy for Endometrial Tissue Interrogation

The development of effective diagnostic, screening and treatment strategies for endometrial cancer finds its basis in a deep understanding of tissue physiological and pathological processes. In particular, to be clinically useful, a new diagnostic or screening tool should be able to accurately distinguish healthy patients from those with disease.

ATR-FtIR and Raman spectroscopy were used to categorise disease and identify cancer or intra-epithelial neoplasia in a number of excised tissues, such as prostate [97–101], gastrointestinal tract [102–106], brain [95,107,108], breast [109–116], lung [117,118] and skin [119–121]. Gynaecological applications include studies of cervical cytology and histopathology [122–127], ovarian cancer [128,129] and vulvar disease [130].

With regards to endometrial tissue, vibrational biospectroscopy was successfully applied in preliminary research to the study of its structural architecture [94,131], as well as the classification of cancerous lesions [10,132,133], specific cancer subtypes [132,133] and the identification of cell phenotypes with different drug sensitivity [134] (see Table 1). There is, however, a paucity of literature, and specifically of large studies, compared with other types of diseases.

Table 1. Spectroscopy of uterine tissues/cells, studies included in this Review.

Author	Year	Sample	No of Patients	Sample Preparation	Spectroscopy Method	Spectral Findings
Theophilou et al. [131]	2018	Benign uterine tissue	3 Multiple tissue sections per patient	Paraffin-fixed sections	Synchrotron FtIR (SR-FtIR) and globar focal plane array-based (FPA) FtIR	Identification of endometrial stem cell putative location with SR-FtIR: changes of stretching vibration in DNA, RNA, nucleic acids and protein secondary structure
						Differentiation between functionalis and basalis epithelial layers with FPA-FtIR: variation in protein secondary structure
Patel et al. [94]	2011	Benign and malignant endometrial tissues	4	Freshly-thawed frozen sections	Raman	Identification of tissue architecture: - high content of DNA and RNA in glandular epithelium - high protein content in collagenous stroma and myometrium
Kelly et al. [132]	2009	Benign and malignant endometrial tissues	26 Non-tamoxifen associated n = 15, tamoxifen-associated n = 8	Paraffin-fixed sections	Synchrotron FtIR (SR-FtIR)	Endometrioid carcinoma vs. benign: variations in protein content and secondary structure
						Serous papillary and malignant mixed mullerian tumours vs. benign: variations in RNA and DNA regions
						Group separation based on tamoxifen usage improved cancer vs. benign classification. Spectral changes were observed in protein secondary structure

Table 1. *Cont.*

Author	Year	Sample	No of Patients	Sample Preparation	Spectroscopy Method	Spectral Findings
Taylor et al. [133]	2011	Benign and malignant endometrial tissues	76	Ethanol-based fixed sections (SurePath^{TM})	ATR-FtIR	Stages of mentrual cycle classification: variation of lipid, Amide I, Amide II and asymmetric phosphate stretching vibration regions Cancer vs. benign: increased content of lipids and proteins Classification of tumour subtypes: - highest lipid content in grades I and III endometrioid, in clear cell tumours, adenosarcomas and carcinosarcomas - variation of protein secondary structure in endometrioid cancers
Depciuch et al. [9]	2021	Benign, pre-cancerous and malignant endometrial tissues	16 patients, 59 serial tissue samples	Paraffin-fixed sections	Raman and ATR-FtIR	Raman - cancer vs. benign: higher content of lipids and proteins and decreased collagen vibrations ATR-FtIR - cancer vs. benign: higher protein content and variations in protein secondary structure
Barnas et al. [10]	2020	Benign, pre-cancerous and malignant endometrial tissues	45	Paraffin-fixed sections	Raman and ATR-FtIR	Raman: - atypical hyperplasia vs. benign: higher nucleic acids, shift in protein and lipid peaks - cancer vs. benign: shift of nucleic acids, protein and lipid peaks ATR-FtIR: - atypical hyperplasia vs. benign: changes in carbohydrates, collagen and protein peak - cancer vs. control: changes in carbohydrates, collagen and protein peaks
Krishna et al. [134]	2005	Chemo-sensitive and multidrug resistant uterine sarcoma cell lines	15 samples	Cell culture	Raman and FtIR	Raman - multidrug resistant phenotype vs. sensitive cell line: changes in protein secondary structure and DNA vibrations FtIR - multidrug resistant phenotype vs. sensitive cell line: changes in protein secondary structure and lipid content

Key: ATR = attenuated total reflectance; DNA = deoxyribonucleic acid; FPA = focal plane array; FtIR = Fourier transform infrared; RNA = ribonucleic acid; SR = synchrotron; SurePath^{TM} is owned by TriPath Imaging.

4.3.1. Assessment of Endometrial Structure

The endometrium is a highly complex regenerative tissue, with proliferation and shedding cycles strictly regulated by the hormones oestrogen and progesterone. The endometrial architecture and cellular turnover are also influenced by the activity of adult endometrial stem cells [135], whose pathological changes may be involved in uterine carcinogenesis and in the development of conditions such as adenomyosis and endometriosis [135,136]. The study of endometrial tissue structure, including stem cell physiology and function, may help further understanding of endometrial proliferative disorders and guide future treatment strategies.

In vivo and in vitro assays were already developed to isolate and characterise endometrial stem cells [137–139]; a novel approach is to scrutinise endometrial stem cells with biospectroscopy.

Theophilou et al. [131] examined endometrial glands, using synchrotron- and focal-plane detector array-based FtIR spectroscopy. Vibrations of Amide I and II and PO_2 in DNA and RNA nucleic acid were the main factors that allowed the segregation between the glands' functionalis and basalis regions and the identification of putative stem cells within the deepest/terminal portion of the endometrial glands in the basalis layer. Furthermore, the putative stem cells were identified in two separate locations of the gland bases, prompting speculation that they might have different functions: one set being dormant and one in active differentiation [131].

The endometrial architecture was also explored by Patel et al. [94], who analysed benign and malignant uterine tissues with Raman microspectroscopy and were able to identify features of myometrium, glandular components and uterine epithelium. Again, different computer algorithms were tested to analyse the spectra obtained. The wavenumbers responsible for achieving the highest contrast between cellular structures were: 1234 cm^{-1} (Amide III), 1390 cm^{-1} (CH_3 bend), 1675 cm^{-1} (Amide I/lipid), 1275 cm^{-1} (Amide III), 918 cm^{-1} (proline) and 936 cm^{-1} (proline, valine and proteins) [96]. By translating the spectral information obtained into spectra-derived contrast images, they were able to reconstruct an image of the tissue structure and match it with the correspondent Haematoxylin and Eosin (H&E) stained architecture.

The use of spectroscopy may, therefore, help to uncover structural and biological functions at a cellular level, leading a variety of potential clinical applications. Indeed, the identification of specific cell types, such as endometrial stem cells, and the observation of their biochemical activity, may correlate with clinical behaviours, such as chemotherapy resistance or tendency to metastasise, thus potentially informing the development of target treatments. The characterisation of the endometrial tissue structure, matched with corresponding histological appearance, may be used as a complementary and objective tool in support of standard histopathological diagnosis, with the aim of reducing inter-observer variability and human errors.

4.3.2. Endometrial Cancer Diagnosis

The ability of FtIR spectroscopy to differentiate benign from malignant endometrial tissue, and to discriminate between different cancer subtypes, has been investigated in the quest for novel approaches to endometrial cancer diagnosis.

In a proof of concept study, Kelly et al. [132] used synchrotron-based Fourier-transform infrared microspectroscopy to examine de-waxed paraffin-embedded blocks of excised uterine tissue. They were able to discriminate between different endometrial cancer subtypes and benign tissues, as well as to distinguish between tamoxifen- and non-tamoxifen-associated cases. Interestingly, the cancer subtypes explored in the study exhibited different discriminating spectral patterns, which may reflect differences in biological characteristics: endometrioid types manifested discriminating wavenumbers, particularly in the protein region (1800–1480 cm^{-1}), whereas those contributing the most to serous papillary or malignant mixed Mullerian tumours segregation were primarily in the DNA/RNA region (1425–900 cm^{-1}) of the vibrational spectrum [132].

While the use of de-paraffinisation procedures allows access to large tissue banks, essential for retrospective and novel exploratory studies, these methods require tissue preparation or electronic spectral manipulation in order to exclude the contribution of paraffin to the tissue spectra [140].

An alternative method, proposed by Taylor et al. [133], is to immerse endometrial samples in ethanol-based liquid fixative (SurePath^{TM}) and then perform sequential distilled H_2O washes prior to spectral collection. In their study, instead of a synchrotron source, endometrial tissue was examined using a blackbody source in an ATR-FtIR instrument. The different phases of the menstrual cycle, could be clearly differentiated, exhibiting large differences in absorption bands in the lipid, Amide I, Amide II and asymmetric phosphate-stretching vibration regions. The authors also evaluated differences between benign and malignant endometrium. Multivariate analysis with principal component analysis (PCA), followed by linear discriminant analysis (LDA), obtained ~80% separation between benign and malignant spectra [133]. Among discriminating wavenumbers, most differences between benign and malignant tissues were consistent with previous data [132] and identified in the Amide I/II (1624, 1750, 1516 cm^{-1}) regions, as well as in the protein band (1477 cm^{-1}), asymmetrical PO_2^- (1230 cm^{-1}), RNA/carbohydrate (1168 cm^{-1}) and phosphorylated proteins (968 cm^{-1}). In addition, the lipid region (1735 cm^{-1}) also appeared to have a significant discriminating role.

More recently, Depciuch et al. [9] identified the chemical changes that occur during the carcinogenesis process in endometrial tissues with both Raman and FtIR spectroscopy and reported classification accuracy ranging between 62.71 and 96.61%; the research group further combined Raman and FtIR spectroscopy to differentiate endometrial cancer, atypical hyperplasia and controls [10]. Alterations in nucleic acid, Amide I, lipids and collagen were seen in the Raman spectra, while changes in carbohydrates and amide vibrations were shown in the FtIR spectra. The shifts in wavenumber reported for tissues at different stages of carcinogenesis were consistent with those previously reported by other authors [94,133,141].

Tables 2 and 3, illustrate a full list of typical FtIR absorption band assignment and Raman shifts, respectively, identified in endometrial cancers.

Table 2. Peak assignment of typical FtIR spectra absorption bands observed in endometrioid and non-endometrioid endometrial cancers [9,131,133,142,143].

Endometrioid Wavenumber/(cm^{-1})	Non-Endometrioid Wavenumber/(cm^{-1})	Peak Assignment
1735	1736	Ester carbonyl of lipids
1682, 1624	1624, 1601	Amide I group in peptide linkages of proteins
1570, 1516	1570, 1516	Amide II group in peptide linkages of proteins
1535		C-N stretching contribution to Amide II
1477, 1462, 1450	1477	CH_2 group scissoring modes in proteins (collagen)
1373		C-O-O symmetric stretching of fatty acids, and amino acid side chains
1340	1340	CH_2 wagging of proline in amino acids and collagen
1240		Amide III-N-H bending, C-N stretch, C-C stretch of proteins, DNA, phospholipids
1234, 1230	1231	Asymmetric PO_2^- stretching in RNA and DNA
1169	1173, 1142	C-O-C and C-O-P stretching and ring vibrations, symmetric C-O stretching coupled to C-O-H bending of carbohydrates
1088	1061	Symmetric PO_2^- stretching in RNA and DNA
1066		C-O stretching mode of C–OH groups of serine, threonine, and tyrosine of protein
1034	1003	Symmetric C-O-C/C-O stretching of Glycogen
964	968	Phosphorylated proteins

Table 3. Raman shift, with corresponding vibrations described in endometrioid adenocarcinoma [9,10,94,144].

Raman Shift/(cm^{-1})	Vibration Assignment
853, 821	Proline, hydroxyproline, tyrosine, PO_2 stretching from nucleic acids
880, 876	C-C stretching from proline and hydroxyproline
1066, 935	Proline, valine, PO_2 stretching from nucleic acids
1299	Phosphodiester groups in nucleic acids
1302	Amide III (collagen assignment)
1335	Adenine
1376, 1374	Tryptophan
1447	CH_2 bending from lipids and proteins
1561	C-C, tryptophan (protein assignment)
1685, 1660	Amide I
1792, 1723	C-O stretching from lipids
2795, 2758	CH_3 stretching from lipids
2873	CH_2 stretching from lipids

The ability of biospectroscopy to discriminate between normal and abnormal samples, rapidly providing label-free information on their biomolecular features bypassing the need for extensive tissue processing, has important clinical implications.

From a risk stratification point of view, biospectroscopy could be employed as a triage tool: the interrogation of endometrial biopsies at the bedside providing information on the likelihood of disease could help streamlining fast-track pathways. Prioritising conventional tissue diagnosis for those patients with the highest risk of cancer would not only speed up treatment, but also reduce patient anxiety that arises from waiting for test results.

Additionally, biospectroscopy could be developed into an intra-operative tool, to assess metastatic lymph node involvement or the nature of peritoneal deposits, to assist surgeons with precious real-time information, thus allowing them to individualise the treatment provided.

4.3.3. Treatment and Surveillance

Beside discriminating normal and malignant tissues, optical spectroscopy techniques were explored for potential application in cancer therapy and disease surveillance. Multi-drug resistance is one of the greatest hurdles to chemotherapy and the ability to discriminate between sensitive/resistant cancer cell lines may improve treatment effectiveness and allow the development of individualised treatment plans.

Krishna et al. [134] reported on the identification by spectroscopy of specific phenotypes in uterine cancer cell lines. By applying PCA analysis to spectral data obtained with Raman and FtIR spectroscopes, multi-drug resistant (MDR) phenotypes of uterine sarcoma cell lines were distinguished from their drug sensitive counterparts, showing promising differentiation of cell types and the identification of their biological functions [134]. Discriminating spectral features included changes in protein secondary structure, DNA vibrations and lipid content.

Similarly, pilot studies have also examined pre- and post-chemotherapy tissue spectra of colon, breast and ovarian cancers, as follows.

Kaznowska et al. [145] applied PCA-LDA to FtIR spectra of cancerous and healthy colon tissues to detect pre- and post-chemotherapy spectral differences. Interestingly, although the baseline structure of healthy tissue was not restored after chemotherapy, the spectra of healthy and post-chemotherapy colon displayed a high degree of similarity, which could be used as a marker of treatment effectiveness: the higher the spectral similarities, the greater the treatment response [145]. Depciuch et al. [115] reported similar spectral findings in breast cancer tissue pre- and post-chemotherapy, where the resemblance of the post-chemotherapy tissue FtIR spectrum to the healthy tissue FtIR spectrum correlated with the clinical response to treatment.

Finally, Zendehdel et al. [146] studied chemotherapy resistance patterns of ovarian cancer cell lines with FtIR spectroscopy coupled with PCA analysis. The authors identified alterations in secondary protein structures and a shift toward the high wavenumbers of the CH_2 stretching vibration to 2920 and 2852 cm^{-1}.

The potential ability of spectroscopy methods to detect chemotherapy resistance and to assess spectral changes in tissues after treatment, could be exploited to guide the choice of cancer therapy and could be further developed in the context of disease monitoring.

4.4. Biospectroscopy of Biofluids: Screening and Cancer Diagnosis

With the endometrial cancer global disease burden expected to rise, disease screening, early detection and treatment monitoring will benefit from the development of more cost-effective, rapid, non-invasive and label-free techniques. Biological fluids, being readily accessible with low-cost procedures, represent the ideal sample target. Indeed, the study of biofluids with spectroscopy is becoming a rapidly emerging field and a number of pilot studies have now been published, focusing on oncological applications, as well as on a broad range of acute and chronic medical conditions [83,88,129,147–153].

With regards to endometrial cancer, biospectroscopy was proposed as a novel approach to test blood, urine and saliva (Table 4). The development of such techniques is particularly relevant as currently available methods, such as radiological imaging and blood biomarkers, lack the required sensitivity, specificity and accuracy to be used as effective screening tools. Interestingly, while Raman spectroscopy of blood serum was recently investigated for the first time as a non-invasive technique to diagnose endometriosis [154], no studies were found on the application of Raman spectroscopy to biofluids for EC diagnosis.

Similar to experiments performed with endometrial tissue, the sample manipulations and chemometric analyses used to test biofluids vary between studies. The first pilot research, by Gajjar et al. [149], investigated the potential role of ATR-FtIR for cancer diagnosis using blood samples and, with the development of "machine classifiers", reported classification rates of endometrial cancer versus controls up to 77.08% and 81.67% for serum and plasma, respectively. The same spectral data were more recently re-analysed by the research group [143], to evaluate the performance of alternative data-processing methods and classifier tools. Furthermore, the authors focused on the water-free sub-section of the spectrum (1430 cm^{-1} to 900 cm^{-1}), in contrast with the more extended bio-fingerprint region (1800 cm^{-1} to 900 cm^{-1}) used in the original paper. The researchers assessed four types of classifiers and reported high discrimination rates for both plasma (sensitivity of 0.865 ± 0.043 and specificity of 0.895 ± 0.023 with k-Nearest Neighbours algorithm) and serum (sensitivity 0.899 ± 0.023, specificity 0.763 ± 0.048 for LDA). This approach demonstrated for the first time the possibility of overcoming the dominant effect of water seen in the analysis of hydrated samples with MIR spectroscopy, which would support future applications of MIR spectroscopy in vivo to cancer diagnosis and screening [143].

Paraskevaidi et al. [155,156] used ATR-FtIR spectroscopy with PCA followed by support vector machine (PCA-SVM) to analyse blood plasma and serum from women with endometrial cancer and age-matched healthy controls. Test performance was described as high as 100% sensitivity and 85% specificity (98% accuracy) and changes in the bands associated with proteins and lipids were consistently responsible for the discrimination between blood plasma and serum from endometrial cancer and healthy samples. Traditionally, low-emissivity (low-E) slides have been used as ATR-FtIR substrates to support samples; however, their high cost may limit their use in large-scale studies and implementation in routine analysis. Aluminium foil substrate may represent an equivalent, cheaper, sample-support alternative for the detection of endometrial cancer with blood plasma and serum [156]. Indeed, Paraskevaidi et al. [156] showed that aluminium foil substrate was able to differentiate blood plasma and serum from patients with endometrial cancer and controls, with sensitivities and specificities comparable with the traditional low-E slides. The cost-effectiveness and ease of use, if validated in larger datasets, would greatly facilitate clinical application.

Table 4. Spectroscopy of biofluids: endometrial cancer studies included in this Review.

Author	Year	Sample	No of Patients	Preparation	Spectroscopy Method	Spectral Findings
Gajjar et al. [149]	2013	Endometrial cancer and healthy blood plasma and serum	60	Dried samples	ATR-FTIR	Plasma cancer vs. healthy: changes of stretching vibration in glycogen, RNA, fatty acids, amino acids and lower levels of lipids
						Serum cancer vs. healthy: changes of stretching vibration in DNA, RNA and lower levels of lipids
Paraskevaidi et al. [155]	2017	Endometrial cancer and healthy blood plasma and serum	89	Dried samples	ATR-FTIR	Discrimination between cancer subtypes for both plasma and serum due to protein and lipid alterations
Paraskevaidi et al. [156]	2018	Endometrial cancer and healthy blood plasma and serum	85 for plasma, 75 for serum	Dried samples	ATR-FTIR	Aluminium foil: - plasma cancer vs. healthy: changes in protein secondary structure and lipids - serum cancer vs. healthy: changes in glycogen, phosphate, fatty acid and amino acids
						Low-E slides: - plasma cancer vs. healthy: changes in protein secondary structure and lipids - serum cancer vs. healthy: changes in protein secondary structure and nucleic acids
Paraskevaidi et al. [152]	2018	Endometrial cancer and healthy urine	20	Dried samples	ATR-FTIR	Cancer vs. healthy: increased proteins and nucleic acids, decreased lipid content and alterations in protein secondary structure
Bel'skaya et al. [153]	2019	Endometrial cancer and controls saliva	55	Lipid extraction with Folch solution	FTIR	Cancer vs. controls: decreased lipid content
Paraskevaidi et al. [141]	2020	Endometrial cancer, atypical hyperplasia, and healthy blood plasma	652	Dried samples	ATR-FTIR	Cancer vs. healthy: increased lipids and decreased content of carbohydrates and fatty acids
						Hyperplasia vs. healthy: higher nucleic acids, collagen and stretching vibration in DNA and RNA
						Type I vs. Type II cancers: changes in protein secondary structure
Mabwa et al. [143]	2021	Endometrial cancer and healthy blood plasma and serum	60	Dried samples	ATR-FTIR	- Plasma and Serum (bio-fingerprint region 1430 cm^{-1} to 900 cm^{-1})—cancer vs. healthy: changes of stretching vibration in DNA, RNA, changes in fatty acid, amino acid and protein content

More recently, the same group, Paraskevaidi et al. [141], examined plasma from women with endometrial cancer, atypical hyperplasia and controls with ATR-FtIR in the largest diagnostic cross-sectional study to date (total n = 652). The study identified the six most discriminatory peaks for each subgroup analysis, suggesting these features could be developed in a panel of diagnostic spectral markers. Endometrial cancers and controls were differentiated with 87% sensitivity and 78% specificity (overall accuracy of 83%); further analysis of cancer subtypes achieved disease discrimination with sensitivities of 71–100% and specificities of 81–88%. In addition, the authors accounted for potential confounding factors, such as age, body mass index (BMI), diabetes, fasting status and blood pressure and found no impact on spectral classification after applying the MANOVA test (multivariate analysis of variance) to the spectral wavenumbers.

Discrimination between endometrial cancer and controls appears consistently superior for plasma over serum samples. It has been speculated that this might be due to the more complex and heterogeneous composition of the plasma; however, the cause of superior diagnostic results, being a panel of biomolecules or the presence of higher concentration of cell-free DNA, still remains to be determined [149,156].

In addition to plasma and serum, pilot research has also applied biospectroscopy techniques for endometrial cancer diagnosis to urine and saliva analysis.

Urine spectra obtained by Paraskevaidi et al. [152] yielded high levels of diagnostic accuracy after the application of multivariate analysis and classification algorithms (95% sensitivity and 100% specificity, 95% accuracy). The classification methods used included: partial least squares discriminant analysis (PLS-DA), PCA-SVM and genetic algorithm with LDA (GA-LDA.) The majority of discriminating wavelengths were once more located in the lipid, protein and acid nucleic infrared regions.

Saliva spectral analysis by Bel'skaya et al. [153] showed interesting alterations in the lipid regions of patients with endometrial and ovarian cancer; in particular, the ratio of the intensity of the absorption bands 2923/2957 cm^{-1} appeared to consistently decrease in cancer samples compared with controls, leading the authors to propose its use as a new diagnostic criterion.

Spectral differences in lipid absorption bands of healthy and diseased samples, also documented in prostate [157] and breast cancer [158], may be explained by tumour-mediated changes of the lipid metabolism and may warrant further investigations in the context of non-invasive endometrial cancer biomarker development.

5. State of Play, Present Challenges and Future Perspectives

We have reviewed here the progress made to date on the application of vibrational biospectroscopy methods to endometrial cancer studies and their role in the analysis of endometrial tissue and biofluids.

Vibrational biospectroscopy appears to be able to provide detailed information on endometrial architecture, and cell function. Furthermore, it was successfully used to discriminate between normal and cancerous tissues, to identify different cancer subtypes and to differentiate between drug-resistant and drug-sensitive cell phenotypes.

The qualitative and quantitative spectral changes associated with specific groups of biomolecules, such as proteins, lipids, amino acids and carbohydrates, can be exploited to develop disease-specific biomarkers.

The flexibility of the methods allows a variety of fresh, fixed, dry or wet specimens to be processed. An additional advantage is the ability of interrogating, simultaneously, the entirety of the molecules in a sample, in contrast with time-consuming individual biochemical mapping. With the development of computer software able to provide spectral data analysis and display in real time [159], there is an opportunity to conceptualise a novel bedside tool, or a point-of-care test for endometrial cancer. Such a tool, in conjunction with already established diagnostic tests, could facilitate patient triage and selection for additional more invasive procedures, and could be used intra-operatively to assess suspicious peritoneal lesions or metastatic lymph node involvement. Finally, the analysis of biofluids with spectroscopy might support the development of non-invasive screening pathways, as well as strategies for disease monitoring and assessment of response to treatment. Focus on the practical applications of biospectroscopy techniques to endometrial cancer diagnosis, screening and monitoring, therefore, fit nicely in the context of current endometrial cancer research priorities.

While vibrational spectroscopy development made significant progress over the last few decades, several challenges were, however, identified that still stand in the way of clinical translation; indeed, the techniques are still at the experimental stage and not yet in general use in clinics. More work is required, with adequately powered studies: from the adjustment of analytical instrumentation to be suitable for use in a clinical context, such a theatre room, a clinic or a laboratory, to the adaptation and standardisation of protocols for sample collection, processing and data analysis of different biological materials, as well as the essential need for spectral panel validation, reproducibility of results and design of analytical systems to fit well with clinical implementation [12].

These general considerations can also be extended to the application of biospectroscopy in Gynaecology. With regards to endometrial cancer, the first large study was published on the use of blood biospectroscopy as an early detection tool [141]; however, most of the other current evidence here-described consists of pilot and feasibility studies, with the inherent limitation of a small sample size. Patient characteristics, such as hormonal status, body mass index, age and co-morbidities, should be considered and patient groups appropriately matched to allow meaningful comparison between groups. Biofluids are ideal candidates for developing minimally invasive diagnostic/screening tests but, while preliminary studies have shown promising results with regards to the biospectroscopy diagnostic and screening potential for EC, current studies have not yet compared discrimination between EC and other cancers. The identification of disease-specific spectral features would be useful and should be explored prior to clinical implementation.

In addition, the majority of the experimental work in endometrial cancer has so far been conducted on preserved specimens, but to develop a point-of-care or intraoperative test, data are needed on the feasibility of performing reliable spectral analysis of fresh samples. In particular, the adverse impact on the quality of the data that may arise from the presence of water and blood components in the samples and the inherent heterogeneity of endometrial tissues are barriers that must be overcome in order to obtain results that are useful for clinical application. Furthermore, the variability of the sample-processing techniques used in different studies, even in the context of dry and fixed tissues, and the different approaches with regards to chemometric analysis and data extraction, make it challenging to draw direct comparison between datasets to date.

The limitations and challenges highlighted in this paper have important implications for future research. Validation and reproducibility of results should be assessed in larger scale, adequately powered, clinical trials. Future study designs should ensure patient groups are adequately matched. Spectral investigation of endometrial fresh tissue should be explored and results compared to dry analysis and "Gold Standard" histopathology. Finally, sample collection, processing and storage, as well as spectral acquisition and data extraction, should follow standardised protocols in order to minimise bias and guarantee the quality and reproducibility of results. Clearly, there is still a long way to go before clinical implementation, but vibrational biospectroscopy is proving to be an evolving technology with promising applications in endometrial cancer, which certainly warrants further exploration.

Author Contributions: Conceptualization, R.S.; writing—original draft preparation, R.S. and A.B.S., writing—review and editing, R.S., D.F., S.P., A.B.S., W.A. and K.B.G. All authors have read and agreed to the published version of the manuscript.

Funding: This research received funding from EPSRC (EP/T010762/1).

Acknowledgments: The Authors thank the Engineering and Physical Sciences Research Council, UK, [EP/T010762/1] Project: SHAPE (Ceramic SHaping: extrusion of glAss Preforms for new fibres in hEalthcare) for funding support of D. Furniss and A.B. Seddon.

Conflicts of Interest: The authors declare no conflict of interest. The funders had no role in the design of the study; in the collection, analyses, or interpretation of data; in the writing of the manuscript, or in the decision to publish the results.

References

1. Bray, F.; Ferlay, J.; Soerjomataram, I.; Siegel, R.L.; Torre, L.A.; Jemal, A. Global cancer statistics 2018: GLOBOCAN estimates of incidence and mortality worldwide for 36 cancers in 185 countries. *CA Cancer J. Clin.* **2018**, *68*, 394–424. [CrossRef] [PubMed]
2. Zhang, S.; Gong, T.-T.; Liu, F.-H.; Jiang, Y.-T.; Sun, H.; Ma, X.-X.; Zhao, Y.-H.; Wu, Q.-J. Global, Regional, and National Burden of Endometrial Cancer, 1990–2017: Results from the Global Burden of Disease Study, 2017. *Front. Oncol.* **2019**, *9*, 1440. [CrossRef] [PubMed]
3. Crosbie, E.; Morrison, J. The emerging epidemic of endometrial cancer: Time to take action. *Cochrane Database Syst. Rev.* **2014**, ED000095. [CrossRef]
4. Brüggmann, D.; Ouassou, K.; Klingelhöfer, D.; Bohlmann, M.K.; Jaque, J.; Groneberg, D.A. Endometrial cancer: Mapping the global landscape of research. *J. Transl. Med.* **2020**, *18*, 386. [CrossRef]
5. James Lind Alliance. Available online: https://www.jla.nihr.ac.uk/priority-setting-partnerships/womb-cancer/top-10-priorities.htm (accessed on 1 May 2021).
6. Wan, Y.L.; Beverley-Stevenson, R.; Carlisle, D.; Clarke, S.; Edmondson, R.J.; Glover, S.; Holland, J.; Hughes, C.; Kitchener, H.C.; Kitson, S.; et al. Working together to shape the endometrial cancer research agenda: The top ten unanswered research questions. *Gynecol. Oncol.* **2016**, *143*, 287–293. [CrossRef]
7. Pahlow, S.; Weber, K.; Popp, J.; Wood, B.R.; Kochan, K.; Rüther, A.; Perez-Guaita, D.; Heraud, P.; Stone, N.; Dudgeon, A.; et al. Application of Vibrational Spectroscopy and Imaging to Point-of-Care Medicine: A Review. *Appl. Spectrosc.* **2018**, *72*, 52–84. [CrossRef]
8. Kendall, C.; Isabelle, M.; Bazant-Hegemark, F.; Hutchings, J.; Orr, L.; Babrah, J.; Baker, R.; Stone, N. Vibrational spectroscopy: A clinical tool for cancer diagnostics. *Anal.* **2009**, *134*, 1029–1045. [CrossRef]
9. Depciuch, J.; Barnaś, E.; Skręt-Magierło, J.; Skręt, A.; Kaznowska, E.; Łach, K.; Jakubczyk, P.; Cebulski, J. Spectroscopic evaluation of carcinogenesis in endometrial cancer. *Sci. Rep.* **2021**, *11*, 9079. [CrossRef]
10. Barnas, E.; Skret-Magierlo, J.; Skret, A.; Kaznowska, E.; Depciuch, J.; Szmuc, K.; Łach, K.; Krawczyk-Marć, I.; Cebulski, J. Simultaneous FTIR and Raman Spectroscopy in Endometrial Atypical Hyperplasia and Cancer. *Int. J. Mol. Sci.* **2020**, *21*, 4828. [CrossRef]
11. Auner, G.W.; Koya, S.K.; Huang, C.; Broadbent, B.; Trexler, M.; Auner, Z.; Elias, A.; Mehne, K.C.; Brusatori, M.A. Applications of Raman spectroscopy in cancer diagnosis. *Cancer Metastasis Rev.* **2018**, *37*, 691–717. [CrossRef]
12. Baker, M.J.; Byrne, H.J.; Chalmers, J.; Gardner, P.; Goodacre, R.; Henderson, A.; Kazarian, S.G.; Martin, F.L.; Moger, J.; Stone, N.; et al. Clinical applications of infrared and Raman spectroscopy: State of play and future challenges. *Analyst* **2018**, *143*, 1735–1757. [CrossRef]
13. Butler, H.J.; Ashton, L.; Bird, B.; Cinque, G.; Curtis, K.; Dorney, J.; Esmonde-White, K.; Fullwood, N.J.; Gardner, B.; Martin-Hirsch, P.L.; et al. Using Raman spectroscopy to characterize biological materials. *Nat. Protoc.* **2016**, *11*, 664–687. [CrossRef]
14. Cui, S.; Zhang, S.; Yue, S. Raman Spectroscopy and Imaging for Cancer Diagnosis. *J. Health Eng.* **2018**, *2018*, 8619342. [CrossRef]
15. Kong, K.; Kendall, C.; Stone, N.; Notingher, I. Raman spectroscopy for medical diagnostics—From in-vitro biofluid assays to in-vivo cancer detection. *Adv. Drug Deliv. Rev.* **2015**, *89*, 121–134. [CrossRef]
16. Nijssen, A.; Koljenović, S.; Schut, T.C.B.; Caspers, P.J.; Puppels, G.J. Towards oncological application of Raman spectroscopy. *J. Biophotonics* **2009**, *2*, 29–36. [CrossRef]
17. Kazarian, S.G.; Chan, K.L.A. ATR-FTIR spectroscopic imaging: Recent advances and applications to biological systems. *Analyst* **2013**, *138*, 1940–1951. [CrossRef]
18. Santos, F.; Magalhaes, S.; Henriques, M.C.; Fardilha, M.; Nunes, A. Spectroscopic Features of Cancer Cells: FTIR Spectroscopy as a Tool for Early Diagnosis. *Curr. Metab.* **2018**, *6*, 103–111. [CrossRef]
19. Siqueira, L.F.S.; Lima, K.M.G. MIR-biospectroscopy coupled with chemometrics in cancer studies. *Analyst* **2016**, *141*, 4833–4847. [CrossRef]

20. Purandare, N.C.; Trevisan, J.; I Patel, I.; Gajjar, K.; Mitchell, A.L.; Theophilou, G.; Valasoulis, G.; Martin, M.; von Bünau, G.; Kyrgiou, M.; et al. Exploiting biospectroscopy as a novel screening tool for cervical cancer: Towards a framework to validate its accuracy in a routine clinical setting. *Bioanalysis* **2013**, *5*, 2697–2711. [CrossRef]
21. National Cancer Institute. Endometrial Cancer Treatment Physician Data Query (PDQ). Available online: http://www.cancer.gov/cancertopics/pdq/treatment/endometrial/healthprofessional (accessed on 1 May 2021).
22. Cancer Research UK. Available online: https://www.cancerresearchuk.org/health-professional/cancer-statistics/statistics-by-cancer-type/uterine-cancer/incidence (accessed on 15 January 2022).
23. Office for National Statistics. Cancer Survival by Stage at Diagnosis for England (Experimental Statistics)—Office for National Statistics. Available online: https://www.ons.gov.uk/peoplepopulationandcommunity/healthandsocialcare/conditionsanddiseases/datasets/cancersurvivalratescancersurvivalinenglandadultsdiagnosed (accessed on 15 January 2022).
24. Lortet-Tieulent, J.; Ferlay, J.; Bray, F.; Jemal, A. International Patterns and Trends in Endometrial Cancer Incidence, 1978–2013. *JNCI J. Natl. Cancer Inst.* **2017**, *110*, 354–361. [CrossRef]
25. Crosbie, E.; Zwahlen, M.; Kitchener, H.C.; Egger, M.; Renehan, A. Body Mass Index, Hormone Replacement Therapy, and Endometrial Cancer Risk: A Meta-Analysis. *Cancer Epidemiol. Biomark. Prev.* **2010**, *19*, 3119–3130. [CrossRef] [PubMed]
26. Hecht, J.L.; Mutter, G.L. Molecular and Pathologic Aspects of Endometrial Carcinogenesis. *J. Clin. Oncol.* **2006**, *24*, 4783–4791. [CrossRef] [PubMed]
27. Bokhman, J.V. Two pathogenetic types of endometrial carcinoma. *Gynecol. Oncol.* **1983**, *15*, 10–17. [CrossRef]
28. Sherman, M.E. Theories of Endometrial Carcinogenesis: A Multidisciplinary Approach. *Mod. Pathol.* **2000**, *13*, 295–308. [CrossRef] [PubMed]
29. Amant, F.; Moerman, P.; Neven, P.; Timmerman, D.; Van Limbergen, E.; Vergote, I. Endometrial cancer. *Lancet* **2005**, *366*, 491–505. [CrossRef]
30. Daley-Brown, D.; Oprea-Ilies, G.M.; Lee, R.; Pattillo, R.; Gonzalez-Perez, R.R. Molecular cues on obesity signals, tumor markers and endometrial cancer. *Horm. Mol. Biol. Clin. Investig.* **2015**, *21*, 89–106. [CrossRef] [PubMed]
31. Creasman, W.T.; Odicino, F.; Maisonneuve, P.; Quinn, M.A.; Beller, U.; Benedet, J.L.; Heintz, A.; Ngan, H.Y.S.; Pecorelli, S. Carcinoma of the Corpus Uteri. FIGO 26th Annual Report on the Results of Treatment in Gynecological Cancer. *Int. J. Gynecol. Obstet.* **2006**, *95*, S105–S143. [CrossRef]
32. Murali, R.; Soslow, R.; Weigelt, B. Classification of endometrial carcinoma: More than two types. *Lancet Oncol.* **2014**, *15*, e268–e278. [CrossRef]
33. Talhouk, A.; McAlpine, J.N. New classification of endometrial cancers: The development and potential applications of genomic-based classification in research and clinical care. *Gynecol. Oncol. Res. Pract.* **2016**, *3*, 14. [CrossRef]
34. Kandoth, C.; Schultz, N.; Cherniack, A.D.; Akbani, R.; Liu, Y.; Shen, H.; Robertson, A.G.; Pashtan, I.; Shen, R.; Benz, C.C.; et al. Integrated genomic characterization of endometrial carcinoma. *Nature* **2013**, *497*, 67–73. [CrossRef]
35. Church, D.N.; Stelloo, E.; Nout, R.A.; Valtcheva, N.; Depreeuw, J.; Ter Haar, N.; Noske, A.; Amant, F.; Tomlinson, I.P.M.; Wild, P.J.; et al. Prognostic Significance of POLE Proofreading Mutations in Endometrial Cancer. *JNCI J. Natl. Cancer Inst.* **2015**, *107*, 402. [CrossRef]
36. Concin, N.; Matias-Guiu, X.; Vergote, I.; Cibula, D.; Mirza, M.R.; Marnitz, S.; Ledermann, J.; Bosse, T.; Chargari, C.; Fagotti, A.; et al. ESGO/ESTRO/ESP guidelines for the management of patients with endometrial carcinoma. *Int. J. Gynecol. Cancer* **2021**, *31*, 12–39. [CrossRef]
37. Luomaranta, A.; Leminen, A.; Loukovaara, M. Magnetic Resonance Imaging in the Assessment of High-Risk Features of Endometrial Carcinoma: A Meta-Analysis. *Int. J. Gynecol. Cancer* **2015**, *25*, 837–842. [CrossRef]
38. Andreano, A.; Rechichi, G.; Rebora, P.; Sironi, S.; Valsecchi, M.G.; Galimberti, S. MR diffusion imaging for preoperative staging of myometrial invasion in patients with endometrial cancer: A systematic review and meta-analysis. *Eur. Radiol.* **2014**, *24*, 1327–1338. [CrossRef]
39. Cignini, P.; Vitale, S.G.; Laganà, A.S.; Biondi, A.; La Rosa, V.L.; Cutillo, G. Preoperative work-up for definition of lymph node risk involvement in early stage endometrial cancer: 5-year follow-up. *Updat. Surg.* **2017**, *69*, 75–82. [CrossRef]
40. Koplay, M.; Dogan, N.U.; Erdoğan, H.; Sivri, M.; Erol, C.; Nayman, A.; Karabagli, P.; Paksoy, Y.; Çelik, C. Diagnostic efficacy of diffusion-weighted MRI for pre-operative assessment of myometrial and cervical invasion and pelvic lymph node metastasis in endometrial carcinoma. *J. Med. Imaging Radiat. Oncol.* **2014**, *58*, 538–546. [CrossRef]
41. Wouk, N.; Helton, M. Abnormal Uterine Bleeding in Premenopausal Women. *Am. Fam. Physician* **2019**, *99*, 435–443.
42. Gill, K.A. Chapter 1: Introduction to Diagnostic Ultrasound. In *Ultrasound in Obstetrics and Gynaecology*; Davies Publishing, Inc.: Pasadena, CA, USA, 2014; pp. 1–2.
43. Cusimano, M.C.; Simpson, A.N.; Han, A.; Hayeems, R.; Bernardini, M.Q.; Robertson, D.; Kives, S.L.; Satkunaratnam, A.; Baxter, N.N.; Ferguson, S.E. Barriers to care for women with low-grade endometrial cancer and morbid obesity: A qualitative study. *BMJ Open* **2019**, *9*, e026872. [CrossRef]
44. Bosch, T.V.D. Ultrasound in the diagnosis of endometrial and intracavitary pathology: An update. *Australas. J. Ultrasound Med.* **2012**, *15*, 7–12. [CrossRef]
45. Clarke, M.A.; Long, B.J.; Del Mar Morillo, A.; Arbyn, M.; Bakkum-Gamez, J.N.; Wentzensen, N. Association of Endometrial Cancer Risk With Postmenopausal Bleeding in Women: A Systematic Review and Meta-analysis. *JAMA Intern. Med.* **2018**, *178*, 1210–1222. [CrossRef]

46. Timmermans, A.; Opmeer, B.C.; Khan, K.S.; Bachmann, L.M.; Epstein, E.; Clark, T.J.; Gupta, J.K.; Bakour, S.H.; Bosch, T.V.D.; van Doorn, H.C.; et al. Endometrial Thickness Measurement for Detecting Endometrial Cancer in Women With Postmenopausal Bleeding. *Obstet. Gynecol.* **2010**, *116*, 160–167. [CrossRef] [PubMed]
47. Smith-Bindman, R.; Kerlikowske, K.; Feldstein, V.A.; Subak, L.L.; Scheidler, J.; Segal, M.R.; Brand, R.; Grady, D. Endovaginal Ultrasound to Exclude Endometrial Cancer and Other Endometrial Abnormalities. *JAMA* **1998**, *280*, 1510–1517. [CrossRef] [PubMed]
48. Getpook, C.; Wattanakumtornkul, S. Endometrial thickness screening in premenopausal women with abnormal uterine bleeding. *J. Obstet. Gynaecol. Res.* **2006**, *32*, 588–592. [CrossRef] [PubMed]
49. Niculä, R.; Costin, N. Management of endometrial modifications in perimenopausal women. *Clujul Med.* **2015**, *88*, 101–110. [CrossRef]
50. Bignardi, T.; Van den Bosch, T.; Condous, G. Abnormal uterine and post-menopausal bleeding in the acute gynaecology unit. *Best Pract. Res. Clin. Obstet. Gynaecol.* **2009**, *23*, 595–607. [CrossRef]
51. Valle, R.F. Operative Hysteroscopy. *Glob. Libr. Women's Med.* **2008**. [CrossRef]
52. British Gynaecological Cancer Society. *BGCS Uterine Cancer Guidelines: Recommendations for Practice*; British Gynaecological Cancer Society: London, UK, 2017.
53. Bakour, S.H.; Jones, S.E.; O'Donovan, P. Ambulatory hysteroscopy: Evidence-based guide to diagnosis and therapy. *Best Pract. Res. Clin. Obstet. Gynaecol.* **2006**, *20*, 953–975. [CrossRef]
54. Kremer, C.; Barik, S.; Duffy, S. Flexible outpatient hysteroscopy without anaesthesia: A safe, successful and well tolerated procedure. *Br. J. Obstet. Gynaecol.* **1998**, *105*, 672–676. [CrossRef]
55. Clark, T.J.; Voit, D.; Gupta, J.K.; Hyde, C.; Song, F.; Khan, K.S. Accuracy of Hysteroscopy in the Diagnosis of Endometrial Cancer and Hyperplasia. *JAMA* **2002**, *288*, 1610–1621. [CrossRef]
56. Royal College of Obstetricians and Gynaecologists. Diagnostic Hysteroscopy under General Anaesthesia. *Consent Advice 1*. Available online: www.rcog.org.uk/guidance/browse-all-guidance/consent-advice/diagnostic-hysteroscopy-under-general-anaesthesia-consent-advice-no-1 (accessed on 28 April 2021).
57. Relph, S.; Lawton, T.; Broadbent, M.; Karoshi, M. Failed hysteroscopy and further management strategies. *Obstet. Gynaecol.* **2016**, *18*, 65–68. [CrossRef]
58. Farquhar, C.; Ekeroma, A.; Furness, S.; Arroll, B. A systematic review of transvaginal ultrasonography, sonohysterography and hysteroscopy for the investigation of abnormal uterine bleeding in premenopausal women. *Acta Obstet. Gynecol. Scand.* **2003**, *82*, 493–504. [CrossRef]
59. Van Dongen, H.; De Kroon, C.D.; Jacobi, C.E.; Trimbos, J.B.; Jansen, F.W. Diagnostic hysteroscopy in abnormal uterine bleeding: A systematic review and meta-analysis. *BJOG Int. J. Obstet. Gynaecol.* **2007**, *114*, 664–675. [CrossRef]
60. National Institute for Health and Care Excellence. Heavy Menstrual Bleeding: Assessment and Management. *NICE Guideline [NG88]*. Available online: https://www.nice.org.uk/guidance/ng88/chapter/Recommendations-investigations-for-the-cause-of-hmb (accessed on 1 October 2021).
61. National Institute for Health and Care Excellence. Heavy Menstrual Bleeding (Update). A: Evidence Reviews for Diagnostic Test Accuracy in Investigation for Women Presenting with Heavy Menstrual Bleeding. Available online: https://www.nice.org.uk/guidance/ng88/evidence/a-diagnostic-test-accuracy-pdf-4782293101 (accessed on 1 October 2021).
62. Slaoui, M.; Fiette, L. Histopathology procedures: From tissue sampling to histopathological evaluation. *Methods Mol. Biol.* **2011**, *691*, 69–82. [CrossRef]
63. Morelli, P.; Porazzi, E.; Ruspini, M.; Restelli, U.; Banfi, G. Analysis of errors in histology by root cause analysis: A pilot study. *J. Prev. Med. Hyg.* **2013**, *54*, 90. [CrossRef]
64. Phillips, V. Results of a questionnaire regarding criteria for adequacy of endometrial biopsies. *J. Clin. Pathol.* **2005**, *58*, 417–419. [CrossRef]
65. Breijer, M.C.; Visser, N.C.M.; van Hanegem, N.; van der Wurff, A.A.; Opmeer, B.C.; van Doorn, H.C.; Mol, B.W.J.; Pijnenborg, J.M.A.; Timmermans, A. A Structured Assessment to Decrease the Amount of Inconclusive Endometrial Biopsies in Women with Postmenopausal Bleeding. *Int. J. Surg. Oncol.* **2016**, *2016*, 3039261. [CrossRef]
66. van Hanegem, N.; Prins, M.M.; Bongers, M.Y.; Opmeer, B.C.; Sahota, D.; Mol, B.W.J.; Timmermans, A. The accuracy of endometrial sampling in women with postmenopausal bleeding: A systematic review and meta-analysis. *Eur. J. Obstet. Gynecol. Reprod. Biol.* **2016**, *197*, 147–155. [CrossRef]
67. Narice, B.F.; Delaney, B.; Dickson, J.M. Endometrial sampling in low-risk patients with abnormal uterine bleeding: A systematic review and meta-synthesis. *BMC Fam. Pract.* **2018**, *19*, 135. [CrossRef]
68. Gordon, S.; Westgate, J. The incidence and management of failed Pipelle sampling in a general outpatient clinic. *Aust. N. Z. J. Obstet. Gynaecol.* **1999**, *39*, 115–118. [CrossRef]
69. Visser, N.C.; Breijer, M.C.; Herman, M.C.; Bekkers, R.L.; Veersema, S.; Opmeer, B.C.; Mol, B.W.; Timmermans, A.; Pijnenborg, J.M. Factors attributing to the failure of endometrial sampling in women with postmenopausal bleeding. *Acta Obstet. Gynecol. Scand.* **2013**, *92*, 1216–1222. [CrossRef]
70. Williams, A.R.W.; Brechin, S.; Porter, A.J.L.; Warner, P.; Critchley, H.O.D. Factors affecting adequacy of Pipelle and Tao Brush endometrial sampling. *BJOG Int. J. Obstet. Gynaecol.* **2008**, *115*, 1028–1036. [CrossRef] [PubMed]

71. Van Doorn, H.C.; Opmeer, B.C.; Burger, C.W.; Duk, M.J.; Kooi, G.S.; Mol, B.W.J. Dutch Study in Postmenopausal Bleeding (DUPOMEB) Inadequate office endometrial sample requires further evaluation in women with postmenopausal bleeding and abnormal ultrasound results. *Int. J. Gynecol. Obstet.* **2007**, *99*, 100–104. [CrossRef] [PubMed]
72. Mohanlal, R.D. Endometrial sampling at an academic hospital in South Africa: Histological findings, lessons learnt and interesting surprises. *Afr. J. Lab. Med.* **2020**, *9*, 7. [CrossRef] [PubMed]
73. Du, J.; Li, Y.; Lv, S.; Wang, Q.; Sun, C.; Dong, X.; He, M.; Ulain, Q.; Yuan, Y.; Tuo, X.; et al. Endometrial sampling devices for early diagnosis of endometrial lesions. *J. Cancer Res. Clin. Oncol.* **2016**, *142*, 2515–2522. [CrossRef] [PubMed]
74. Clark, T.J.; Mann, C.H.; Shah, N.; Khan, K.S.; Song, F.; Gupta, J.K. Accuracy of outpatient endometrial biopsy in the diagnosis of endometrial cancer: A systematic quantitative review. *BJOG: Int. J. Obstet. Gynaecol.* **2002**, *109*, 313–321. [CrossRef]
75. Dijkhuizen, F.P.; Mol, B.W.; Brölmann, H.A.; Heintz, A.P. The accuracy of endometrial sampling in the diagnosis of patients with endometrial carcinoma and hyperplasia: A meta-analysis. *Cancer* **2000**, *89*, 1765–1772. [CrossRef]
76. Feldman, S.; Berkowitz, R.S.; Tosteson, A.N. Cost-effectiveness of strategies to evaluate postmenopausal bleeding. *Obstet. Gynecol.* **1993**, *81*, 968–975. [CrossRef]
77. Clark, T.J.; Mann, C.H.; Shah, N.; Khan, K.S.; Song, F.; Gupta, J.K. Accuracy of outpatient endometrial biopsy in the diagnosis of endometrial hyperplasia. *Acta Obstet. Gynecol. Scand.* **2001**, *80*, 784–793. [CrossRef]
78. World Health Organisation. Screening. Available online: https://www.who.int/cancer/prevention/diagnosis-screening/screening/en/ (accessed on 15 May 2021).
79. Møller, P.; Seppälä, T.; Bernstein, I.; Holinski-Feder, E.; Sala, P.; Evans, D.G.; Lindblom, A.; Macrae, F.; Blanco, I.; Sijmons, R.; et al. Cancer incidence and survival in Lynch syndrome patients receiving colonoscopic and gynaecological surveillance: First report from the prospective Lynch syndrome database. *Gut* **2017**, *66*, 464–472. [CrossRef]
80. Jacobs, I.; Gentry-Maharaj, A.; Burnell, M.; Manchanda, R.; Singh, N.; Sharma, A.; Ryan, A.; Seif, M.W.; Amso, N.; Turner, G.; et al. Sensitivity of transvaginal ultrasound screening for endometrial cancer in postmenopausal women: A case-control study within the UKCTOCS cohort. *Lancet Oncol.* **2010**, *12*, 38–48. [CrossRef]
81. Clark, T.; Barton, P.; Coomarasamy, A.; Gupta, J.; Khan, K. Gynaecological oncology: Investigating postmenopausal bleeding for endometrial cancer: Cost-effectiveness of initial diagnostic strategies. *BJOG Int. J. Obstet. Gynaecol.* **2006**, *113*, 502–510. [CrossRef]
82. Jacobs, I.J.; Menon, U.; Ryan, A.; Gentry-Maharaj, A.; Burnell, M.; Kalsi, J.K.; Amso, N.; Apostolidou, S.; Benjamin, E.; Cruickshank, D.; et al. Ovarian cancer screening and mortality in the UK Collaborative Trial of Ovarian Cancer Screening (UKCTOCS): A randomised controlled trial. *Lancet* **2016**, *387*, 945–956. [CrossRef]
83. Baker, M.J.; Hussain, S.R.; Lovergne, L.; Untereiner, V.; Hughes, C.; Lukaszewski, R.A.; Thiéfin, G.; Sockalingum, G.D. Developing and understanding biofluid vibrational spectroscopy: A critical review. *Chem. Soc. Rev.* **2016**, *45*, 1803–1818. [CrossRef]
84. Byrne, H.J.; Baranska, M.; Puppels, G.J.; Stone, N.; Wood, B.; Gough, K.M.; Lasch, P.; Heraud, P.; Sulé-Suso, J.; Sockalingum, G.D. Spectropathology for the next generation: Quo vadis? *Analyst* **2015**, *140*, 2066–2073. [CrossRef]
85. Seddon, A.B.; Napier, B.; Lindsay, I.; Lamrini, S.; Moselund, P.M.; Stone, N.; Bang, O.; Farries, M. Prospective on using fibre mid-infrared supercontinuum laser sources for in vivo spectral discrimination of disease. *Analyst* **2018**, *143*, 5874–5887. [CrossRef]
86. Mantsch, H.H. Foreword. In *Biomedical Applications of Synchrotron Infrared Microspectroscopy*; Analytical Spectroscopy Series; Moss, D., Ed.; The Royal Society of Chemistry: London, UK, 2011.
87. Baker, M.J.; Trevisan, J.; Bassan, P.; Bhargava, R.; Butler, H.J.; Dorling, K.M.; Fielden, P.R.; Fogarty, S.W.; Fullwood, N.J.; Heys, K.A.; et al. Using Fourier transform IR spectroscopy to analyze biological materials. *Nat. Protoc.* **2014**, *9*, 1771–1791. [CrossRef]
88. Mitchell, A.L.; Gajjar, K.B.; Theophilou, G.; Martin, F.L.; Martin-Hirsch, P.L. Vibrational spectroscopy of biofluids for disease screening or diagnosis: Translation from the laboratory to a clinical setting. *J. Biophotonics* **2014**, *7*, 153–165. [CrossRef]
89. The International Organisation for Standardisation. ISO 20473: 2007 Optics and Photonics. Spectral Bands 2007, 10. Available online: www.iso.org/standard/39482.html (accessed on 1 May 2021).
90. Stuart, B. *Infrared Spectroscopy: Fundamentals and Applications*; John Wiley & Sons Ltd.: Chichester, UK, 2004.
91. Filik, J.; Frogley, M.D.; Pijanka, J.K.; Wehbe, K.; Cinque, G. Electric field standing wave artefacts in FTIR micro-spectroscopy of biological materials. *Analyst* **2012**, *137*, 853–861. [CrossRef]
92. Bassan, P.; Lee, J.; Sachdeva, A.; Pissardini, J.; Dorling, K.M.; Fletcher, J.S.; Henderson, A.; Gardner, P. The inherent problem of transflection-mode infrared spectroscopic microscopy and the ramifications for biomedical single point and imaging applications. *Analyst* **2013**, *138*, 144–157. [CrossRef]
93. Raman, C.V.; Krishnan, K.S. A New Type of Secondary Radiation. *Nature* **1928**, *121*, 501–502. [CrossRef]
94. Patel, I.I.; Trevisan, J.; Evans, G.; Llabjani, V.; Martin-Hirsch, P.L.; Stringfellow, H.F.; Martin, F.L. High contrast images of uterine tissue derived using Raman microspectroscopy with the empty modelling approach of multivariate curve resolution-alternating least squares. *Analyst* **2011**, *136*, 4950–4959. [CrossRef] [PubMed]
95. Kirsch, M.; Schackert, G.; Salzer, R.; Krafft, C. Raman spectroscopic imaging for in vivo detection of cerebral brain metastases. *Anal. Bioanal. Chem.* **2010**, *398*, 1707–1713. [CrossRef] [PubMed]
96. Cui, L.; Butler, H.J.; Martin-Hirsch, P.L.; Martin, F.L. Aluminium foil as a potential substrate for ATR-FTIR, transflection FTIR or Raman spectrochemical analysis of biological specimens. *Anal. Methods* **2016**, *8*, 481–487. [CrossRef]
97. Crow, P.; Stone, N.; Kendall, C.A.; Uff, J.S.; Farmer, J.A.M.; Barr, H.; Wright, M.P.J. The use of Raman spectroscopy to identify and grade prostatic adenocarcinoma in vitro. *Br. J. Cancer* **2003**, *89*, 106–108. [CrossRef] [PubMed]

98. Theophilou, G.; Lima, K.M.G.; Briggs, M.; Martin-Hirsch, P.L.; Stringfellow, H.F.; Martin, F.L. A biospectroscopic analysis of human prostate tissue obtained from different time periods points to a trans-generational alteration in spectral phenotype. *Sci. Rep.* **2015**, *5*, 13465. [CrossRef] [PubMed]
99. Patel. I.I.; Trevisan, J.; Singh, P.B.; Nicholson, C.M.; Krishnan, R.K.G.; Matanhelia, S.S.; Martin, F.L. Segregation of human prostate tissues classified high-risk (UK) versus low-risk (India) for adenocarcinoma using Fourier-transform infrared or Raman microspectroscopy coupled with discriminant analysis. *Anal. Bioanal. Chem.* **2011**, *401*, 969–982. [CrossRef]
100. Güler, G.; Guven, U.; Oktem, G. Characterization of CD133$^+$/CD44$^+$ human prostate cancer stem cells with ATR-FTIR spectroscopy. *Analyst* **2019**, *144*, 2138–2149. [CrossRef]
101. German, M.; Hammiche, A.; Ragavan, N.; Tobin, M.; Cooper, L.J.; Matanhelia, S.S.; Hindley, A.C.; Nicholson, C.M.; Fullwood, N.J.; Pollock, H.M.; et al. Infrared Spectroscopy with Multivariate Analysis Potentially Facilitates the Segregation of Different Types of Prostate Cell. *Biophys. J.* **2006**, *90*, 3783–3795. [CrossRef]
102. Bergholt, M.S.; Zheng, W.; Lin, K.; Ho, K.Y.; Teh, M.; Yeoh, K.G.; So, J.; Huang, Z. In Vivo Diagnosis of Esophageal Cancer Using Image-Guided Raman Endoscopy and Biomolecular Modeling. *Technol. Cancer Res. Treat.* **2011**, *10*, 103–112. [CrossRef]
103. Yao, H.-W.; Liu, Y.-Q.; Fu, W.; Shi, X.-Y.; Zhang, Y.-F.; Xu, Y.-Z. Initial research on Fourier transform infrared spectroscopy for the diagnosis of colon neoplasms. *Guang Pu Xue Yu Guang Pu Fen Xi Guang Pu* **2011**, *31*, 297–301.
104. Sahu, R.K.; Argov, S.; Walfisch, S.; Bogomolny, E.; Moreh, R.; Mordechai, S. Prediction potential of IR-micro spectroscopy for colon cancer relapse. *Analyst* **2010**, *135*, 538–544. [CrossRef]
105. Nallala, J.; Piot, O.; Diebold, M.-D.; Gobinet, C.; Bouché, O.; Manfait, M.; Sockalingum, G.D. Infrared and Raman Imaging for Characterizing Complex Biological Materials: A Comparative Morpho-Spectroscopic Study of Colon Tissue. *Appl. Spectrosc.* **2014**, *68*, 57–68. [CrossRef]
106. Maziak, D.E.; Do, M.T.; Shamji, F.M.; Sundaresan, S.R.; Perkins, D.G.; Wong, P.T. Fourier-transform infrared spectroscopic study of characteristic molecular structure in cancer cells of esophagus: An exploratory study. *Cancer Detect. Prev.* **2007**, *31*, 244–253. [CrossRef]
107. Cameron, J.M.; Butler, H.J.; Smith, B.R.; Hegarty, M.G.; Jenkinson, M.D.; Syed, K.; Brennan, P.M.; Ashton, K.; Dawson, T.; Palmer, D.S.; et al. Developing infrared spectroscopic detection for stratifying brain tumour patients: Glioblastoma multiforme vs. lymphoma. *Analyst* **2019**, *144*, 6736–6750. [CrossRef]
108. Bury, D.; Morais, C.L.M.; Martin, F.L.; Lima, K.M.G.; Ashton, K.M.; Baker, M.J.; Dawson, T.P. Discrimination of fresh frozen non-tumour and tumour brain tissue using spectrochemical analyses and a classification model. *Br. J. Neurosurg.* **2020**, *34*, 40–45. [CrossRef]
109. Kong, K.; Zaabar, F.; Rakha, E.; Ellis, I.; Koloydenko, A.; Notingher, I. Towards intra-operative diagnosis of tumours during breast conserving surgery by selective-sampling Raman micro-spectroscopy. *Phys. Med. Biol.* **2014**, *59*, 6141–6152. [CrossRef]
110. Zúñiga, W.C.; Jones, V.; Anderson, S.M.; Echevarria, A.; Miller, N.L.; Stashko, C.; Schmolze, D.; Cha, P.D.; Kothari, R.; Fong, Y.; et al. Raman Spectroscopy for Rapid Evaluation of Surgical Margins during Breast Cancer Lumpectomy. *Sci. Rep.* **2019**, *9*, 14639. [CrossRef]
111. Nargis, H.; Nawaz, H.; Ditta, A.; Mahmood, T.; Majeed, M.; Rashid, N.; Muddassar, M.; Bhatti, H.; Saleem, M.; Jilani, K.; et al. Raman spectroscopy of blood plasma samples from breast cancer patients at different stages. *Spectrochim. Acta Part A Mol. Biomol. Spectrosc.* **2019**, *222*, 117210. [CrossRef]
112. Lyng, F.M.; Traynor, D.; Nguyen, T.N.Q.; Meade, A.; Rakib, F.; Al-Saady, R.; Goormaghtigh, E.; Al-Saad, K.; Ali, M.H. Discrimination of breast cancer from benign tumours using Raman spectroscopy. *PLoS ONE* **2019**, *14*, e0212376. [CrossRef]
113. Walsh, M.J.; Holton, S.E.; Kajdacsy-Balla, A.; Bhargava, R. Attenuated total reflectance Fourier-transform infrared spectroscopic imaging for breast histopathology. *Vib. Spectrosc.* **2012**, *60*, 23–28. [CrossRef]
114. Tian, P.; Zhang, W.; Zhao, H.; Lei, Y.; Cui, L.; Wang, W.; Li, Q.; Zhu, Q.; Zhang, Y.; Xu, Z. Intraoperative diagnosis of benign and malignant breast tissues by fourier transform infrared spectroscopy and support vector machine classification. *Int. J. Clin. Exp. Med.* **2015**, *8*, 972–981. [PubMed]
115. Depciuch, J.; Kaznowska, E.; Zawlik, I.; Wojnarowska-Nowak, R.; Cholewa, M.; Heraud, P.; Cebulski, J. Application of Raman Spectroscopy and Infrared Spectroscopy in the Identification of Breast Cancer. *Appl. Spectrosc.* **2016**, *70*, 251–263. [CrossRef] [PubMed]
116. Depciuch, J.; Kaznowska, E.; Golowski, S.; Koziorowska, A.; Zawlik, I.; Cholewa, M.; Szmuc, K.; Cebulski, J. Monitoring breast cancer treatment using a Fourier transform infrared spectroscopy-based computational model. *J. Pharm. Biomed. Anal.* **2017**, *143*, 261–268. [CrossRef] [PubMed]
117. Huang, Z.; McWilliams, A.; Lui, H.; McLean, D.I.; Lam, S.; Zeng, H. Near-infrared Raman spectroscopy for optical diagnosis of lung cancer. *Int. J. Cancer* **2003**, *107*, 1047–1052. [CrossRef] [PubMed]
118. Sun, X.; Xu, Y.; Wu, J.; Zhang, Y.; Sun, K. Detection of lung cancer tissue by attenuated total reflection–Fourier transform infrared spectroscopy—A pilot study of 60 samples. *J. Surg. Res.* **2013**, *179*, 33–38. [CrossRef] [PubMed]
119. Lui, H.; Zhao, J.; McLean, D.; Zeng, H. Real-time Raman Spectroscopy for In Vivo Skin Cancer Diagnosis. *Cancer Res.* **2012**, *72*, 2491–2500. [CrossRef] [PubMed]
120. Lima, C.A.; Goulart, V.P.; Côrrea, L.; Pereira, T.M.; Zezell, D.M. ATR-FTIR Spectroscopy for the Assessment of Biochemical Changes in Skin Due to Cutaneous Squamous Cell Carcinoma. *Int. J. Mol. Sci.* **2015**, *16*, 6621–6630. [CrossRef]

121. Lima, C.A.; Goulart, V.P.; Correa, L.; Zezell, D.M. Using Fourier transform infrared spectroscopy to evaluate biological effects induced by photodynamic therapy. *Lasers Surg. Med.* **2016**, *48*, 538–545. [CrossRef]
122. Walsh, M.J.; Singh, M.N.; Pollock, H.M.; Cooper, L.J.; German, M.J.; Stringfellow, H.F.; Fullwood, N.J.; Paraskevaidis, E.; Martin-Hirsch, P.L.; Martin, F.L. ATR microspectroscopy with multivariate analysis segregates grades of exfoliative cervical cytology. *Biochem. Biophys. Res. Commun.* **2007**, *352*, 213–219. [CrossRef]
123. Kelly, J.G.; Angelov, P.P.; Trevisan, J.; Vlachopoulou, A.; Paraskevaidis, E.; Martin-Hirsch, P.L.; Martin, F.L. Robust classification of low-grade cervical cytology following analysis with ATR-FTIR spectroscopy and subsequent application of self-learning classifier eClass. *Anal. Bioanal. Chem.* **2010**, *398*, 2191–2201. [CrossRef]
124. Kelly, J.G.; Cheung, K.T.; Martin, C.; O'Leary, J.J.; Prendiville, W.; Martin-Hirsch, P.L.; Martin, F.L. A spectral phenotype of oncogenic human papillomavirus-infected exfoliative cervical cytology distinguishes women based on age. *Clin. Chim. Acta* **2010**, *411*, 1027–1033. [CrossRef]
125. Halliwell, D.E.; Morais, C.L.; Lima, K.M.; Trevisan, J.; Siggel-King, M.R.; Craig, T.; Ingham, J.; Martin, D.S.; Heys, K.; Kyrgiou, M.; et al. An imaging dataset of cervical cells using scanning near-field optical microscopy coupled to an infrared free electron laser. *Sci. Data* **2017**, *4*, 170084. [CrossRef]
126. Ramos, I.R.; Meade, A.D.; Ibrahim, O.; Byrne, H.J.; McMenamin, M.; McKenna, M.; Malkin, A.; Lyng, F.M. Raman spectroscopy for cytopathology of exfoliated cervical cells. *Faraday Discuss.* **2016**, *187*, 187–198. [CrossRef]
127. Krishna, C.M.; Prathima, N.B.; Malini, R.; Vadhiraja, B.M.; Bhatt, R.A.; Fernandes, D.J.; Kushtagi, P.; Vidyasagar, M.S.; Kartha, V.B. Raman spectroscopy studies for diagnosis of cancers in human uterine cervix. *Vib. Spectrosc.* **2006**, *41*, 136–141. [CrossRef]
128. Paraskevaidi, M.; Ashton, K.M.; Stringfellow, H.F.; Wood, N.J.; Keating, P.J.; Rowbottom, A.W.; Martin-Hirsch, P.L.; Martin, F.L. Raman spectroscopic techniques to detect ovarian cancer biomarkers in blood plasma. *Talanta* **2018**, *189*, 281–288. [CrossRef]
129. Lima, K.M.G.; Gajjar, K.B.; Martin-Hirsch, P.L.; Martin, F.L. Segregation of ovarian cancer stage exploiting spectral biomarkers derived from blood plasma or serum analysis: ATR-FTIR spectroscopy coupled with variable selection methods. *Biotechnol. Prog.* **2015**, *31*, 832–839. [CrossRef]
130. Frost, J.; Ludeman, L.; Hillaby, K.; Gornall, R.; Lloyd, G.; Kendall, C.; Shore, A.C.; Stone, N. Raman spectroscopy and multivariate analysis for the non invasive diagnosis of clinically inconclusive vulval lichen sclerosus. *Analyst* **2016**, *142*, 1200–1206. [CrossRef]
131. Theophilou, G.; Morais, C.L.M.; Halliwell, D.E.; Lima, K.M.G.; Drury, J.; Martin-Hirsch, P.L.; Stringfellow, H.F.; Hapangama, D.K.; Martin, F.L. Synchrotron- and focal plane array-based Fourier-transform infrared spectroscopy differentiates the basalis and functionalis epithelial endometrial regions and identifies putative stem cell regions of human endometrial glands. *Anal. Bioanal. Chem.* **2018**, *410*, 4541–4554. [CrossRef]
132. Kelly, J.G.; Singh, M.; Stringfellow, H.F.; Walsh, M.J.; Nicholson, J.M.; Bahrami, F.; Ashton, K.M.; Pitt, M.A.; Martin-Hirsch, P.L.; Martin, F.L. Derivation of a subtype-specific biochemical signature of endometrial carcinoma using synchrotron-based Fourier-transform infrared microspectroscopy. *Cancer Lett.* **2009**, *274*, 208–217. [CrossRef]
133. Taylor, S.E.; Cheung, K.T.; Patel, I.I.; Trevisan, J.; Stringfellow, H.F.; Ashton, K.M.; Wood, N.J.; Keating, P.J.; Martin-Hirsch, P.L.; Martin, F.L. Infrared spectroscopy with multivariate analysis to interrogate endometrial tissue: A novel and objective diagnostic approach. *Br. J. Cancer* **2011**, *104*, 790–797. [CrossRef]
134. Murali Krishna, C.; Kegelaer, G.; Adt, I.; Rubin, S.; Kartha, V.; Manfait, M.; Sockalingum, G. Characterisation of uterine sarcoma cell lines exhibiting MDR phenotype by vibrational spectroscopy. *Biochim. Biophys. Acta (BBA)—Gen. Subj.* **2005**, *1726*, 160–167. [CrossRef] [PubMed]
135. Figueira, P.G.M.; Abrão, M.S.; Krikun, G.; Taylor, H.S. Stem cells in endometrium and their role in the pathogenesis of endometriosis. *Ann. N. Y. Acad. Sci.* **2011**, *1221*, 10–17. [CrossRef] [PubMed]
136. Rosen, J.M.; Jordan, C.T. The Increasing Complexity of the Cancer Stem Cell Paradigm. *Science* **2009**, *324*, 1670–1673. [CrossRef] [PubMed]
137. Schwab, K.E.; Gargett, C.E. Co-expression of two perivascular cell markers isolates mesenchymal stem-like cells from human endometrium. *Hum. Reprod.* **2007**, *22*, 2903–2911. [CrossRef]
138. Wolff, E.F.; Wolff, A.B.; Du, H.; Taylor, H.S. Demonstration of Multipotent Stem Cells in the Adult Human Endometrium by In Vitro Chondrogenesis. *Reprod. Sci.* **2007**, *14*, 524–533. [CrossRef]
139. Gargett, C.E.; Chan, R.; Schwab, K.E. Endometrial stem cells. *Curr. Opin. Obstet. Gynecol.* **2007**, *19*, 377–383. [CrossRef]
140. Nallala, J.; Lloyd, G.R.; Stone, N. Evaluation of different tissue de-paraffinization procedures for infrared spectral imaging. *Analyst* **2015**, *140*, 2369–2375. [CrossRef]
141. Paraskevaidi, M.; Morais, C.L.M.; Ashton, K.M.; Stringfellow, H.F.; McVey, R.J.; Ryan, N.A.J.; O'Flynn, H.; Sivalingam, V.N.; Kitson, S.J.; Mackintosh, M.L.; et al. Detecting Endometrial Cancer by Blood Spectroscopy: A Diagnostic Cross-Sectional Study. *Cancers* **2020**, *12*, 1256. [CrossRef]
142. Krishna, C.M.; Kegelaer, G.; Adt, I.; Rubin, S.; Kartha, V.B.; Manfait, M.; Sockalingum, G.D. Combined Fourier transform infrared and Raman spectroscopic approach for identification of multidrug resistance phenotype in cancer cell lines. *Biopolymers* **2006**, *82*, 462–470. [CrossRef]
143. Mabwa, D.; Gajjar, K.; Furniss, D.; Schiemer, R.; Crane, R.; Fallaize, C.; Martin-Hirsch, P.L.; Martin, F.L.; Kypraios, T.; Seddon, A.B.; et al. Mid-infrared spectral classification of endometrial cancer compared to benign controls in serum or plasma samples. *Analyst* **2021**, *146*, 5631–5642. [CrossRef]

144. Movasaghi, Z.; Rehman, S.; Rehman, I.U. Raman Spectroscopy of Biological Tissues. *Appl. Spectrosc. Rev.* **2007**, *42*, 493–541. [CrossRef]
145. Kaznowska, E.; Depciuch, J.; Szmuc, K.; Cebulski, J. Use of FTIR spectroscopy and PCA-LDC analysis to identify cancerous lesions within the human colon. *J. Pharm. Biomed. Anal.* **2017**, *134*, 259–268. [CrossRef]
146. Zendehdel, R.; Masoudi-Nejad, A.; Mohammadzadeh, J.; Shirazi, F.H. Cisplatin Resistant Patterns in Ovarian Cell Line Using FTIR and Principle Component Analysis. *Iran J. Pharm. Res.* **2012**, *11*, 235–240. [CrossRef]
147. Owens, G.; Gajjar, K.; Trevisan, J.; Fogarty, S.W.; Taylor, S.E.; Da Gama-Rose, B.; Martin-Hirsch, P.L.; Martin, F.L. Vibrational biospectroscopy coupled with multivariate analysis extracts potentially diagnostic features in blood plasma/serum of ovarian cancer patients. *J. Biophotonics* **2014**, *7*, 200–209. [CrossRef]
148. Taleb, I.; Thiéfin, G.; Gobinet, C.; Untereiner, V.; Bernard-Chabert, B.; Heurgué, A.; Truntzer, C.; Hillon, P.; Manfait, M.; Ducoroy, P.; et al. Diagnosis of hepatocellular carcinoma in cirrhotic patients: A proof-of-concept study using serum micro-Raman spectroscopy. *Analyst* **2013**, *138*, 4006–4014. [CrossRef]
149. Gajjar, K.; Trevisan, J.; Owens, G.; Keating, P.J.; Wood, N.J.; Stringfellow, H.F.; Martin-Hirsch, P.L.; Martin, F.L. Fourier-transform infrared spectroscopy coupled with a classification machine for the analysis of blood plasma or serum: A novel diagnostic approach for ovarian cancer. *Analyst* **2013**, *138*, 3917–3926. [CrossRef]
150. Hands, J.R.; Clemens, G.; Stables, R.; Ashton, K.; Brodbelt, A.; Davis, C.; Dawson, T.P.; Jenkinson, M.D.; Lea, R.W.; Walker, C.; et al. Brain tumour differentiation: Rapid stratified serum diagnostics via attenuated total reflection Fourier-transform infrared spectroscopy. *J. Neuro-Oncol.* **2016**, *127*, 463–472. [CrossRef]
151. Yu, M.-C.; Rich, P.; Foreman, L.; Smith, J.; Tanna, A.; Dibbur, V.; Unwin, R.; Tam, F.W.K. Label Free Detection of Sensitive Mid-Infrared Biomarkers of Glomerulonephritis in Urine Using Fourier Transform Infrared Spectroscopy. *Sci. Rep.* **2017**, *7*, 4601. [CrossRef]
152. Paraskevaidi, M.; Morais, C.L.M.; Lima, K.M.G.; Ashton, K.M.; Stringfellow, H.F.; Martin-Hirsch, P.L.; Martin, F.L. Potential of mid-infrared spectroscopy as a non-invasive diagnostic test in urine for endometrial or ovarian cancer. *Analyst* **2018**, *143*, 3156–3163. [CrossRef]
153. Bel'Skaya, L.V.; Sarf, E.A.; Solomatin, D.V.; Kosenok, V.K. Analysis of the lipid profile of saliva in ovarian and endometrial cancer by IR fourier spectroscopy. *Vib. Spectrosc.* **2019**, *104*, 102944. [CrossRef]
154. Parlatan, U.; Inanc, M.T.; Ozgor, B.Y.; Oral, E.; Bastu, E.; Unlu, M.B.; Basar, G. Raman spectroscopy as a non-invasive diagnostic technique for endometriosis. *Sci. Rep.* **2019**, *9*, 19795. [CrossRef] [PubMed]
155. Paraskevaidi, M.; Morais, C.; Raglan, O.; Lima, K.M.G.; Martin-Hirsch, P.L.; Paraskevaidis, E.; Kyrgiou, M.; Martin, F.L. Spectroscopy of blood samples for the diagnosis of endometrial cancer and classification of its different subtypes. *J. Clin. Oncol.* **2017**, *35*, 5596. [CrossRef]
156. Paraskevaidi, M.; Morais, C.L.M.; Raglan, O.; Lima, K.M.G.; Paraskevaidis, E.; Martin-Hirsch, P.L.; Kyrgiou, M.; Martin, F.L. Aluminium foil as an alternative substrate for the spectroscopic interrogation of endometrial cancer. *J. Biophotonics* **2018**, *11*, e201700372. [CrossRef] [PubMed]
157. Baker, M.J.; Gazi, E.; Brown, M.D.; Shanks, J.H.; Gardner, P.; Clarke, N.W. FTIR-based spectroscopic analysis in the identification of clinically aggressive prostate cancer. *Br. J. Cancer* **2008**, *99*, 1859–1866. [CrossRef] [PubMed]
158. Fabian, H.; Thi, N.A.N.; Eiden, M.; Lasch, P.; Schmitt, J.; Naumann, D. Diagnosing benign and malignant lesions in breast tissue sections by using IR-microspectroscopy. *Biochim. Biophys. Acta (BBA)—Biomembr.* **2006**, *1758*, 874–882. [CrossRef] [PubMed]
159. Trevisan, J.; Angelov, P.; Scott, A.D.; Carmichael, P.L.; Martin, F.L. IRootLab: A free and open-source MATLAB toolbox for vibrational biospectroscopy data analysis. *Bioinformatics* **2013**, *29*, 1095–1097. [CrossRef] [PubMed]

International Journal of
Molecular Sciences

MDPI

Article

Sera Protein Signatures of Endometrial Cancer Lymph Node Metastases

Doris Mangiaracina Benbrook [1,*], James Randolph Sanders Hocker [2,*,†], Katherine Marie Moxley [3] and Jay S. Hanas [2]

1 Gynecologic Oncology Section, Department of Obstetrics and Gynecology, Stephenson Cancer Center, University of Oklahoma Health Sciences Center, Oklahoma City, OK 73104, USA
2 Department of Biochemistry and Molecular Biology, University of Oklahoma Health Sciences Center, Oklahoma City, OK 73104, USA; jay-hanas@ouhsc.edu
3 Department of Obstetrics and Gynecology, Rogel Cancer Center, University of Michigan Health System, Ann Arbor, MI 48109, USA; kmoxley@med.umich.edu
* Correspondence: doris-benbrook@ouhsc.edu (D.M.B.); james-hocker@ouhsc.edu (J.R.S.H.); Tel.: +1-405-271-5523 (D.M.B.)
† Current Address: Microbiology and Immunology Department, University of Oklahoma Health Sciences Center, Oklahoma City, OK 73104, USA.

Abstract: The presence of lymph node metastases in endometrial cancer patients is a critical factor guiding treatment decisions; however, surgical and imaging methods for their detection are limited by morbidity and inaccuracy. To determine if sera can predict the presence of positive lymph nodes, sera collected from endometrial cancer patients with or without lymph node metastases, and benign gynecology surgical patients (N = 20 per group) were subjected to electron spray ionization mass spectrometry (ES-MS). Peaks that were significantly different among the groups were evaluated by leave one out cross validation (LOOCV) for their ability to differentiation between the groups. Proteins in the peaks were identified by MS/MS of five specimens in each group. Ingenuity Pathway Analysis was used to predict pathways regulated by the protein profiles. LOOCV of sera protein discriminated between each of the group comparisons and predicted positive lymph nodes. Pathways implicated in metastases included loss of PTEN activation and PI3K, AKT and PKA activation, leading to calcium signaling, oxidative phosphorylation and estrogen receptor-induced transcription, leading to platelet activation, epithelial-to-mesenchymal transition and senescence. Upstream activators implicated in these events included neurostimulation and inflammation, activation of G-Protein-Coupled Receptor Gβγ, loss of HER-2 activation and upregulation of the insulin receptor.

Keywords: endometrial cancer; lymph node; metastasis; epithelial-to-mesenchymal transition; leave one out cross validation; mass spectrometry; signaling pathways

Citation: Benbrook, D.M.; Hocker, J.R.S.; Moxley, K.M.; Hanas, J.S. Sera Protein Signatures of Endometrial Cancer Lymph Node Metastases. *Int. J. Mol. Sci.* **2022**, *23*, 3277. https://doi.org/10.3390/ijms23063277

Academic Editor: Laura Paleari

Received: 25 February 2022
Accepted: 16 March 2022
Published: 18 March 2022

1. Introduction

Endometrial cancer is one of the ten major cancers in women and its incidence and mortality are increasing worldwide [1,2]. The presence of lymph node metastases in endometrial cancer patients is key to the clinical prediction of disease recurrence and overall survival [3–5]. Treatment decisions are based on surgical staging with stages I and II confined to the uterus, while stages III and IV have spread beyond the uterus. Metastasis to the locoregional (pelvic and para-aortic) lymph node basins are classified as stage IIIC1 and IIIC2, respectively. Currently, the method and extent of evaluation for lymph node metastases at the time of surgery is highly controversial [6]. Removal and evaluation of lymph nodes (lymphadenectomy) has the benefit of reducing the potential of hematogenous and lymphatic spread of the cancer to distant sites, but at the cost of increased patient morbidity due to surgical complications such as lymphedema, lymphocyst formation, cellulitis, endothelial and neurovascular injury [6,7]. These issues are especially pertinent to obese and morbidly obese patients in whom surgical resection of the lymph nodes is complicated by poor surgical exposure and increased risk of surgical complications. Additionally, the co-morbidities that result from long-standing obesity often make these women poor surgical candidates [7,8]. Thus, the decision to perform a lymphadenectomy remains a topic of debate among gynecologic oncology surgeons and especially in patients with low grade endometrioid histology and multiple medical co-morbidities who are at low risk of extrauterine disease and high risk of surgical morbidity [9,10].

A less morbid alternative to complete pelvic and aortocaval lymphadenectomy is lymphatic mapping (via imaging following cervical injection with dye) followed by directed sampling of the first lymph node(s) that directly drain the uterus (sentinel lymph nodes); however, the utility of sentinel lymph node mapping especially in high-risk endometrial cancer histologies remains controversial [10,11]. Sentinel lymph node mapping increases the overall detection rate for metastases, but has <5% false negative rate [10] when performed with a complete lymphadenectomy. In clinical trials evaluating the predictive value of sentinel lymph node sampling for endometrial cancer, 67% of gynecologic oncologists performed a back-up lymphadenectomy most notably observed when staging patients at high risk for recurrence [12]. Retrospective, multi-institutional studies found that back-up lymphadenectomy did not improve disease-free or overall survival of high-risk or obese endometrial cancer patients [8,13], but did connote clinical benefit by directing the use of adjuvant therapy in patients with identified positive sentinel lymph nodes. The addition of this therapy, chemotherapy and/or radiation therapy was shown to reduce endometrial cancer recurrence in the pelvic sidewall [14], and expanded the use of surgical staging in high morbidity, obese and medically compromised patients.

Based on the importance of knowing the lymph node status in dictating adjuvant therapy and predicting disease-specific survival from endometrial cancer, the current limitations of radiographic assessments in predicting nodal metastases and the controversies surrounding utilization of sentinel lymph node mapping and biopsy versus complete pelvic and aortocaval lymphadenectomy, additional means to identify and understand the biology of metastatic lymph nodes in endometrial cancer patients are needed. Analysis of peripheral blood represents a novel, low morbidity approach to evaluating tumor characteristics at the time of diagnosis that could provide cancer-specific characteristics that can guide surgical management and adjuvant treatment decisions. The objective of this study was to determine if sera proteomic profiling can prediction the presence of lymph node metastases in endometrial cancer patients and to gain insight into the tumor biology associated with lymphatic dissemination of endometrial cancer. Protein profiles present at significantly different levels in sera collected from endometrial cancer patient with or without lymph node metastases and surgical patients with benign gynecologic conditions were identified and found to predict the presence of metastases.

2. Results

Twenty sera samples from each of the following three groups of patients were evaluated by electron-spray ionization MS (ESI-MS), Stages I (N-19) and II (N = 1) endometrial cancer patients (no lymph node metastases), Stage IIIC (N = 19) and IVB (N = 1) endometrial cancer patients and Benign Gynecology surgical patients. All of the endometrial cancer histologies included in the study were endometrioid. There were no significant differences in patient age, body mass indices (BMIs), race, use of tobacco, alcohol, NSAIDs, aspirin, metformin or insulin, or diagnosis of Type 2 diabetes, hypertension, cardiovascular disease or arthritis between the groups (Table 1).

Table 1. Comparison of patient characteristics between groups.

Characteristic	Benign	Stage I or II	Stage IIIC or IVB	p-Value *
Age, y, median (range)	58 (50–64)	57 (52–63)	62 (32–77)	0.43
BMI, mean (SD)	35.4 (7.9)	35.3 (7.5)	34.7 (6.5)	0.95
Race:				0.56
White	10	17	15	
Black	1	2	4	
American Indian	0	1	0	
Asian	0	0	1	
Unknown/Other	9	0	0	
Endometrioid Histology:				
No	N/A			
Yes	N/A			
Tobacco Use:				
No	15	13	13	>0.99
Yes	1	1	2	
Second-hand	3	6	4	
Unknown	1	0	0	
Alcohol Use:				0.78
No	14	17	16	
Yes	5	3	3	
Unknown	1	0	0	
NSAIDs Use:				>0.99
No	17	14	12	
Yes	3	6	7	
Aspirin Use:				>0.99
No	18	16	16	
Yes	2	4	3	
Metformin Use:				>0.99
No	20	19	18	
Yes	0	1	1	
Insulin Use:				>0.99
No	20	20	18	
Yes	0	0	1	
Type 2 Diabetes:				>0.99
No	19	17	16	
Yes	1	3	3	
Hypertension:				>0.99
No	10	13	10	
Yes	10	7	9	
Cardiovascular Disease:				>0.99
No	19	19	19	
Yes	1	1	0	
Arthritis:				>0.99
No	19	18	14	
Yes	1	2	5	

* Ordinary one-way ANOVA used for age and BMI, Friedman test use for all other characteristics.

Mass spectra of diluted serum samples were compared between the groups using leave one out cross validation (LOOCV) (Figure 1A). Individual peaks in the electron spray ionization MS (ESI-MS) that exhibited significantly different normalized areas between the different groups were identified (Figure 1B). The area of each peak represents the average amount of observable components detected in the prepared sera at each indicated centroided m/Z for the specific group of patients indicated. The LOOCV analysis of the data were able to discriminate sera of endometrial cancer patients from Benign Gynecology controls (Figure 1C, $p = 6.6 \times 10^{-20}$). As a validation of the analysis, LOOCV was performed on groups consisting of patients randomly assigned into the groups while matching for age and pathologic staging in each group. If the identification of patients to their randomly assigned group fails (as it did in this situation), this would suggest that the pathology originally defining the patient samples is the major factor producing the group separations. This analysis demonstrated no significant differences in the sera profiles of RND groups (Figure 1D). The overlap of, or lack of differentiation between, the groups when the specimens were randomly assigned to being cancer or benign, provides a negative control for the significant differentiation observed between the groups when they are assigned to their true pathology group of cancer or benign, and suggests that the pathology is the major factor producing the distinction between groups.

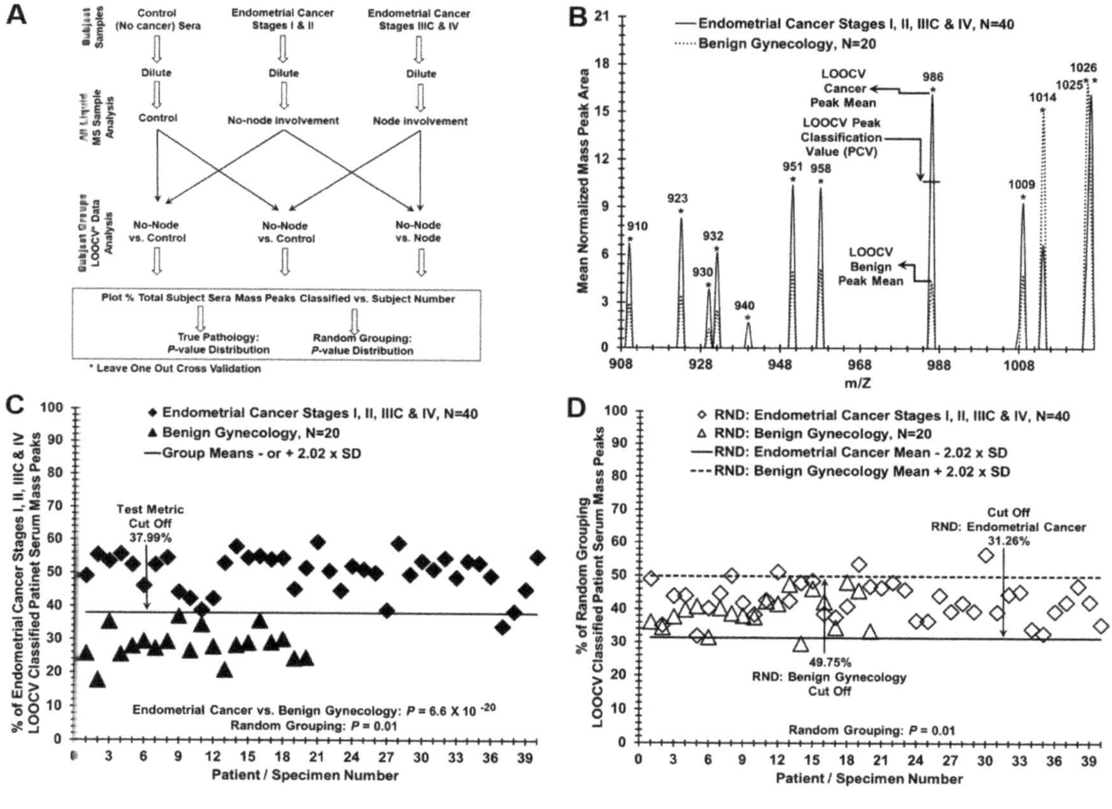

Figure 1. LOOCV of ES-MS data to identify peaks that are significantly different in cancer compared to control subjects. (**A**) Flowchart of approach. (**B**) Plot of peaks (an "*" is above each peak and peak m/z's are specified) with significantly different areas under the curve/amount of protein detected in cancer and control sera. (**C**) LOOCV of all endometrial cancer stages compared to benign. (**D**) LOOCV of all endometrial cancer stages randomized compare to benign gynecology groups when the specimens are randomly assigned to the groups (RND).

LOOCV analysis of the data was also able to differentiate between each combination of two of the three groups (Figure 2A: Stages I and II compared to Benign, Figure 2B: Stages IIC and IV Compared to Benign, Figure 2C: Stages I and II compared to Stages IIIC and IV). Lack of LOOCV analysis differentiation of the groups when specimens were randomized to the different groups instead of being assigned to their true groups documented the lack of arbitrary results (Supplemental Figure S1. LOOCV analysis of randomized groups). This approach was also able to distinguish sera from ovarian cancer compared to endometrial cancer using ESI-MS data from our previous publication [15] that distinguished sera from ovarian cancer from benign controls, and ovarian cancer without (Stages I and II) and with (Stages III and IV) positive lymph nodes (Figure 2D and Supplementary SA1: Author comments concerning the Ovarian data previously published). Again, there were no differentiation between the groups when LOOCV analysis was performed on samples randomly assigned to groups instead of being assigned to their true groups (Supplemental Figure S1). LOOCV analysis was also able to distinguish endometrial cancer from ovarian cancer when evaluated by individual stages of I and III (Supplemental Figure S2). Detailed LOOCV data are provided in Supplementary SB (File SB1: Sera as used in LOOCV Figures; File SB2: Master peak areas for each patient De-identified; File SB3. LOOCV Significant peaks by test and figure; and File SB4: LOOCV results of sera MS peaks).

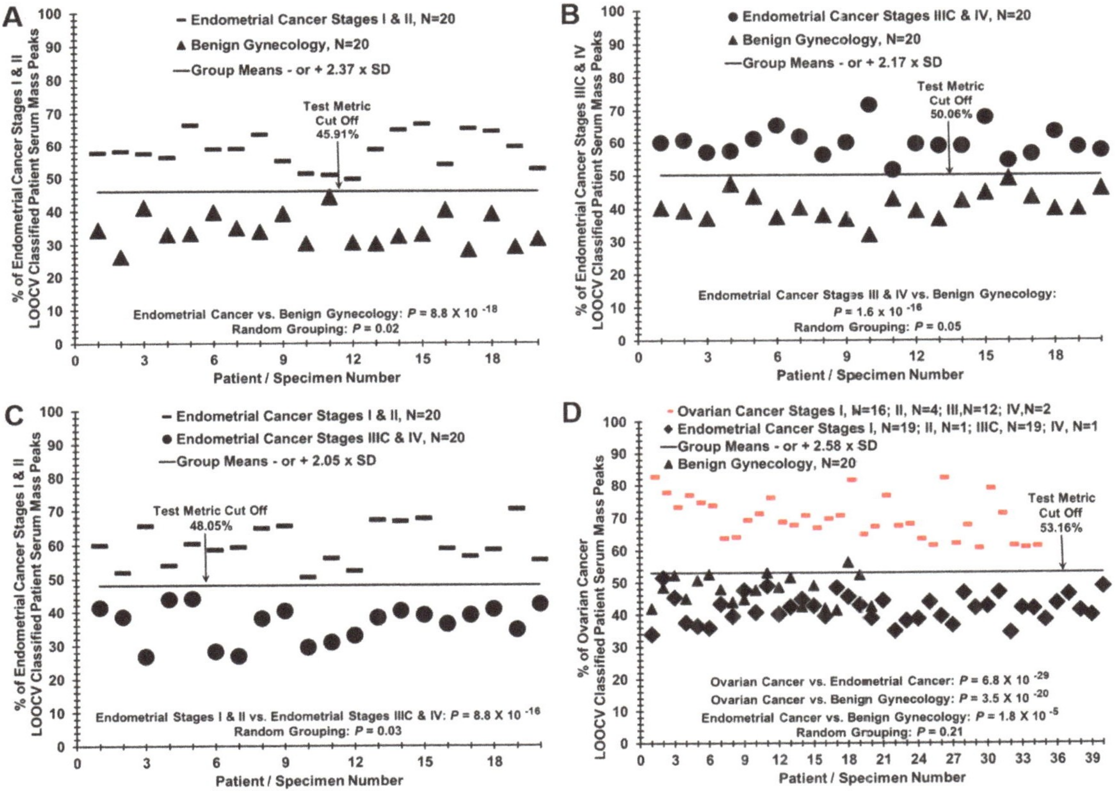

Figure 2. LOOCV analysis comparison of individual groups and ovarian cancer sera. (**A**) Endometrial Cancer Stages I and II compared to Benign; (**B**) Endometrial Cancer Stages IIC and IV compared to Benign; (**C**) Endometrial Cancer Stages I and I compared to Stages IIIC and IV; (**D**) Endometrial Cancer Stages I, II, IIIC and IV endometrial cancer compared to Ovarian Cancer Stages I, II, III and IV.

The proteins in these significant peaks were then identified by subjecting five samples from each group (Supplementary SA. File SA2: Patient samples utilized for MS/MS) to tandem MS/MS mass peak peptide/protein structure identification. Proteins that are overexpressed in diseased cells can often be detected in blood, considered as biomarkers of the disease conditions, and provide insight into the disease mechanism. Therefore, we evaluated the sera protein profiles of the different groups for their involvement with specific canonical signaling pathways which might be driving the endometrial cancer biology. Ingenuity pathway analysis (IPA) of the differential protein profiles in the endometrial cancer groups predicted multiple canonical pathways to be upregulated or downregulated (Table 2 and Figure 3). Comparison of the Stage I and II and Benign groups predicted repression (negative z-scores) of pathways involved in epithelial-to-mesenchymal transition (EMT), fibrosis, HER-2 signaling, neuroinflammation, GP6 and STAT3; and activation (positive z-scores) of Th2 immune cells, PTEN Signaling and Protein Kinase A (PKA) signaling. Comparison of the Stages IIIC and IV and Benign groups identified activation of multiple pathways involved in neuronal signaling, estrogen receptor (ER) signaling, senescence, oxidative phosphorylation, protein kinase A (PKA), G Coupled Protein Receptor (GCPR) $G_{\beta\gamma}$ Signaling and phosphoinositide 3-kinase (PI3K) in B Lymphocytes and Integrin Signaling. No pathways were identified to be significantly repressed in this comparison. The osteoarthritis pathway was the only activated pathway found to be significantly different in the comparison of Stages IIIC and IV and Stages I and II groups.

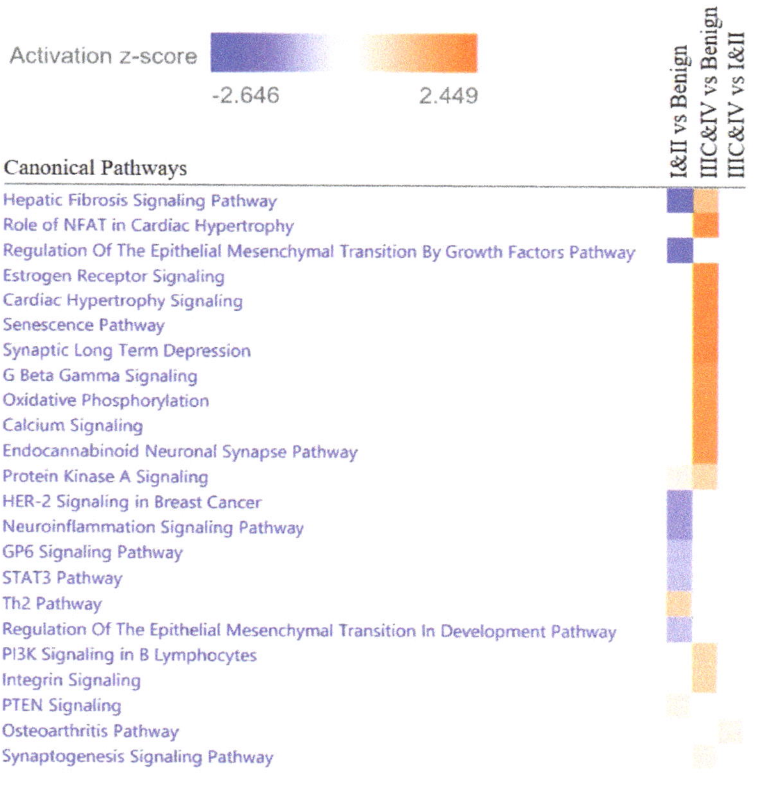

Figure 3. Comparison of z-scores of canonical pathways identified by IPA analysis to be involved in proteins differentially present in sera of groups.

Table 2. Canonical pathways involved with significantly different proteins between groups.

Group Comparison	Pathway	z-Score	*p*-Value	Molecules *
Stages I and II vs. Benign	Hepatic Fibrosis Signaling	−2.646	0.019	FGFR1,GLI1, ITGB2, NOX1,SOS2,TGFBR2,TTN,WNT6
	Regulation of EMT by Growth Factors	−2.236	0.018	EGF,FGFR1,GSC,SOS2,TGFBR2
	HER−2 Signaling in Breast Cancer	−1.342	0.019	EGF, ITGB2, MT-CO1, NRG1, SOS2
	Neuroinflammation Signaling	−1.342	0.040	CASP8, ELP1, GRIN3B, NOX1, PPP3R1, TGFBR2
	GP6 Signaling	−1	0.018	COL4A3, LAMA2, LAMA5, NOX1
	Regulation of EMT in Development	−1	0.005	GLI1, GSC, JAG2, WNT6
	STAT3	−1	0.022	EGF, FGFR1, INSR, TGFBR2
	Th2	1.0	0.0005	ITGB2, JAG2, NOTCH3, NOTCH4,TGFBR2
	PTEN Signaling	0.447	0.0002	ELP1, FGFR1. INSR, ITGB2, MAGI2, SOS2, TGFBR2
	PKA Signaling	0.447	0.041	AKAP13, PPP3R1, PTPN22, PTPRS, RYR2, TGFBR2, TTN
Stages IIIC and IV vs. Benign	Cardiac Hypertrophy Signaling	2.236	0.01	CACNA1A, CACNA1E, KRAS, NFATC4, PLCG1
	Synaptic Long-term Depression	2.236	0.003	CACNA1A,CACNA1E,GRIA4,KRAS, PLCG1
	Role of NFAT in Cardiac Hypertrophy	2.236	0.006	CACNA1A, CACNA1E, KRAS, NFATC4, PLCG1
	ER Signaling	2.236	0.005	CACNA1A,CACNA1E,KRAS, MT-CYB, MT-ND1, MT-ND5, PLCG1
	Senescence	2.236	0.019	CACNA1A, CACNA1E, EP400, KRAS, NFATC4
	Oxidative Phosphorylation	2	0.003	MT-CO2,MT-CYB,MT-ND1,MT-ND5
	G Beta Gamma Signaling	2	0.005	CACNA1A,CACNA1E,KRAS, PLCG1
	Endocannabinoid Neuronal Synapse	2	0.007	CACNA1A,CACNA1E, GRIA4,PLCG1
	Ca^{2+} Signaling	2	0.026	CACNA1A, CACNA1E, GRIA4, NFATC4
	PKA Signaling	1	0.018	FLNB, NFATC4, PDE3B, PLCG1,PTPN5, TTN
	PI3K Signaling in B Lymphocytes	1	0.007	CBL, KRAS, NFATC4, PLCG1
	Integrin Signaling	1	0.025	KRAS, PLCG1, TLN2, TTN
	Synaptogenesis Signaling	0.447	0.023	CDH23, GRIA4, KRAS,PLCG1, RELN
	Osteoarthritis	0.447	0.0006	FN1, ITGB6, LRP1, NOTCH1, PPARD, TNFRSF1B
Stages IIIC and IV vs. Stages II and III	Osteoarthritis	0.447	0.0006	FN1, ITGB6, LRP1, NOTCH1, PPARD, TNFRSF1B

* Abbreviations defined in Supplemental Table S1: Gene IDs and Names of Proteins. Direction of difference between first and second group listed in agreement with pathway activation (green) or repression (red), or no anticipated direction of difference (black).

3. Discussion

The results of this study identify proteins that are differentially present in sera from patients undergoing surgery for Stages I and II Endometrial Cancer, Stages IIIC and IV Endometrial Cancer or Benign Gynecologic Conditions. LOOCV analyses of the protein profiles in these groups were able to distinguish the cancer from the controls or ovarian cancer patients, as well as the two different stage groups from each other. These results suggest that sera and novel biomarkers in sera can be used for early diagnosis and management of endometrial cancer. The ability to predict Stages IIIC and IV compared to Stages I and II offers promise for developing a blood-based assay that can be added to the clinical and surgical endpoints currently used to make treatment decisions. These decisions include potentially avoiding lymphadenectomy or sentinel lymph node mapping in patients who are poor surgical candidates. This preliminary finding provides proof-of-principle for the capability of blood-based assays to predict the presence of lymph node metastases in endometrial cancer patients. Further studies with improved mass spectrometry technology and validation of the differential levels of the identified proteins among the three groups of patients are needed to translate these findings toward clinical application.

The results of this study were not likely affected by patient demographics, which were not significantly different between the groups, or by the endometrial histology type, because only the endometrioid type of endometrial cancer histology was included in this study. This is an important consideration as histology type has been classically used to categorize the aggressiveness of endometrial cancer. Before the advent of tumor molecular characterization, endometrial cancer was categorized into a less-aggressive Type I associated with diabetes, obesity, high estrogen exposure and estrogen receptor α expression, and the more-aggressive Type II, which has a higher risk of metastases and occurs more often in post-menopausal non-obese women [16]. Type I endometrial cancers are primarily endometrioid histology, while the more aggressive histologic types of serous, mucinous and clear cell are categorized as Type II endometrial cancers. The Cancer Genome Atlas (TCGA) analysis of endometrial cancer identified four risk categories based on tumor DNA mutation profile subtypes [17]. Although endometrioid histology is not associated with the most aggressive TCGA molecular subtype, it was chosen for this study because it is the most common endometrial cancer histology and allowed availability of a sufficient number of samples within both early- and late-stage patients across a single histologic type. In general, the Stages I and II endometrial cancer group in this study could be considered to be categorized as the less aggressive Type I, while the Stages IIIC and IV could be categorized as the more aggressive Type II endometrial cancer.

The differential proteins identified in this study offer an opportunity to study the similarities and differences of early and late stages of endometrial cancer biological processes. The presence of proteins in blood can reflect altered levels of these proteins within diseased cells. Therefore, the differential proteins in sera of patients with and without positive lymph nodes can provide insight into the biology of the endometrial cancer metastatic process (Figure 4).

In the comparison of the early Stages I and II Endometrial Cancer with the Benign Gynecology Control Group, several processes commonly deregulated in both endometrial cancer and diabetes were identified: EMT, PTEN/PI3K, and STAT3. EMT is a process that drives endometrial cancer metastases conferring on cells the capability to migrate, invade and resist apoptosis, and is considered the first step in endometrial cancer metastases [18]. Animal models have shown that diabetes increases the EMT of cancer cells [19,20]. Consistent with this association, EMT/fibrosis signaling pathways were found in this study to be repressed in Stages I and II, but not in Stages IIIC and IV endometrial cancer. This loss of EMT/fibrosis prevention is concordant with the presence of positive lymph nodes in Stages IIIC and IV, but not in Stages I and II. The pathways that regulate the EMT in cancer cells were also identified to be associated with differential sera proteins identified in this study.

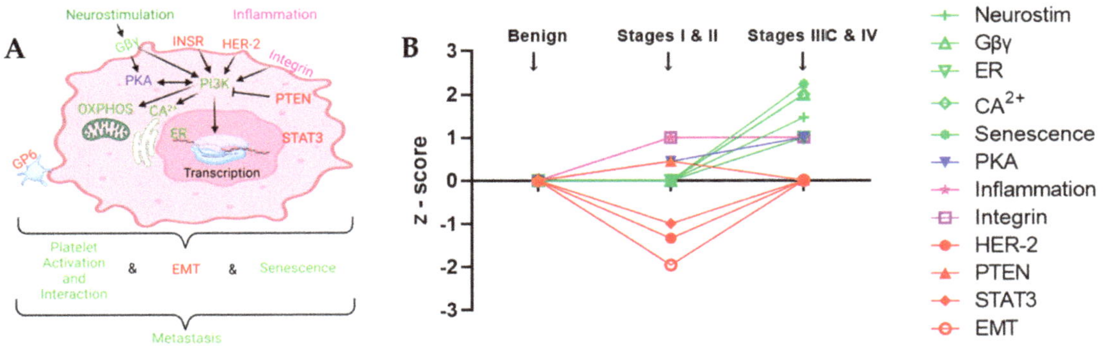

Figure 4. Evolution of canonical pathways in progression from Benign to Early Endometrial Cancer Stages I and II to Metastatic Endometrial Cancer Stages IIIC and IV. (**A**) Illustration of the cellular context of pathway components implicated in endometrial cancer metastases. (**B**) The z-scores for the association of the differential sera proteins in the individual pathways plotted for the different groups. Positive z-scores indicate activation, negative z-scores indicate repression of the pathways. EMT/Fibrosis is the average of hepatic fibrosis signaling, regulation of EMT by growth factors, GP6 signaling and regulation of EMT in development pathways in the Stages I and II group. Neurostimulation (Neurostim) represents the average of cardiac hypertrophy signaling, synaptic long-term depression, role of nfat in cardiac hypertrophy, endocannabinoid neuronal synapse and synaptogenesis signaling in the stages IIIC and IV group. Inflammation refers to Th2 signaling in the stages I and II group and PI3K signaling in B lymphocytes in stages IIIC and IV. (**A**,**B**) In both panels, red indicates loss of pathway regulation is associated with metastatic progression; purple indicates activation of pathway occurs in early endometrial cancer; blue indicates progressively increased activation with metastatic progression; green indicates that pathway regulation is associated with metastatic progression. Subfigure (**A**) created with BioRender.com.

EMT is repressed by PTEN and induced PI3K/Akt pathways [18]. PTEN acts as a tumor suppressor gene and represses activation of the PI3K/Akt pathway [21–23]. PI3K and/or PTEN mutations are present in over 90% of endometrioid endometrial cancers [17]. PTEN expression is commonly lost in endometrial pre-cancer and cancer due to genetic and epigenetic causes; however, its association with patient outcome is controversial and its prognostic significance may be modulated by obesity [24]. Upregulation of the PI3K/Akt pathway by hyperinsulinemia has been shown to drive endometrial cancer development [25]. On the other hand, PTEN expression is elevated in diabetes. Elevated serum PTEN levels in women with gestational diabetes has been associated with increased insulin resistance [26]. In this study, activation of PTEN signaling identified in Stages I and II, but not Stages IIIC and IV compared to the Benign groups may be due to the higher association of diabetes with early stage or Type I endometrial cancer compared to late stage or Type II endometrial cancer. Furthermore, the activation of PTEN signaling in the Stages I and II group is only partial, because it is associated with downregulation of three proteins, FGFR1, SOS2 and TGFBR2, which are decreased in PTEN activation, and is counteracted with the upregulation of INSR and MAGI2, which are increased with PTEN signaling.

The PI3K pathway is a major player in diabetes through its induction by insulin and effects on metabolism [27]. PI3K activation and lack of PTEN Signaling activation in the Stages IIIC and IV Group, compared to each of the other groups, is concordant with a loss of PTEN repression of PI3K as endometrial cancer cells undergo EMT and metastasize. The PI3K pathway identified was specific to B lymphocytes and may reflect activation of an immune response. This lack of PTEN activation and presence of PI3K signaling activation in Stages IIIC and IV compared to Benign groups are logical upstream mediators of the activation of oxidative phosphorylation, CA^{2+} signaling pathways and senescence.

PTEN inhibits, while PI3K stimulates, oxidative phosphorylation in cancer cells [28]. PI3K activation leads to downstream induction of CA^{2+} signaling through effects on lipids and PKA, which modulate CA^{2+} ion channels [29]. Senescence is induced in cancer cells by several oncogenic events, including loss of PTEN and activation of PI3K signaling, while other alterations mutations compromise this oncogene-induced senescence to allow the cancer to thrive [30].

A pathway common to both Stages I and II and Stages IIIC and IV compared to Benign groups is PKA signaling. The z-score for PKA activation increased from the earlier to the later stages, while only the TTN kinase was present in both the early and late stage PKA pathway proteins. Little is known about the role of PKA signaling in endometrial cancer except for a report that luteinizing hormone increases invasion of endometrial cancer cell lines through activation of PKA [31].

The signaling pathways identified in this study (PTEN, PI3K, PKA, CA^{2+}) are known to interact and influence each other's activities and to mediate the downstream activation of the $G_{\beta\gamma}$ GPCR [32]. GPCRs have been shown to be activated by neurotransmitters in ovarian cancer [33]. Neuronal and neuroinflammation pathways were observed to be repressed in Stages I and II and activated in Stages IIIC and IV compared to Benign groups. Proteins involved in neuronal development and excitability have been reported to be overexpressed in endometrial cancer. Axotrophin, which is involved in neuronal development and immune response, was found to be elevated in endometrial cancer and to cause EMT, migration and invasion in endometrial cancer cell lines [34]. HERG K(+)channels, which regulate neuronal and cardiac excitability, were found to be expressed at higher levels in endometrial cancer tissues compared to normal and hyperplastic endometrial epithelium [35]. Neuronal activation pathways may integrate with the other signaling pathways in this study based on their activation of GCPRs in epithelial cancer cells [33].

The HER-2 membrane receptor is another upstream regulator of PTEN/PI3K and PKA signaling identified in this study. Repression of HER-2 signaling was also noted in Stages I and II, but not in Stages IIIC and IV compared to the Benign group. Reports of HER-2 expression association with prognostic significance and clinicopathologic features of endometrial cancer are inconsistent. Positive HER-2 expression has been found to be an independent prognosticator of worse endometrial cancer patient survival in several studies [36–40], while a study of 315 patients with endometrioid histology endometrial cancer found no significant association of HER-2 expression with survival, stages I through IV cancer or other clinicopathologic features [41]. A study of 79 patients with Stages I through IV endometrial cancer with various histologies found that the number of cancers with positive HER-2 staining was greater in tumors with $\leq$50% compared to $\geq$50% myometrial invasion and in patients with absence compared to presence of positive lymph nodes [42]. Another study of 68 Stages I and II endometrial cancer also reported the number of cancers with positive HER-2 staining to be greater tumors with $\geq$50% myometrial invasion compared to $\leq$50% myometrial invasion [43].

EMT is a downstream consequence of STAT3 activation [44]. In this study, STAT3 signaling was repressed in Stages I and II, but not in Stages IIIC and IV compared to Benign groups. Gain of function STAT3 mutations have been shown to promote diabetes, while inhibition of the JAK kinase upstream of STAT3 has been shown to control lymphoproliferation in a diabetes patient with a STAT3 gain of function mutation [45,46]. STAT3 was reported to be upregulated in early-stage endometrial cancer, and inhibition of its signaling pathway has been shown to repress endometrial cancer stem cells and tumor growth [47,48]. Thus, the STAT3 repression observed in this study is likely counteracted by the significantly elevated levels of the INSR insulin receptor observed in the Stages I and II group compared to benign controls. STAT3 repression in Stages I and II, and the lack of STAT3 repression in Stages IIIC and IV, is concordant with the absence and presence of lymph node metastases, respectively, in these groups.

Platelets are also implicated in the mechanism of metastases in this study based on the repression of GP6 signaling in Stages I and II, and lack of this repression in Stages IIIC and

IV. GP6 is a collagen receptor expressed primarily on platelets where it participates platelet activation [49]. Platelets and cancer cells interact in a pathological feed-forward loop in which the cancer cells activate platelets and platelets support cancer cells. Platelet counts are significantly elevated in endometrial cancer patients, and, in combination with other factors, are prognostic for presence of positive lymph nodes and worse prognosis [50–52]. Blood-based assays including platelet counts have also been shown to differentiate between endometrial cancer from atypical endometrial hyperplasia [53]. PI3K and CA^{2+} signaling, and oxidative phosphorylation, mediate platelet activation downstream of GP6 [54].

The identification of activation of the osteoarthritis canonical pathway with Stage IIIC and IV compared to Stages I and II endometrial cancers is consistent with some preliminary findings associating arthritis with increased risk of endometrial cancer. There are several case reports of endometrial cancer in patients with arthritis [55,56]. Patients with the arthritis-associated Myhre syndrome were found to have a 9% incidence of endometrial cancer [57]. A retrospective cohort study found a significant elevation of endometrial cancer in patients diagnosed with dermatomyositis, but not with other rheumatoid arthritis conditions (95% CI = 3.7 to 110.3) [58].

4. Materials and Methods

4.1. Clinical Specimens

The sera samples were obtained from the Stephenson Cancer Center Biospecimen Bank under a University of Oklahoma Health Sciences Center Internal Review Board-approved protocol. The twenty specimens in the Benign Gynecology group were collected from surgical patients with the following pathology diagnoses: ovarian benign neoplasms (N = 8), endometriosis (N = 3), uterine fibroids (N = 2), both ovarian neoplasms and uterine fibroids (N = 3), ovarian and endometrial fibroids and cervical intraepithelial neoplasia (N = 1), atypical endometrial hyperplasia (N = 1), vulvar intraepithelial neoplasia (N = 1) and unknown benign pathology (N = 1). All of the histologies of the 20 endometrial cancers in the Stages I and II groups were endometrioid. All of the histologies of the 20 endometrial cancers in the Stages IIIC and IV groups were endometrioid.

4.2. ESI-MS

Serum ESI-MS and mass peak analysis were performed as described [15,59–61]. In brief, serum samples were diluted (4 µL sera + 1200 µL of a mixture of 50% methanol, 2% formic acid and 48% water). The ESI-MS was performed on an ADVANTAGE (Thermo Fisher, Waltham, MA, USA) ion-trap mass spectrometer in positive ion mode. The source had been modified with a fused silica tip (Polymicro Technologies: Phoenix, AZ, USA) with a 20 micron inside diameter and a 90 micron external diameter. Voltages were set with 1.75 Kv source voltage and 0.34 µamperes current with a capillary temp of 195 °C, for each injection. The flow rate was 0.23 µL per minute using an Eldex MicroPro HPLC pumping system. Samples were triplicate loop injected for MS spectral data and data were acquired for 15–30 min periods. MS signal data (timed from stable injection peak) were extracted from each file in 1.0 m/z units.

4.3. LOOCV Mass Peak Analysis and Statistics

As previously described [15,59–61], post-acquisition MS spectral data processing was performed by locally normalizing each injection data stream set to a total peak intensity value of 100 within a 10 m/z window along the 350–2000 m/z observed range. Peaks were identified using valley to valley definition and averaged within closest 1 unit m/z values. Data acquired in triplicate for each sample were averaged as a representative spectral peak pattern. LOOCV analysis format was used to assess the similarity and significance of peak patterns between the known individual patient groups, using t-tests (significance designated at $p < 0.05$, one-tailed, unequal variance). LOOCV analysis was used to remove the overfitting potential, which is the over-optimistic bias potentially observed when a sample is tested against a set of test variables constructed from a group of samples that also included the sample being tested. Use of the LOOCV procedure reduces this bias by serially removing one sample from the dataset and creating a set of test variables from the remaining samples. The variable sample database was then only valid for use against that particular left-out sample. The sample data were then replaced into its proper group and the next sample is removed to create a new variable database against which it is tested. This sample removal and data base construction continues until all samples have been tested. The peak area at each m/z was compared to the remaining samples. If the left-out sample peak was above the PCV, then that peak was scored with the pathology of the group with the higher mean at the specific peak; otherwise, the peak was scored with the pathology of the group with the lower mean at that peak. All peaks having been scored, the percentage of peaks scored in each group was plotted on the y axis (see patient scoring distribution figures). Additionally, a database series was created where treatments groups are evenly mixed (referred to as random groups "RND") to assess the potential for identifying random unrelated data patterns using the same methodology and number of significant peaks. Each sera sample was scored against its respective database by performing the above described LOOCV, analyzing each peak between the 400–2000 m/z range. Additional detailed information is provided in Supplementary SA (File SA3: Binary LOOCV test metrics, File SA4: Concerning Files uploaded to JPost repository and File SA5: LCQ-ADVANTAGE Instrument Method for MSMS example).

4.4. Tandem MS/MS and Bioinformatic Analysis

For tandem MS peptide sequence analysis [59], samples from sera were randomly selected (Supplementary SA. File SA2: Patient samples utilized for MSMS) and re-analyzed in the MS ion-trap instrument via selected reaction monitoring MS/MS without chromatographic separation of sera. The mass peaks analyzed were pre-determined in the ESI-MS positive mode and found to be significant from the LOOCV analysis ($p < 0.05$). These peaks were chosen for MS/MS isolation and fragmentation. The significant selected peaks were between a 500–1100 m/z range and are a master set of the peaks represented in figures showing MS trace peaks. Identification of peak proteins was established using SEQUEST Proteome Discoverer 1.0 (Thermo Fisher) employing the "no cleavage" setting on a Human database created through the Discoverer software from an NCBI non-redundant database downloaded on 6 October 2015. Peptides/proteins MS/MS identification from samples involved a cross-correlation value (Xcorr) of 2.0 or better [62]. For Ingenuity® Pathway Analysis (IPA®, QIAGEN, Germantown, MD, USA www.qiagen.com/ingenuity (accessed on 10 January 2022)), associated gene names and the number of identified MS/MS sequences were each imported as base-2 log ratios of untreated tumor sequence "hits" divided by treated tumor "hits". Detected pathways were manually inspected and verified using Medline/PubMed. Additional detailed information is provided in Supplementary SA (File SA5: LCQ-ADVANTAGE Instrument Method for MSMS example and File SA6: LCQ-ADVANTAGE Machine Status Log MSMS example) and Supplementary SB (File SB5: Peak list for MSMS analysis of sera).

4.5. Test Metrics

A test/procedure diagnostic value is determined for each group after analysis of each sample through the LOOCV process and is defined by sensitivity, specificity, predictive value and efficiency [63,64]. The sensitivity of the test was determined from TP/(TP + FN) for each comparison, where TP was the number of true positives for known disease presence, and FN was the number of false negatives for known disease presence. Specificity was calculated from TN/(TN + FP), where TN is the number of true negatives and FP is the number of false positives. Each comparison of groups utilized TP, FP, TN and FN values defined using cutoffs #SD (standard deviations) above and below the mean percentage % of classified mass patient serum peaks. The number of SDs from the mean for each group was set equal to each other so there was one cutoff with an equal number of SD on each side from the respective group mean.

5. Conclusions

Sera protein profiles have the potential to predict the presence of positive lymph nodes in endometrial cancer patients. The canonical pathways associated with proteins differentially present in sera of Benign Gynecology, Stages I and II and Stages IIIC and IV groups of patients provide insight into the biology of endometrial cancer metastases (Figure 4). Several pathways involved in the Stages I and II versus Benign groups comparison would logically inhibit metastases (activation of PTEN and repression of HER-2, STAT3 and EMT), and concordantly, these pathways were not present in the Stage IIIC and IV versus to Benign groups comparison, suggesting that loss of PTEN activation and HER-2, STAT3 and EMT repression are involved in endometrial cancer metastatic progression. Activation of GPCR, HER-2 and integrin receptors, leading to PI3K and PKA activation, leading to oxidative phosphorylation and CA^{2+} signaling, leading to EMT and senescence, are implicated in endometrial cancer metastasis, along with ER-mediated transcription and loss of PTEN activation. Neurostimulation and inflammation are implicated as upstream mediators of these events.

Supplementary Materials: The following supporting information can be downloaded at: https://www.mdpi.com/article/10.3390/ijms23063277/s1.

Author Contributions: Conceptualization, D.M.B., J.S.H. and K.M.M.; methodology, J.S.H. and J.R.S.H.; validation, J.R.S.H.; formal analysis, D.M.B., J.R.S.H. and J.S.H.; investigation, K.M.M.; resources, D.M.B.; data curation, J.R.S.H.; writing—original draft preparation, D.M.B.; writing—review and editing, D.M.B., J.R.S.H., and K.M.M.; visualization, D.M.B. and J.R.S.H.; supervision, J.S.H.; project administration, J.S.H.; funding acquisition, D.M.B. All authors have read and agreed to the published version of the manuscript.

Funding: This research was funded by the Gynecologic Cancer Program of the Stephenson Cancer Center NCI Grant P30CA225520.

Institutional Review Board Statement: The animal study protocol was approved by the Institutional Review Board of the University of Oklahoma Health Sciences Center (protocol code 2105 and 20 September 2014).

Informed Consent Statement: Informed consent was obtained from all subjects involved in the study.

Data Availability Statement: https://repository.jpostdb.org/entry/JPST001476 and ftp://ftp.biosciencedbc.jp/archive/jpostrepos/JPST001476 (last accessed 24 February 2022) The following file types have been uploaded: *.raw mass spectrometric (MS and MS/MS) files from ThermoFinnigan/ThermoFisher LCQ-ADVANTAGE, Excel spreadsheets with search results, peak area for analyzed serum, LOOCV results, example LOOCV database file and summarized search tables.

Conflicts of Interest: The authors declare no conflict of interest.

References

1. Forman, D.; Bray, F.B.; Brewster, D.H.; Gombe Mbalawa, C.; Kohler, B.; Pineros, M.; Steliarova-Foucher, E.; Swaminathan, R.; Ferlay, J. *Cancer Incidence in Five Continents*; IARC Scientific Publications No. 164; IARC Scientific Publications: Lyon, France, 2014; Volume X.
2. Siegel, R.L.; Miller, K.D.; Fuchs, H.E.; Jemal, A. Cancer Statistics, 2021. *CA Cancer J. Clin.* **2021**, *71*, 7–33. [CrossRef] [PubMed]
3. Buldukoglu, O.C.; Turker, A.; Usubutun, A.; Salman, M.C. Relationship of lymph node status with survival and recurrence among women with endometrial cancer. *Int. J. Gynaecol. Obs.* **2020**, *151*, 267–271. [CrossRef] [PubMed]
4. Press, J.Z.; Gotlieb, W.H. Controversies in the Treatment of Early Stage Endometrial Carcinoma. *Obstet. Gynecol. Int.* **2012**, *2012*, 578490. [CrossRef]
5. Matsuo, K.; Garcia-Sayre, J.; Medeiros, F.; Casabar, J.K.; Machida, H.; Moeini, A.; Roman, L.D. Impact of depth and extent of lymphovascular space invasion on lymph node metastasis and recurrence patterns in endometrial cancer. *J. Surg. Oncol.* **2015**, *112*, 669–676. [CrossRef] [PubMed]
6. Zhai, L.; Zhang, X.; Cui, M.; Wang, J. Sentinel Lymph Node Mapping in Endometrial Cancer: A Comprehensive Review. *Front. Oncol.* **2021**, *11*, 701758. [CrossRef] [PubMed]

7. Åkesson, Å.; Wolmesjö, N.; Adok, C.; Milsom, I.; Dahm-Kähler, P. Lymphadenectomy, obesity and open surgery are associated with surgical complications in endometrial cancer. *Eur. J. Surg. Oncol.* **2021**, *47*, 2907–2914. [CrossRef]
8. Matanes, E.; Eisenberg, N.; Amajoud, Z.; Gupta, V.; Yasmeen, A.; Ismail, S.; Racovitan, F.; Raban, O.; Lau, S.; Salvador, S.; et al. Sentinel Lymph Node Sampling as an Alternative to Lymphadenectomy in Patients with Endometrial Cancer and Obesity. *J. Obs. Gynaecol. Can.* **2021**, *43*, 1136–1144.e1131. [CrossRef] [PubMed]
9. Colombo, N.; Creutzberg, C.; Amant, F.; Bosse, T.; González-Martín, A.; Ledermann, J.; Marth, C.; Nout, R.; Querleu, D.; Mirza, M.R. et al. ESMO-ESGO-ESTRO Consensus Conference on Endometrial Cancer: Diagnosis, treatment and follow-up. *Ann. Oncol.* **2016** *27*, 16–41. [CrossRef]
10. Holloway, R.W.; Abu-Rustum, N.R.; Backes, F.J.; Boggess, J.F.; Gotlieb, W.H.; Jeffrey Lowery, W.; Rossi, E.C.; Tanner, E.J.; Wolsky, R.J. Sentinel lymph node mapping and staging in endometrial cancer: A Society of Gynecologic Oncology literature review with consensus recommendations. *Gynecol. Oncol.* **2017**, *146*, 405–415. [CrossRef]
11. Marchocki, Z.; Cusimano, M.C.; Clarfield, L.; Kim, S.R.; Fazelzad, R.; Espin-Garcia, O.; Bouchard-Fortier, G.; Rossi, E.C.; Stewart, K.I.; Soliman, P.T.; et al. Sentinel lymph node biopsy in high-grade endometrial cancer: A systematic review and meta-analysis of performance characteristics. *Am. J. Obstet. Gynecol.* **2021**, *225*, 367.e361–367.e339. [CrossRef]
12. Casarin, J.; Multinu, F.; Abu-Rustum, N.; Cibula, D.; Cliby, W.A.; Ghezzi, F.; Leitao, M.; Konishi, I.; Nam, J.-H.; Querleu, D.; et al. Factors influencing the adoption of the sentinel lymph node technique for endometrial cancer staging: An international survey of gynecologic oncologists. *Int. J. Gynecol. Cancer* **2019**, *29*, 60. [CrossRef] [PubMed]
13. Bogani, G.; Papadia, A.; Buda, A.; Casarin, J.; Di Donato, V.; Gasparri, M.L.; Plotti, F.; Pinelli, C.; Paderno, M.C.; Lopez, S.; et al. Sentinel node mapping vs. sentinel node mapping plus back-up lymphadenectomy in high-risk endometrial cancer patients: Results from a multi-institutional study. *Gynecol. Oncol.* **2021**, *161*, 122–129. [CrossRef] [PubMed]
14. How, J.; Gauthier, C.; Abitbol, J.; Lau, S.; Salvador, S.; Gotlieb, R.; Pelmus, M.; Ferenczy, A.; Probst, S.; Brin, S.; et al. Impact of sentinel lymph node mapping on recurrence patterns in endometrial cancer. *Gynecol. Oncol.* **2017**, *144*, 503–509. [CrossRef] [PubMed]
15. Hocker, J.R.; Bishop, E.A.; Lightfoot, S.A.; Lerner, M.R.; Peyton, M.D.; Brackett, D.J.; Hanas, R.J.; McMeekin, D.S.; Walker, J.L.; Hanas, J.S. Serum profiling to distinguish early- and late-stage ovarian cancer patients from disease-free individuals. *Cancer Investig.* **2012**, *30*, 189–197. [CrossRef] [PubMed]
16. Murali, R.; Soslow, R.A.; Weigelt, B. Classification of endometrial carcinoma: More than two types. *Lancet Oncol.* **2014**, *15*, e268–e278. [CrossRef]
17. Kandoth, C.; Schultz, N.; Cherniack, A.D.; Akbani, R.; Liu, Y.; Shen, H.; Robertson, A.G.; Pashtan, I.; Shen, R.; Benz, C.C.; et al. Integrated genomic characterization of endometrial carcinoma. *Nature* **2013**, *497*, 67–73. [CrossRef] [PubMed]
18. Kent, C.N.; Guttilla Reed, I.K. Regulation of epithelial-mesenchymal transition in endometrial cancer: Connecting PI3K, estrogen signaling, and microRNAs. *Clin. Transl. Oncol.* **2016**, *18*, 1056–1061. [CrossRef] [PubMed]
19. Srivastava, S.P.; Goodwin, J.E. Cancer Biology and Prevention in Diabetes. *Cells* **2020**, *9*, 1380. [CrossRef] [PubMed]
20. Viedma-Rodriguez, R.; Martinez-Hernandez, M.G.; Martinez-Torres, D.I.; Baiza-Gutman, L.A. Epithelial Mesenchymal Transition and Progression of Breast Cancer Promoted by Diabetes Mellitus in Mice Are Associated with Increased Expression of Glycolytic and Proteolytic Enzymes. *Horm. Cancer* **2020**, *11*, 170–181. [CrossRef] [PubMed]
21. Parsons, R. Discovery of the PTEN Tumor Suppressor and Its Connection to the PI3K and AKT Oncogenes. *Cold Spring Harb. Perspect. Med.* **2020**, *10*, a036129. [CrossRef] [PubMed]
22. Papa, A.; Pandolfi, P.P. The PTEN-PI3K Axis in Cancer. *Biomolecules* **2019**, *9*, 153. [CrossRef] [PubMed]
23. Ijuin, T. Phosphoinositide phosphatases in cancer cell dynamics-Beyond PI3K and PTEN. *Semin. Cancer Biol.* **2019**, *59*, 50–65. [CrossRef] [PubMed]
24. Westin, S.N.; Ju, Z.; Broaddus, R.R.; Krakstad, C.; Li, J.; Pal, N.; Lu, K.H.; Coleman, R.L.; Hennessy, B.T.; Klempner, S.J.; et al. PTEN loss is a context-dependent outcome determinant in obese and non-obese endometrioid endometrial cancer patients. *Mol. Oncol.* **2015**, *9*, 1694–1703. [CrossRef] [PubMed]
25. Kyo, S.; Nakayama, K. Endometrial Cancer as a Metabolic Disease with Dysregulated PI3K Signaling: Shedding Light on Novel Therapeutic Strategies. *Int. J. Mol. Sci.* **2020**, *21*, 6073. [CrossRef] [PubMed]
26. Li, Y.-Y.; Xiao, R.; Li, C.-P.; Huangfu, J.; Mao, J.-F. Increased plasma levels of FABP4 and PTEN is associated with more severe insulin resistance in women with gestational diabetes mellitus. *Med. Sci. Monit. Int. Med. J. Exp. Clin. Res.* **2015**, *21*, 426–431. [CrossRef]
27. Benchoula, K.; Parhar, I.S.; Wong, E.H. The crosstalk of hedgehog, PI3K and Wnt pathways in diabetes. *Arch. Biochem. Biophys.* **2021**, *698*, 108743. [CrossRef] [PubMed]
28. Cerniglia, G.J.; Dey, S.; Gallagher-Colombo, S.M.; Daurio, N.A.; Tuttle, S.; Busch, T.M.; Lin, A.; Sun, R.; Esipova, T.V.; Vinogradov, S.A.; et al. The PI3K/Akt Pathway Regulates Oxygen Metabolism via Pyruvate Dehydrogenase (PDH)-E1α Phosphorylation. *Mol. Cancer Ther.* **2015**, *14*, 1928–1938. [CrossRef]
29. Ghigo, A.; Laffargue, M.; Li, M.; Hirsch, E. PI3K and Calcium Signaling in Cardiovascular Disease. *Circ. Res.* **2017**, *121*, 282–292. [CrossRef]
30. Xu, Y.; Li, N.; Xiang, R.; Sun, P. Emerging roles of the p38 MAPK and PI3K/AKT/mTOR pathways in oncogene-induced senescence. *Trends Biochem. Sci.* **2014**, *39*, 268–276. [CrossRef]

31. Dabizzi, S.; Noci, I.; Borri, P.; Borrani, E.; Giachi, M.; Balzi, M.; Taddei, G.L.; Marchionni, M.; Scarselli, G.F.; Arcangeli, A. Luteinizing hormone increases human endometrial cancer cells invasiveness through activation of protein kinase A. *Cancer Res.* **2003**, *63*, 4281–4286.
32. Law, N.C.; White, M.F.; Hunzicker-Dunn, M.E. G protein-coupled receptors (GPCRs) That Signal via Protein Kinase A (PKA) Cross-talk at Insulin Receptor Substrate 1 (IRS1) to Activate the phosphatidylinositol 3-kinase (PI3K)/AKT Pathway. *J. Biol. Chem.* **2016**, *291*, 27160–27169. [CrossRef] [PubMed]
33. Predescu, D.-V.; Crețoiu, S.M.; Crețoiu, D.; Pavelescu, L.A.; Suciu, N.; Radu, B.M.; Voinea, S.-C. G Protein-Coupled Receptors (GPCRs)-Mediated Calcium Signaling in Ovarian Cancer: Focus on GPCRs activated by Neurotransmitters and Inflammation-Associated Molecules. *Int. J. Mol. Sci.* **2019**, *20*, 5568. [CrossRef] [PubMed]
34. Liu, L.; Hu, J.; Yu, T.; You, S.; Zhang, Y.; Hu, L. miR-27b-3p/MARCH7 regulates invasion and metastasis of endometrial cancer cells through Snail-mediated pathway. *Acta Biochim. Biophys. Sin.* **2019**, *51*, 492–500. [CrossRef] [PubMed]
35. Cherubini, A.; Taddei, G.L.; Crociani, O.; Paglierani, M.; Buccoliero, A.M.; Fontana, L.; Noci, I.; Borri, P.; Borrani, E.; Giachi, M.; et al. HERG potassium channels are more frequently expressed in human endometrial cancer as compared to non-cancerous endometrium. *Br. J. Cancer* **2000**, *83*, 1722–1729. [CrossRef] [PubMed]
36. Ryan, A.J.; Susil, B.; Jobling, T.W.; Oehler, M.K. Endometrial cancer. *Cell Tissue Res.* **2005**, *322*, 53–61. [CrossRef] [PubMed]
37. Rolitsky, C.D.; Theil, K.S.; McGaughy, V.R.; Copeland, L.J.; Niemann, T.H. HER-2/neu amplification and overexpression in endometrial carcinoma. *Int. J. Gynecol. Pathol. Off. J. Int. Soc. Gynecol. Pathol.* **1999**, *18*, 138–143. [CrossRef] [PubMed]
38. Benevolo, M.; Vocaturo, A.; Novelli, F.; Mariani, L.; Vocaturo, G.; Cianciulli, A.M.; Marandino, F.; Perrone-Donnorso, R.; Giannarelli, D.; Natali, P.G.; et al. Prognostic value of HER2 and progesterone receptor expression in endometrial carcinoma with positive peritoneal washing. *Anticancer Res.* **2007**, *27*, 2839–2844.
39. Srijaipracharoen, S.; Tangjitgamol, S.; Tanvanich, S.; Manusirivithaya, S.; Khunnarong, J.; Thavaramara, T.; Leelahakorn, S.; Pataradool, K. Expression of ER, PR, and Her-2/neu in endometrial cancer: A clinicopathological study. *Asian Pac. J. Cancer Prev. APJCP* **2010**, *11*, 215–220. [PubMed]
40. Kalogiannidis, I.; Petousis, S.; Bobos, M.; Margioula-Siarkou, C.; Topalidou, M.; Papanikolaou, A.; Vergote, I.; Agorastos, T. HER-2/neu is an independent prognostic factor in type I endometrial adenocarcinoma. *Arch. Gynecol. Obstet.* **2014**, *290*, 1231–1237. [CrossRef]
41. Jongen, V.H.; Briët, J.M.; de Jong, R.A.; Joppe, E.; ten Hoor, K.A.; Boezen, H.M.; Evans, D.B.; Hollema, H.; van der Zee, A.G.; Nijman, H.W. Aromatase, cyclooxygenase 2, HER-2/neu, and p53 as prognostic factors in endometrioid endometrial cancer. *Int. J. Gynecol. Cancer* **2009**, *19*, 670–676. [CrossRef]
42. Gai, Q.Z.; Lv, Y.B.; Li, G.Y.; Zhang, D.Q.; Gao, Z.; Fang, X.H. Value of metabolic parameters of primary lesions examined by 18F-FDG PET/CT for endometrial cancer in preoperative evaluation. *Eur. Rev. Med. Pharmacol. Sci.* **2021**, *25*, 2493–2502. [CrossRef] [PubMed]
43. Buchynska, L.G.; Brieieva, O.V.; Iurchenko, N.P. Assessment of HER-2/neu, c-MYC and CCNE1 gene copy number variations and protein expression in endometrial carcinomas. *Exp. Oncol.* **2019**, *41*, 138–143. [CrossRef] [PubMed]
44. Valdez, I.A.; Dirice, E.; Gupta, M.K.; Shirakawa, J.; Teo, A.K.K.; Kulkarni, R.N. Proinflammatory Cytokines Induce Endocrine Differentiation in Pancreatic Ductal Cells via STAT3-Dependent NGN3 Activation. *Cell Rep.* **2016**, *15*, 460–470. [CrossRef] [PubMed]
45. Wegehaupt, O.; Muckenhaupt, T.; Johnson, M.B.; Schwab, K.O.; Speckmann, C. Ruxolitinib Controls Lymphoproliferation and Diabetes in a STAT3-GOF Patient. *J. Clin. Immunol.* **2020**, *40*, 1207–1210. [CrossRef] [PubMed]
46. Warshauer, J.T.; Belk, J.A.; Chan, A.Y.; Wang, J.; Gupta, A.R.; Shi, Q.; Skartsis, N.; Peng, Y.; Phipps, J.D.; Acenas, D.; et al. A human mutation in STAT3 promotes type 1 diabetes through a defect in CD8+ T cell tolerance. *J. Exp. Med.* **2021**, *218*, e20210759. [CrossRef] [PubMed]
47. Wallbillich, J.J.; Josyula, S.; Saini, U.; Zingarelli, R.A.; Dorayappan, K.D.P.; Riley, M.K.; Wanner, R.A.; Cohn, D.E.; Selvendiran, K. High Glucose-Mediated STAT3 Activation in Endometrial Cancer Is Inhibited by Metformin: Therapeutic Implications for Endometrial Cancer. *PLoS ONE* **2017**, *12*, e0170318. [CrossRef] [PubMed]
48. van der Zee, M.; Sacchetti, A.; Cansoy, M.; Joosten, R.; Teeuwssen, M.; Heijmans-Antonissen, C.; Ewing-Graham, P.C.; Burger, C.W.; Blok, L.J.; Fodde, R. IL6/JAK1/STAT3 Signaling Blockade in Endometrial Cancer Affects the ALDHhi/CD126+ Stem-like Component and Reduces Tumor Burden. *Cancer Res.* **2015**, *75*, 3608–3622. [CrossRef] [PubMed]
49. Nieswandt, B.; Bergmeier, W.; Eckly, A.; Schulte, V.; Ohlmann, P.; Cazenave, J.-P.; Zirngibl, H.; Offermanns, S.; Gachet, C. Evidence for cross-talk between glycoprotein VI and Gi-coupled receptors during collagen-induced platelet aggregation. *Blood* **2001**, *97*, 3829–3835. [CrossRef] [PubMed]
50. Muzykiewicz, K.P.; Iwanska, E.; Janeczek, M.; Glanowska, I.; Karolewski, K.; Blecharz, P. The analysis of the prognostic value of the neutrophil/ lymphocyte ratio and the platelet/lymphocyte ratio among advanced endometrial cancer patients. *Ginekol. Pol.* **2021**, *92*, 16–23. [CrossRef] [PubMed]
51. Cong, R.; Kong, F.; Ma, J.; Li, Q.; Wu, Q.; Ma, X. Combination of preoperative neutrophil-lymphocyte ratio, platelet-lymphocyte ratio and monocyte-lymphocyte ratio: A superior prognostic factor of endometrial cancer. *BMC Cancer* **2020**, *20*, 464. [CrossRef] [PubMed]
52. Chen, H.; Wu, Q.; Zhang, Y.; Li, Q.; Ma, J.; Kong, F.; Ma, X. Nomograms based on the novel platelet index score predict postoperative prognosis in endometrial cancer. *Gynecol. Oncol.* **2020**, *158*, 689–697. [CrossRef]

53. Selen, S.; Kilic, F.; Kimyon Comert, G.; Unsal, M.; Kilic, C.; Karalok, A.; Turkmen, O.; Turan, T. Can preoperative inflammatory markers differentiate endometrial cancer from complex atypical hyperplasia/endometrial intraepithelial neoplasia? *J. Obstet. Gynaecol. Res.* **2020**, *46*, 1148–1156. [CrossRef] [PubMed]
54. Fidler, T.P.; Campbell, R.A.; Funari, T.; Dunne, N.; Balderas Angeles, E.; Middleton, E.A.; Chaudhuri, D.; Weyrich, A.S.; Abel, E.D. Deletion of GLUT1 and GLUT3 Reveals Multiple Roles for Glucose Metabolism in Platelet and Megakaryocyte Function. *Cell Rep.* **2017**, *20*, 881–894. [CrossRef] [PubMed]
55. Androutsopoulos, G.; Adonakis, G.; Terzakis, E.; Geropoulou, E.; Decavalas, G. Endometrial cancer in a patient with rheumatoid arthritis. *Eur. J. Gynaecol. Oncol.* **2015**, *36*, 91–93. [PubMed]
56. Abdalla, H.; Bagchi, A.; Bandagi, S. Rheumatoid Arthritis as a Therapeutic Challenge in a Patient with Lynch Syndrome. *Am. J. Case Rep.* **2015**, *16*, 390–392. [CrossRef] [PubMed]
57. Lin, A.E.; Alali, A.; Starr, L.J.; Shah, N.; Beavis, A.; Pereira, E.M.; Lindsay, M.E.; Klugman, S. Gain-of-function pathogenic variants in SMAD4 are associated with neoplasia in Myhre syndrome. *Am. J. Med. Genet. Part A* **2020**, *182*, 328–337. [CrossRef] [PubMed]
58. Chang, S.H.; Park, J.K.; Lee, Y.J.; Yang, J.A.; Lee, E.Y.; Song, Y.W.; Lee, E.B. Comparison of cancer incidence among patients with rheumatic disease: A retrospective cohort study. *Arthritis Res. Ther.* **2014**, *16*, 428. [CrossRef] [PubMed]
59. Hocker, J.R.; Postier, R.G.; Li, M.; Lerner, M.R.; Lightfoot, S.A.; Peyton, M.D.; Deb, S.J.; Baker, C.M.; Williams, T.L.; Hanas, R.J.; et al. Discriminating patients with early-stage pancreatic cancer or chronic pancreatitis using serum electrospray mass profiling. *Cancer Lett.* **2015**, *359*, 314–324. [CrossRef] [PubMed]
60. Hocker, J.R.; Mohammed, A.; Aston, C.E.; Brewer, M.; Lightfoot, S.A.; Rao, C.V.; Hanas, J.S. Mass profiling of serum to distinguish mice with pancreatic cancer induced by a transgenic Kras mutation. *Int. J. Cancer* **2013**, *133*, 2662–2671. [CrossRef] [PubMed]
61. Hocker, J.R.; Peyton, M.D.; Lerner, M.R.; Mitchell, S.L.; Lightfoot, S.A.; Lander, T.J.; Bates-Albers, L.M.; Vu, N.T.; Hanas, R.J.; Kupiec, T.C.; et al. Serum discrimination of early-stage lung cancer patients using electrospray-ionization mass spectrometry. *Lung Cancer* **2011**, *74*, 206–211. [CrossRef]
62. Dittmar, G.A.; Wilkinson, C.R.; Jedrzejewski, P.T.; Finley, D. Role of a ubiquitin-like modification in polarized morphogenesis. *Science* **2002**, *295*, 2442–2446. [CrossRef] [PubMed]
63. Altman, D.G.; Bland, J.M. Diagnostic tests. 1: Sensitivity and specificity. *BMJ* **1994**, *308*, 1552. [CrossRef] [PubMed]
64. Altman, D.G.; Bland, J.M. Statistics Notes: Diagnostic tests 2: Predictive values. *BMJ* **1994**, *309*, 102. [CrossRef] [PubMed]

International Journal of
Molecular Sciences

Article

Gel-Based Proteomic Identification of Suprabasin as a Potential New Candidate Biomarker in Endometrial Cancer

Fulvio Celsi [1], Lorenzo Monasta [1], Giorgio Arrigoni [2,3,4,*], Ilaria Battisti [2,3], Danilo Licastro [5], Michelangelo Aloisio [1], Giovanni Di Lorenzo [1], Federico Romano [1], Giuseppe Ricci [1,6] and Blendi Ura [1,*]

1 Institute for Maternal and Child Health—IRCCS Burlo Garofolo, 65/1 Via dell'Istria, 34137 Trieste, Italy; fulvio.celsi@burlo.trieste.it (F.C.); lorenzo.monasta@burlo.trieste.it (L.M.); michelangelo.aloisio@burlo.trieste.it (M.A.); giovanni.dilorenzo@burlo.trieste.it (G.D.L.); federico.romano@burlo.trieste.it (F.R.); giuseppe.ricci@burlo.trieste.it (G.R.)
2 Department of Biomedical Sciences, University of Padova, 35131 Padova, Italy; ilaria.battisti@studenti.unipd.it
3 Proteomics Centre, University of Padova and Azienda Ospedaliera di Padova, 35131 Padova, Italy
4 CRIBI Biotechnology Center, University of Padova, 35131 Padova, Italy
5 ARGO Laboratorio Genomica ed Epigenomica, AREA Science Park, Basovizza, 34149 Trieste, Italy; danilo.licastro@cbm.fvg.it
6 Department of Medicine, Surgery and Health Sciences, University of Trieste, 34129 Trieste, Italy
* Correspondence: giorgio.arrigoni@unipd.it (G.A.); blendi.ura@burlo.trieste.it (B.U.)

Abstract: Endometrial cancer (EC) is the most frequent gynaecologic cancer in postmenopausal women. We used 2D-DIGE and mass spectrometry to identify candidate biomarkers in endometrial cancer, analysing the serum protein contents of 10 patients versus 10 control subjects. Using gel-based proteomics, we identified 24 candidate biomarkers, considering only spots with a fold change in volume percentage ≥ 1.5 or intensity change ≤ 0.6, which were significantly different between cases and controls ($p < 0.05$). We used Western blotting analysis both in the serum and tissue of 43 patients for data validation. Among the identified proteins, we selected Suprabasin (SBSN), an oncogene previously associated with poor prognosis in different cancers. SBSN principal isoforms were subjected to Western blotting analysis in serum and surgery-excised tissue: both isoforms were downregulated in the tissue. However, in serum, isoform 1 was upregulated, while isoform 2 was downregulated. Data-mining on the TCGA and GTEx projects, using the GEPIA2.0 interface, indicated a diminished SBSN expression in the Uterine Corpus Endometrial Cancer (UCEC) database compared to normal tissue, confirming proteomic results. These results suggest that SBSN, specifically isoform 2, in tissue or serum, could be a potential novel biomarker in endometrial cancer.

Keywords: endometrial cancer; mass spectrometry; serum proteome; Suprabasin; 2D-DIGE; Western blotting

Citation: Celsi, F.; Monasta, L.; Arrigoni, G.; Battisti, I.; Licastro, D.; Aloisio, M.; Di Lorenzo, G.; Romano, F.; Ricci, G.; Ura, B. Gel-Based Proteomic Identification of Suprabasin as a Potential New Candidate Biomarker in Endometrial Cancer. *Int. J. Mol. Sci.* **2022**, *23*, 2076. https://doi.org/10.3390/ijms23042076

Academic Editor: Laura Paleari

Received: 24 December 2021
Accepted: 7 February 2022
Published: 14 February 2022

Publisher's Note: MDPI stays neutral with regard to jurisdictional claims in published maps and institutional affiliations.

1. Introduction

Endometrial cancer (EC), with an increasing incidence, is the most frequent gynaecologic cancer in postmenopausal women [1]. Most EC cases are in the early stages of the disease [2]. Uterine EC is of two types: type 1 is correlated to oestrogen and comprises 80% of cases, while type 2 is described as an independent oestrogen tumour [3].

Many factors increase the risk of developing EC, such as obesity, age, and type 2 diabetes [4].

At present, no diagnostic test is available for EC screening. Abnormal vaginal bleeding is the most common symptom [5]. Further invasive investigations, such as hysteroscopy [6], are needed to obtain a definitive diagnosis.

A test based on biological fluids can dramatically change the diagnosis and treatment of this disease and contribute to its early detection [7]. In this context, molecular biology techniques are fundamental in the early diagnosis and prediction of a cancer therapy's benefits [8]. Serum proteomic permits the identification of new biomarkers for the diagnosis and prognosis of EC [9]. Several candidate biomarkers, such as FAM83D [10], ITIH4, CLU, C1R, and SERPINC1, have been previously identified with proteomic technology [11].

Many oncogenes have been identified as possible biomarkers in EC, including EGFR, PI3KCA, K-Ras, HER2/neu, and FGFR2 [12]. In physiological conditions, these genes are inactivated, while their activation would lead to an uncontrollable proliferation of cells [13].

Suprabasin (SBSN) is an oncogene and biomarker in several cancers such as lung carcinoma, salivary adenoid cystic carcinoma, and myelodysplastic syndromes (MDS). The physiological role of SBSN is still unknown [14]. The human SBSN gene is localised in chromosome 19 and consists of five exons and four introns, while mRNA produces three isoforms by alternative splicing [15]. The first two proteoforms of SBSN are well defined [16]: isoform 1 has 590 aa, with a predicted mass of 60.541 Da; while isoform 2, a 247 aa long polypeptide with a predicted mass of 25.335 Da, is a proposed oncogene in human malignancies [16].

2D-DIGE is a modified version of 2D-PAGE, which uses up to four fluorescent tags for protein labelling [11]. This technology was successfully employed for the identification of several biomarkers in cancers [17–19] and to characterise new pathways in cancer pathophysiology [20,21].

In this study, for the first time, we quantified the abundance of the two isoforms of SBSN in cancer tissue and serum by using Western blotting and evaluated the possible benefit of this protein as a potential novel biomarker in endometrial cancer.

2. Results

2.1. Proteomic Study

We used 2D-DIGE coupled with MS for the proteomic study to compare the enriched serum proteomic profile of 10 controls (Cys 3) and 10 EC (Cys 5). Proteomweaver software detected more than 2500 (Figure 1) protein spots in both types of the proteome. After software analysis, 24 protein spots (Table 1) showed a significant alteration ($p < 0.05$) of their volume in EC vs. control samples, with a fold change of $\geq$1.5 or $\leq$0.6. Seven of them revealed a fold change $\geq$1.5 (APOC3, APOC2, APOE, SERPINC1, C1R, SERPINA1, A2M), while 17 proteins indicated a fold change $\leq$0.6 (APOA1, APOA1, APCS, APOE, CLU, CD5L, CFHR1, VTN, C9, C8A, ALB, C4BPA, IGHM, ITIH2, C1R, SERPINA1, FLG2, SBSN, APOA4, CPS1). Spots of interest were subjected to in-gel digestion and LC-MS/MS analysis, and proteins were identified by searching the MS/MS data against the human section of the UniProt database. All parameters functional in assessing the quality of peptide and protein identifications have been reported in Supplementary Files S1 and S2.

Table 1. Different abundance of proteins identified by mass spectrometry in EC compared to the abundance in control serum.

Accession Number	Spot Number	Protein Description	Gene Symbol	Protein Score	Fold Change *	*p*-Value
A0A3B3ISR2	28	Complement subcomponent C1r	C1R	164.93	4	0.044
P01009	29	Alpha-1-antitrypsin	SERPINA1	325.07	3.66	0.033
P01023	31	Alpha-2-macroglobulin	A2M	150.15	3	0.022
P10909	10	Clusterin	CLU	398.29	2.5	0.033
P01008	14	Antithrombin-III	SERPINC1	403.15	2.22	0.029
P02655	2A	Apolipoprotein C-II	APOC2	152.70	2	0.044
P02656	1A	Apolipoprotein C-III	APOC3	623.84	1.98	0.033
P02743	7	Serum amyloid P-component	APCS	557.00	0.6	0.049
P02649	9	Apolipoprotein E	APOE	324.58	0.6	0.048
P02768	21	Albumin	ALB	1017.31	0.6	0.041
P02748	17	Complement component C9	C9	261.24	0.54	0.021
P07357	20	Complement component C8 alpha chain	C8A	111.76	0.53	0.045
Q5D862	35	Filaggrin 2	FLG2	105.48	0.45	0.036
Q6UWP8	37	Suprabasin	SBSN	156.66	0.43	0.022
P06727	38	Apolipoprotein A-IV	APOA4	844.88	0.4	0.046
P04004	16	Vitronectin	VTN	368.01	0.4	0.021
B1AKG0	12	Complement factor H-related protein 1	CFHR1	354.47	0.39	0.030

Table 1. *Cont.*

Accession Number	Spot Number	Protein Description	Gene Symbol	Protein Score	Fold Change *	p-Value
P02647	3	Apolipoprotein A-I	APOA1	378.52	0.38	0.028
P31327	33	Carbamoyl-phosphate synthase [ammonia], mitochondrial	CPS1	110.97	0.3	0.036
P02647	5	Apolipoprotein A-I	APOA1	481.70	0.28	0.034
O43866	11	CD5 antigen-like	CD5L	127.16	0.28	0.033
P04003	23	C4b-binding protein alpha chain	C4BPA	373.27	0.24	0.0099

* Fold change is defined as the mean % volume ratio according to the formula: %V = Volume single spot/Volume total spot of EC vs. C.

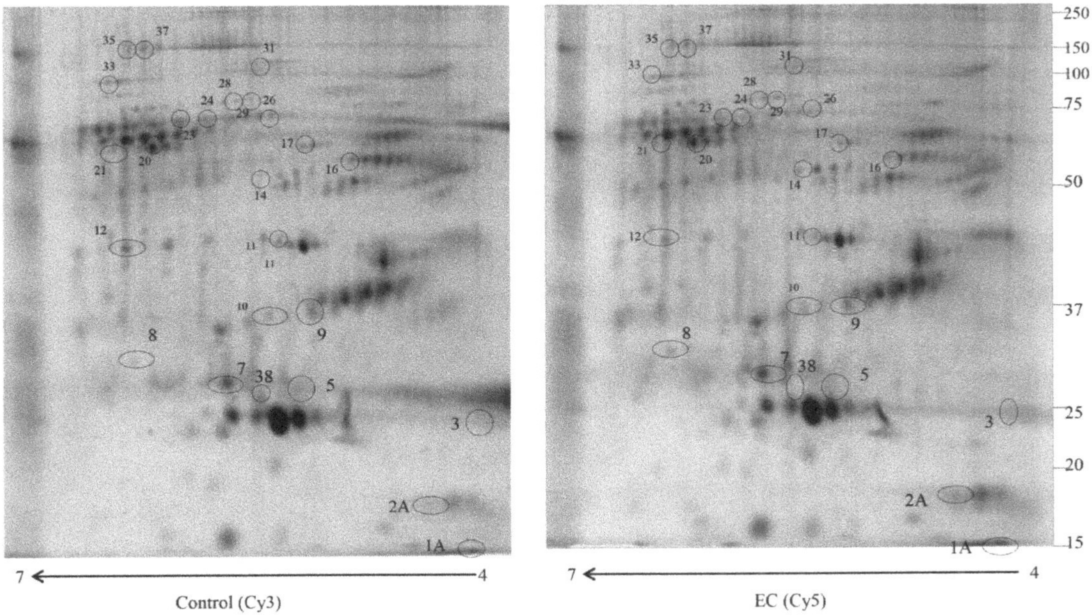

Figure 1. 2D-DIGE map of depleted serum from control serum and endometrial serum. IPG strips 4–7 were used for the first dimension, and 10% SDS-PAGE was used for the second dimension. The numbered circles indicate the differently abundant spots.

2.2. Western Blotting for SBSN Validation

The altered abundance of 2D-DIGE SBSN in depleted serum was validated by Western blotting. SBSN was chosen for proteomic data validation since it has been previously reported either as an oncogene or as a biomarker in other cancers. For both isoforms, the abundance of SBSN in the enriched serum was validated in 30 controls versus 30 EC patients. The abundance of isoform 2 was lower in EC serum than in controls ($p = 0.0005$ and ROC = 0.7544) (Figure 2), while isoform 1 in serum is not significantly higher than in controls ($p = 0.2523$ and ROC = 0.5867).

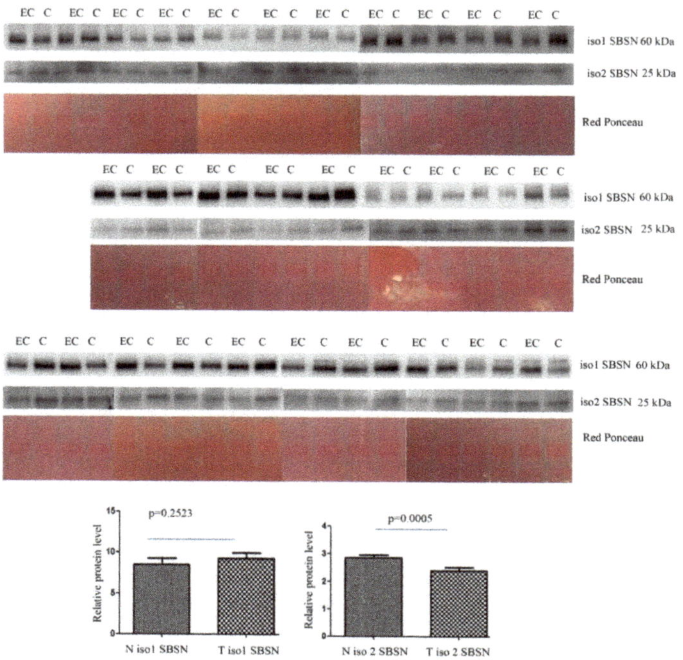

Figure 2. Western blotting analysis of serum isoform 1 and isoform 2 of SBSN in controls (C) and endometrial cancer (EC) patients. The intensity of immunostained bands was normalised against the total protein intensities measured from the same blot stained with Red Ponceau. The graph shows the relative abundance of the two isoforms in control and endometrial cancer serum. Results are shown as a histogram ($p < 0.05$), each bar representing mean ± standard deviation.

The abundance of isoform 1 in tissue was lower in EC than in controls ($p = 0.0001$ and ROC = 0.7928) (Figure 3). Isoform 2 was also lower in EC than in controls ($p = 0.0001$ and ROC = 0.7933).

We calculated the ratio of SBSN-1 between ADK patients and controls, and did the same for SBSN-2. We then calculated the Spearman's rank correlation between the two ratios, both in the serum and in tissue samples. Figures 4 and 5 show the plotted values of the ratios for serum and tissue samples, respectively. In serum samples, the rank correlation between the two ratios was significant (rho = 0.4433, $p = 0.0142$). In tissue samples, the rank correlation between the two ratios was somehow weaker but still significant (rho = 0.3820, $p = 0.0372$).

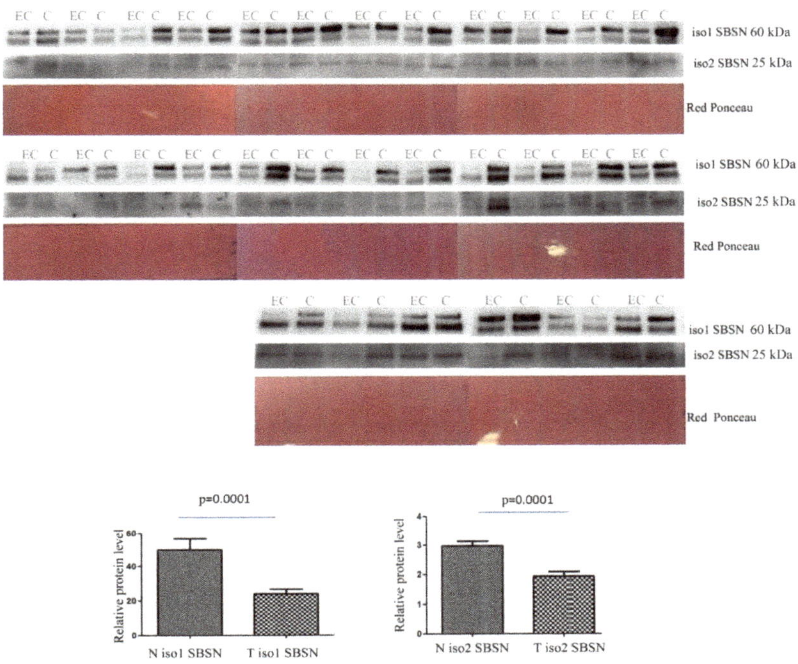

Figure 3. Western blotting analysis of tissue isoform 1 and isoform 2 of SBSN in controls (C) and endometrial cancer (EC) patients. The intensity of immunostained bands was normalised against the total protein intensities measured from the same blot stained with Red Ponceau. The graph shows the relative abundance of the two isoforms in control and endometrial cancer tissue. Results are shown as a histogram ($p < 0.05$), each bar representing mean ± standard deviation.

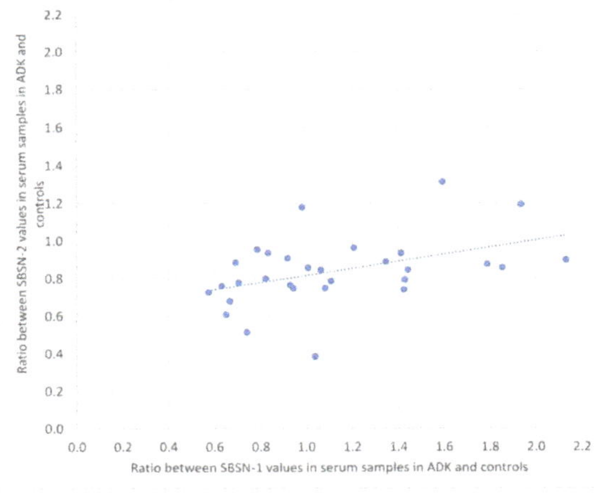

Figure 4. Plotted values of the ratios between SBSN-1 values in ADK vs. controls and the corresponding values for SBSN-2 in serum samples.

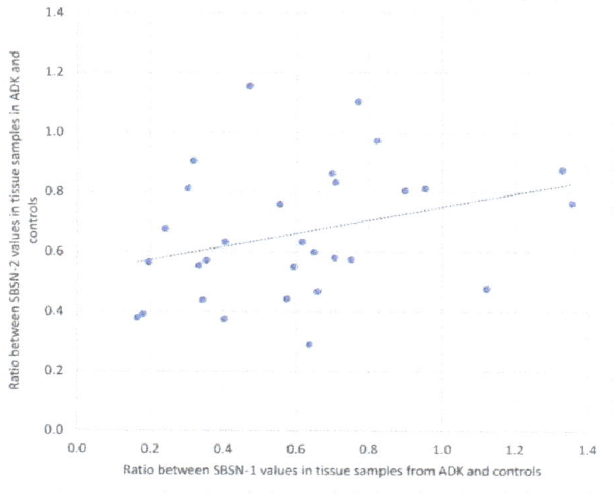

Figure 5. Plotted values of the ratios between SBSN-1 values in ADK vs. controls and the corresponding values for SBSN-2 in tissue samples.

2.3. Bioinformatic Analysis

We aimed to confirm the proteomic results and investigated the TCGA database to assess SBSN expression in EC compared to normal tissue using the GEPIA2 portal. In Figure 6 we reported SBSN mRNA expression in normal versus EC tissues, showing a slight reduction in gene expression in tumours, although not statistically significant ($p = 0.072$).

For enrichment data, we used g: Profiler classification. This tool categorised the identified proteins into groups according to their molecular function, biological processes, and cellular component (Figure 7). Regarding the molecular function, proteins were categorised into phosphatidylcholine-sterol O-acyltransferase, lipoprotein particle receptor binding, enzyme inhibitor activity, and lipase inhibitor activity, while for biological processes, proteins were classified into complement activation, humoral immune response, high-density lipoprotein particle remodelling, and reverse cholesterol transport. Proteins were organised into blood microparticles, extracellular region, extracellular space, and collagen-containing extracellular matrix for cellular components. Pathway enrichment analysis was performed using the REACTOME tool (Figure 7). Proteins were then grouped into six main pathways: plasma lipoprotein remodelling (APOA4, APOE, APOA1, APOC3, APOC2, ALB), plasma lipoprotein assembly, remodelling, and clearance (APOA4, APOE, APOA1, APOC3, APOC2, ALB, A2M), complement cascade (C4BPA, C1R, CFHR1, C9, APCSVTN, C8A, CLU), post-translational protein phosphorylation (APOE, APOA1, ITIH2, SERPINC1, ALB, SERPINA1), plasma lipoprotein assembly (APOA4, APOE, APOA1, APOC3, APOC2, A2M), and chylomicron assembly (APOA4, APOE, APOA1, APOC3, APOC2).

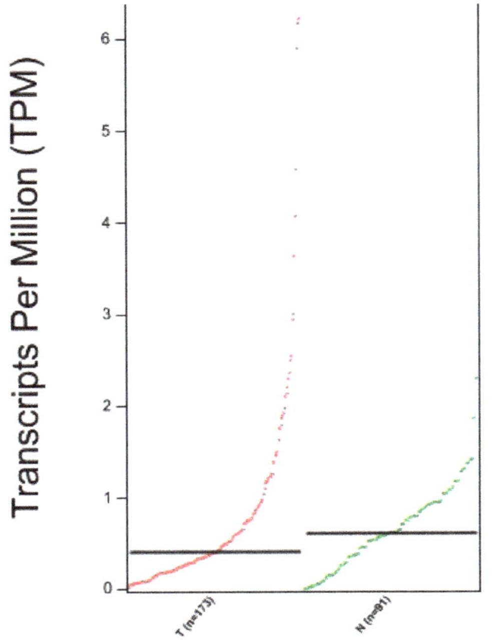

Figure 6. Expression analysis of SBSN mRNA through GEPIA2 interface. T = Tumour tissue, N = Normal Tissue; n = number of samples. Data are expressed as number of transcripts of the gene of interest per million of total expressed genes. (TPM) in logarithmic scale. Black transversal line indicates data median.

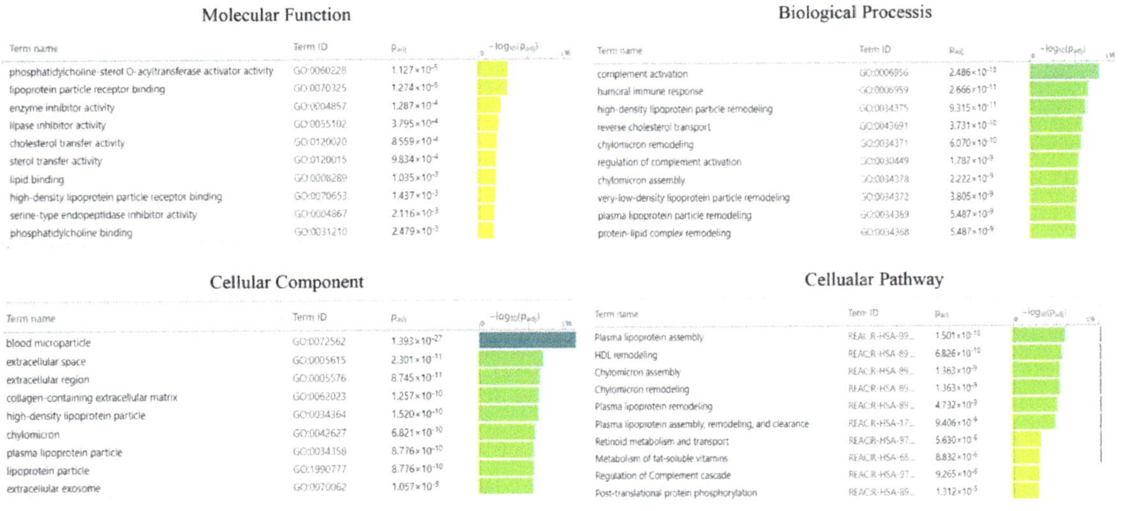

Figure 7. g: Profiler and REACTOME classification of identified proteins in the EC serum according to their molecular function, biological processes, cellular component, and cellular pathway.

The top networks in which these proteins were required corresponded to (Figure 8): (1) cell spreading, (2) cellular infiltration, (3) apoptosis, (4) adhesion of immune cells, (5) metastasis, and (6) migration of cells. Four proteins were implicated in cell spreading: A2M, ALB, VTN, and APCS. Seven proteins were implicated in cellular infiltration: ALB, APCS, APO1, APOE, IGHM, SERPINA1, and SERPINC1. Thirteen proteins were involved in apoptosis: APOA1, APOC3, APOE, CD5L, CLU, IGHM, SERPINA1, SERPINC1, VTN, Phosphate, A2M, ALB, and APCS. Nine proteins were involved in the adhesion of immune cells: A2M, APCS, APOA1, APOA4, APOE, CFHR1, CLU, SERPINA1, and VTN. Seven proteins were implicated in metastasis: ALB, APOA1, C1R, CFHR1, CLU, SERPINA1, and SERPINC1. Twelve were involved in the migration of cells: SERPINC1, VTN, A2M, ALB, APCS, APOA1, APOE, CD5L, CFHR1, CLU, IGHM, and SERPINA1.

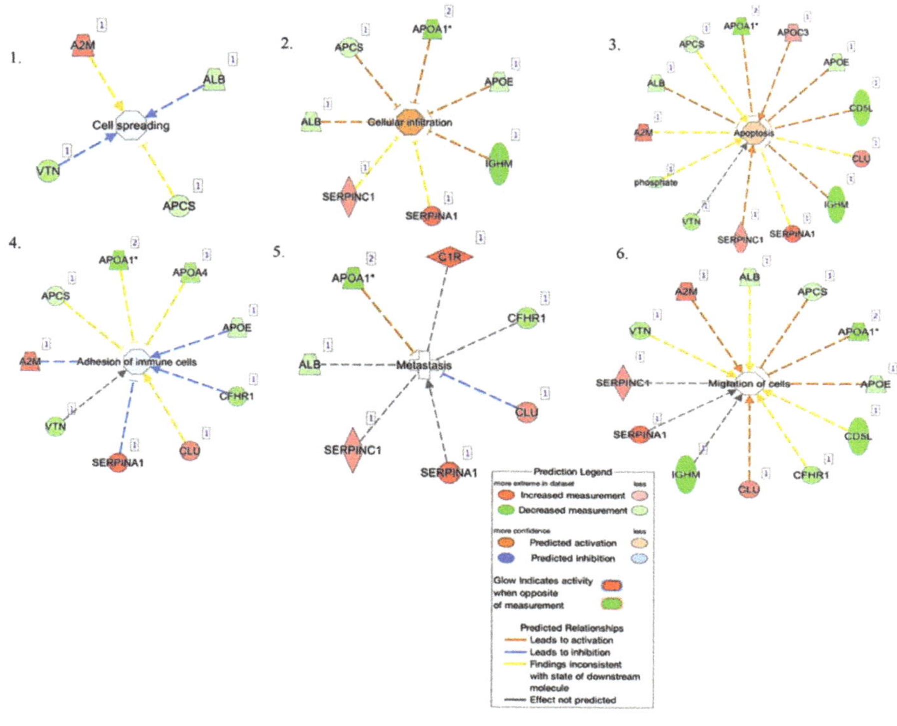

Figure 8. Network build-up from one of the most significant bio-functions: (**1**) cell spreading, (**2**) cellular infiltration, (**3**) apoptosis, (**4**) adhesion of immune cells, (**5**) metastasis, and (**6**) migration of cells.

3. Discussion

Biomarkers are crucial to detecting cancers early, improving treatment, and testing the response of ongoing therapies [22] at later stages. Although there has been a great effort in identifying the biomarkers of EC in recent years, none of them have yet reached the clinical stage. Identification of oncoproteins as biomarkers of EC may help develop new diagnostic and therapeutic approaches [23].

This study combined ProteoMiner, 2D-DIGE, and mass spectrometry to identify 24 proteins with different abundances that could act as candidate biomarkers, among which SBSN was chosen for further evaluation due to its potential role as an oncoprotein [14]. The quantification of the two isoforms of SBSN in serum and tissue was performed by Western blotting. These data proved that SBSN was downregulated in EC tissue, where the two isoforms of the protein had a good AUC (area under the ROC curve) (isoform 1 AUC = 0.7928 and isoform 2 AUC = 0.7933). Conversely, SBSN abundance in the serum behaved differently. Isoform 1 in serum had a low AUC (AUC = 0.5867) and did not appear to be a promising biomarker, while isoform 2 reached a better predictive value (AUC = 0.75). The low AUC of isoform 1 and the slight difference in abundance measured for isoform 2 in serum, which affected its reliability as a candidate biomarker, could be related to the release/leakage of the protein from other tissues in addition to the endometrium.

Our data pointed to a higher specificity of both SBSN isoforms in tissues than in serum: again, this could be related to the possibility that other cells or tissues could release SBSN into the bloodstream, thus reducing the differences in abundance observed for these proteins in serum. This occurrence eventually led to a lower sensitivity of SBSN as a putative biomarker when measured in serum compared to its tissue levels.

Further studies are needed to evaluate the performance of SBSN as a biomarker, combined with other serum biomarkers for EC. Taken together, our data in tissue indicated a downregulation of all SBSN isoforms in EC. To further confirm these observations, we performed data mining on the TCGA and GTEx databases, finding that SBSN mRNA expression on EC tissue was lower than in normal uterine tissue; although not statistically significant, this analysis further strengthened our results.

SBSN physiological functions have not yet been entirely ascertained. This protein was originally described as a component of the cornified envelope, which is expressed by corneocytes. Isoform 1 possessed structural features classified as a structural protein, while isoform 2 (and 3) lacked this signature. Post-translational modification has been proposed for both isoforms but demonstrated only for isoform 2. Different shreds of evidence have supported the role of SBSN in the pathogenesis of various kinds of cancer such as ovarian, cervical, and breast carcinomas [24]. In oesophageal squamous cell carcinoma (ESCC), SBSN was proposed as a potential biomarker [25]. In ESCC cell lines, the overexpression of isoform 2 promoted cell growth and proliferation, probably through the WNT/β-catenin signalling pathway [26]. In colorectal cancer, WNT/β-catenin and RAS/ERK signalling pathways interacted with active GSK3β as a mediator [27].

Oncogenes were not necessarily upregulated in carcinogenesis. For example, Cyclin D1 oncogene [28] was downregulated in breast cancer, increasing cell migration. PML, a proto-oncogene, was downregulated in prostate cancer, leading to the downregulation of the cell surface HLA class I molecule and immune escape [29].

Furthermore, in this study, we identified several proteins associated with the adhesion of immune cells. This mechanism played a key role in the recruitment and activation of immune T-cells [30], which are crucial in tumour development.

SERPINA1 was an inhibitor of serine proteases [31]. This protein, in some cases, could act like a tumour-promoting factor, leading to the activation of a variety of oncogenic pathways [32,33]. The upregulation of this protein led to a loss of its immune surveillance function, thus promoting tumour progression [34].

A2M was a plasma protein that acted as an antiprotease, inactivating several proteinases [35]. This activity was associated with cell adhesion modulation, contributing to cancer resistance [36].

Cell spreading was the key mechanism that permitted the cancer cell to invade the other parts of the body [37]. VTN was a cell adhesion and spreading factor found in serum and tissues [38].

T-cell infiltration was associated with a good prognosis in patients in early-stage EC [39]. APOE was a protein associated with lipid particles, carrying lipids between organs via the interstitial fluids and plasma [40]. Pancreatic cancer was characterised by an inflammatory environment that included abundant infiltrating immune cells [41]. In pancreatic cancer, APOE involved the expression of *Cxcl1* and *Cxcl5*, known immunosuppressive factors, leading to immunosuppression [42].

Our data, thus, suggested a possible association of the identified protein with metastatization. IPA analysis correlated the inhibition of metastasis by APOA1, following the literature, which suggested that AIBP, in combination with APOA1, had an anticancer effect on colorectal cancer [43].

SBSN is a secreted protein and, as such, it is very probable that a glycosylated form might be responsible for the apparent high MW observed in the 2-DE map. The fact that the 150 kDa form cannot be detected in our blots may be related to the inability of the commercial antibodies to recognize the protein when heavily glycosylated.

In conclusion, our results quantified the abundance of SBSN in EC, both in tissue and serum. Furthermore, our findings indicated that isoform 2, either in tissue or serum, could be used as a potential novel biomarker in EC. In our opinion, isoform 2 of SBSN should be combined with other biomarkers to reach the validation phase.

4. Materials and Methods

4.1. Patients

A total of 103 patients (60 non-EC controls and 44 EC patients) were recruited at the Institute for Maternal and Child Health—IRCCS "Burlo Garofolo" (Trieste, Italy)—from 2018 to 2021. All procedures complied with the Declaration of Helsinki and were approved by the Institute's Technical and Scientific Committee. All patients signed informed consent forms. The median age of patients was 45 years, ranging from 33 to 56 years. As controls, endometrial tissue samples from 30 patients who underwent hysterectomy for symptomatic uterine leiomyomas were obtained. For serum analysis, we used another 30 controls with normal endometrium and whose median age was 42 years, ranging from 32 to 77 years.

The clinical and pathological characteristics of the patients enrolled in this study are described in Supplementary Table S1. Controls were chosen by excluding oncologic patients, Human immunodeficiency virus (HIV), Hepatitis B virus (HBV), Hepatitis C virus (HCV) seropositive subjects, and patients with leiomyomas or adenomyosis. EC cases were also selected, ruling out women with other oncologic pathologies, Human immunodeficiency virus (HIV), Hepatitis B virus (HBV), Hepatitis C virus (HCV) seropositive patients, and patients with leiomyomas or adenomyosis.

4.2. Serum Sample Collection and Enrichment

To separate serum, blood was centrifugated at $5000\times g \times 5$ min. After centrifugation, serum was stored at −80 °C. Serum enrichment of low abundance proteins was achieved using a ProteoMiner column (Bio-Rad Laboratories, Inc., Hercules, CA, USA). In brief, 1 mL of serum was incubated for 2 h at room temperature with ProteoMiner beads. After three cycles of washing with PBS, protein elution was performed from the column with TUC buffer: 7 M urea, 2 M thiourea, 4% CHAPS, and 50 mM Tris pH = 8.5. Subsequently, a second elution was conducted with 4% SDS, 100 mM beta-mercaptoethanol, and the sample was precipitated in methanol and chloroform. The pellets were dissolved in TUC buffer and reunited with the first fraction, and the protein content was determined using the Bradford assay.

4.3. Sample Preparation for 2D-DIGE and Gel Image Analysis

For 2D-DIGE analysis, 50 µg of protein of the enriched serum from endometrial cancer patients and controls were labelled with 400 pmol of either Cy5 or Cy3. For internal standards, the samples were pooled and labelled with Cy2. The chemical reaction for protein labelling was carried out by incubating the samples on ice for 30 min in the dark. 1 µL of 10 mM lysine was added to stop the reaction. Following that, proteins were diluted to a final volume of 320 µL in the rehydration buffer: 7 M urea, 2 M thiourea, 2% (*w*/*v*) CHAPS, 65 mM DTT, and 0.24% Bio-Lyte (3–10) (Bio-Rad Laboratories, Inc., Hercules, CA, USA). For 2-DE analysis [44], 4–7 18 cm immobilised pH gradient (IPG) strips (Bio-Rad Laboratories, Inc., Hercules, CA, USA) were rehydrated at 50 V for 12 h at 20 °C, and isoelectric focusing (IEF) was performed in a PROTEAN IEF Cell (Bio-Rad Laboratories, Inc., Hercules, CA, USA) as detailed in [44]. After IEF, IPG strip equilibration was executed with two incubations: the first equilibration in 6 M urea, 2% SDS, 50 mM Tris-HCl (pH 8.8) and 30% glycerol for 5 min, and a second equilibration step performed in 4% iodoacetamide for 10 min. Proteins were separated by SDS-PAGE at a constant voltage of 100 V for 10 h

After electrophoresis, 2-DE gels were scanned with a Molecular Imager PharosFX System (Bio-Rad Laboratories, Inc., Hercules, CA, USA). Molecular weights were determined by Precision Plus Protein Prestained Standards (Bio-Rad Laboratories, Inc., Hercules, CA, USA), covering a molecular weight range from 10 to 250 kDa. Two experimental replicates were performed. Gel analysis was conducted using the MFA (multi fluorescence analysis) module of Proteomweaver 4.0 software (both from Bio-Rad Laboratories, Inc., Hercules, CA, USA) to normalise and quantify protein spots.

4.4. Western Blotting

Western blotting was used for SBSN data validation in enriched serum (i.e., treated with ProteoMiner beads, as detailed above) and tissue, as previously described [45]. The control and EC tissues were lysed with 1% NP-40, 50 mM Tris-HCl (pH 8.0), NaCl 150 mM with Phosphatase Inhibitor Cocktail Set II 1× (Millipore, Burlington, VT, USA), 2 mM phenylmethylsulphonyl fluoride (PMSF), and 1 mM benzamidine.

In this study, 30 μg of protein from the tissue and the enriched serum were loaded on a 4–20% precast gel (Bio-Rad) and then transferred to a nitrocellulose membrane. The membrane was blocked with 5% defatted milk in TBS-tween 20 after protein transfer and incubated overnight at 4 °C with 1:1000 diluted primary rabbit polyclonal antibody against SBSN (Abcam). After primary antibody incubation, membranes were washed three times with TBS-Tween 0.05% and incubated with HRP-conjugated anti-rabbit IgG and anti-mouse IgG (1:3000, Sigma-Aldrich; Merck Kagan, Darmstadt, Germany). The protein band signal was visualised using SuperSignal West Pico Chemiluminescent (Thermo Fisher Scientific Inc., Ottawa, ON, Canada). The intensities of the immunostained bands were normalised with the total protein intensities measured by staining the membranes from the same blot with a Red Ponceau solution (Sigma-Aldrich, St. Louis, MO, USA).

4.5. Trypsin Digestion and MS Analysis

A preparative 2-DE gel (300 μg of loaded proteins) was run and stained with Coomassie colloidal blue for protein visualisation. After gel decolouration, the spots of interest from 2-DE were digested and analysed by mass spectrometry, as previously described by Ura and colleagues [46]. The spots excised from the gel were washed four times with 50 mM ammonium bicarbonate (AB) and acetonitrile (ACN) (Sigma-Aldrich, St. Louis, MO, USA) and dried under vacuum in a SpeedVac system. For spot digestion, 3 μL of 12.5 ng/μL sequencing grade modified trypsin (Promega, Madison, WI, USA) in 50 mM AB were added. Samples were digested overnight at 37 °C. After digestion, peptide extraction was conducted with three changes in extraction by 50% ACN/0.1% formic acid (FA) (Fluka, Ammerbuch, Germany), and samples were dried under vacuum and stored at −20 °C until mass spectrometry (MS) analysis was performed. Samples were dissolved in 12 μL of 3% ACN/0.1% FA and peptides were separated in a 10 cm pico-frit column (75 μm ID, 15 μm Tip; New Objective) packed in-house with C18 material (Aeris Peptide 3.6 μm XB-C18, Phenomenex) using a nano-HPLC system (Ultimate 3000, Dionex—Thermo Fisher Scientific) coupled with an LTQ-Orbitrap XL mass spectrometer (Thermo Fisher Scientific). H_2O/FA 0.1% and ACN/FA 0.1% were used as eluents A and B, respectively, and chromatographic separation of peptides were performed at a flow rate of 0.25 μL/min using a linear gradient of eluent B from 3% to 40% in 20 min. A Data Dependent Acquisition (DDA) method was employed: a full scan between 300 and 1700 Da was conducted at high resolution (60,000) on the Orbitrap, and the 10 most intense ions were selected for CID fragmentation and MS/MS data acquisition at low resolution in the linear ion trap. Raw data files were analysed with the software package Proteome Discoverer 1.4 (Thermo Fisher Scientific) interfaced with the Mascot Search Engine (version 2.2.4, Matrix Science, London, UK). MS/MS spectra were searched against the human section of the UniProt database (version September 2020, 75,074 entries) using the following parameters: enzyme specificity was set on trypsin with one missed cleavage allowed; precursor and fragment ion tolerance were 10 ppm and 0.6 Da, respectively. Carbamidomethylcysteine and methionine oxidation were formulated as fixed and variable modifications, respectively. The Percolator algorithm was used to assess the False Discovery Rate (FDR) at the protein and peptide level. Proteins identified with at least three unique peptides with high confidence (FDR < 1%) were considered positive hits.

4.6. Bioinformatic Analysis

Gene Expression Profiling Interactive Analysis (GEPIA2) was employed to assess SBSN RNA expression in EC. This tool permits RNA expression analysis from a total of 9736 tumours and 8587 standard samples expunged from the TCGA and GTEx projects. The TCGA-UCEC (The Cancer Genome Atlas Uterine Corpus Endometrial Carcinoma) dataset was explored to assess SBSN expression, and the results were compared with those from the uterus dataset from the GTEx (Genotype-Tissue Expression (GTEx) repository.

Proteins identified by MS were analysed by g: Profiler classification systems and categorised according to their molecular function involvement, biological processes, and protein class. For pathway enrichment, the REACTOME tool was used. We employed the Ingenuity Pathway Analysis (IPA) to generate bio-functions [47]. We considered $p < 0.01$ a statistically significant value in IPA. For the filter summary, we only considered associations where confidence was high (predicted) or that had been observed experimentally.

4.7. Statistical Analysis

Differences were considered significant between patients and controls when spots showed a fold change $\pm$ 1.5 and satisfied the Mann–Whitney sum rank test ($p < 0.05$). All analyses were conducted with Stata/IC 16.1 for Windows (StataCorp LP, College Station, TX, USA).

Supplementary Materials: The following supporting information can be downloaded at: https://www.mdpi.com/article/10.3390/ijms23042076/s1.

Author Contributions: Conceptualization, B.U.; methodology, B.U., F.C., I.B. and G.A.; software, L.M., D.L. and M.A.; validation, B.U.; formal analysis, G.A. and L.M.; investigation, G.D.L. and F.R.; resources, G.R.; data curation, G.A., L.M. and F.C.; writing—original draft preparation, B.U., G.A., L.M. and F.C.; writing—review and editing, L.M., G.D.L., G.A. and F.C.; visualization, L.M.; supervision, B.U.; project administration, L.M. and G.R.; funding acquisition, G.R. All authors have read and agreed to the published version of the manuscript.

Funding: This research was funded by the Italian Health Ministry, grant number RC 18/19. The APC was funded by the Italian Health Ministry, grant number RC 18/19.

Institutional Review Board Statement: The study was conducted according to the guidelines of the Declaration of Helsinki and approved by the Institutional Review Board of IRCCS Burlo Garofolo and Regional Ethics Committee (protocol code RC18/19 approved in 2019 and CEUR-2020-Os-030).

Informed Consent Statement: Informed consent was obtained from all subjects involved in the study.

Data Availability Statement: Raw data are available upon request to the corresponding author.

Acknowledgments: The authors thank Martina Bradaschia for the English revision of the manuscript.

Conflicts of Interest: The authors declare no conflict of interest.

References

1. Shaw, E.; Farris, M.; McNeil, J.; Friedenreich, C. Obesity and endometrial cancer. *Recent Results Cancer Res.* **2016**, *208*, 107–136. [PubMed]
2. Kim, S.-I.; Kim, J.-W. Endometrial Cancer. *N. Engl. J. Med.* **2021**, *384*, 586. [PubMed]
3. Onstad, M.A.; Schmandt, R.E.; Lu, K.H. Addressing the Role of Obesity in Endometrial Cancer Risk, Prevention, and Treatment. *J. Clin. Oncol.* **2016**, *34*, 4225–4230. [CrossRef] [PubMed]
4. Sjögren, L.L.; Mørch, L.S.; Løkkegaard, E. Hormone replacement therapy and the risk of endometrial cancer: A systematic review. *Maturitas* **2016**, *91*, 25–35. [CrossRef] [PubMed]
5. Paleari, L.; Pesce, S.; Rutigliani, M.; Greppi, M.; Obino, V.; Gorlero, F.; Vellone, V.G.; Marcenaro, E. New Insights into Endometrial Cancer. *Cancers* **2021**, *13*, 1496. [CrossRef]

6. Paulino, E.; De Melo, A.C. Adjuvant treatment of endometrial cancer in molecular era: Are we ready to move on? *Crit. Rev. Oncol. Hematol.* **2020**, *153*, 103016. [CrossRef]
7. O'Flynn, H.; Ryan, N.A.J.; Narine, N.; Shelton, D.; Rana, D.; Crosbie, E.J. Diagnostic accuracy of cytology for the detection of endometrial cancer in urine and vaginal samples. *Nat. Commun.* **2021**, *12*, 952. [CrossRef]
8. Talhouk, A.; McConechy, M.K.; Leung, S.; Li-Chang, H.H.; Kwon, J.S.; Melnyk, N.; Yang, W.; Senz, J.; Boyd, N.F.; Karnezis, A.N.; et al. A clinically applicable molecular-based classification for endometrial cancers. *Br. J. Cancer* **2015**, *113*, 299–310. [CrossRef]
9. Fleming, G.F.; Emens, L.A.; Eder, J.P.; Hamilton, E.P.; Liu, J.F.; Liu, B.; Molinero, L.; Fasso, M.; O'Hear, C.; Braiteh, F.S. Clinical activity, safety and biomarker results from a phase Ia study of atezolizumab (atezo) in advanced/recurrent endometrial cancer (rEC). *J. Clin. Oncol.* **2017**, *35*, 5585. [CrossRef]
10. Uyar, D.S.; Huang, Y.-W.; Chesnik, M.A.; Doan, N.B.; Mirza, S.P. Comprehensive serum proteomic analysis in early endometrial cancer. *J. Proteom.* **2021**, *234*, 104099. [CrossRef]
11. Ura, B.; Biffi, S.; Monasta, L.; Arrigoni, G.; Battisti, I.; Di Lorenzo, G.; Romano, F.; Aloisio, M.; Celsi, F.; Addobbati, R.; et al. Two Dimensional-Difference in Gel Electrophoresis (2D-DIGE) Proteomic Approach for the Identification of Biomarkers in Endometrial Cancer Serum. *Cancers* **2021**, *13*, 3639. [CrossRef] [PubMed]
12. Banno, K.; Kisu, I.; Yanokura, M.; Tsuji, K.; Masuda, K.; Ueki, A.; Kobayashi, Y.; Yamagami, W.; Nomura, H.; Tominaga, E.; et al. Biomarkers in endometrial cancer: Possible clinical applications (Review). *Oncol. Lett.* **2012**, *3*, 1175–1180. [CrossRef] [PubMed]
13. Caduff, R.F.; Johnston, C.M.; Frank, T.S. Mutations of the Ki-ras oncogene in carcinoma of the endometrium. *Am. J. Pathol.* **1995**, *146*, 182–188. [PubMed]
14. Pribyl, M.; Hodny, Z.; Kubikova, I. Suprabasin—A Review. *Genes* **2021**, *12*, 108. [CrossRef] [PubMed]
15. Clark, H.F.; Gurney, A.L.; Abaya, E.; Baker, K.; Baldwin, D.; Brush, J.; Chen, J.; Chow, B.; Chui, C.; Crowley, C.; et al. The Secreted Protein Discovery Initiative (SPDI), a Large-Scale Effort to Identify Novel Human Secreted and Transmembrane Proteins: A Bioinformatics Assessment. *Genome Res.* **2003**, *13*, 2265–2270. [CrossRef]
16. Consortium, T.U. Activities at the Universal Protein Resource (UniProt). *Nucleic Acids Res.* **2014**, *42*, D191–D198.
17. Kondo, T. Cancer biomarker development and two-dimensional difference gel electrophoresis (2D-DIGE). *Biochim. Biophys. Acta (BBA)—Proteins Proteom.* **2018**, *1867*, 2–8. [CrossRef]
18. Ichikawa, H.; Kanda, T.; Kosugi, S.-I.; Kawachi, Y.; Sasaki, H.; Wakai, T.; Kondo, T. Laser Microdissection and Two-Dimensional Difference Gel Electrophoresis Reveal the Role of a Novel Macrophage-Capping Protein in Lymph Node Metastasis in Gastric Cancer. *J. Proteome Res.* **2013**, *12*, 3780–3791. [CrossRef]
19. Andersen, J.D.; Boylan, K.L.M.; Xue, F.S.; Anderson, L.B.; Witthuhn, B.A.; Markowski, T.W.; Higgins, L.; Skubitz, A. Identification of candidate biomarkers in ovarian cancer serum by depletion of highly abundant proteins and differential in-gel electrophoresis. *Electrophoresis* **2010**, *31*, 599–610. [CrossRef]
20. Ummanni, R.; Mundt, F.; Pospisil, H.; Venz, S.; Scharf, C.; Barett, C.; Fälth, M.; Köllermann, J.; Walther, R.; Schlomm, T.; et al. Identification of Clinically Relevant Protein Targets in Prostate Cancer with 2D-DIGE Coupled Mass Spectrometry and Systems Biology Network Platform. *PLoS ONE* **2011**, *6*, e16833. [CrossRef]
21. Poli, G.; Ceni, E.; Armignacco, R.; Ercolino, T.; Canu, L.; Baroni, G.; Nesi, G.; Galli, A.; Mannelli, M.; Luconi, M. 2D-DIGE proteomic analysis identifies new potential therapeutic targets for adrenocortical carcinoma. *Oncotarget* **2015**, *6*, 5695–5706. [CrossRef] [PubMed]
22. Henry, L.N.; Hayes, D.F. Cancer biomarkers. *Mol. Oncol.* **2012**, *6*, 140–146. [CrossRef] [PubMed]
23. Vogelstein, B.; Kinzler, K.W. Cancer genes and the pathways they control. *Nat. Med.* **2004**, *10*, 789–799. [CrossRef] [PubMed]
24. Connelly, J.T.; Gautrot, J.E.; Trappmann, B.; Tan, D.W.-M.; Donati, G.; Huck, W.T.; Watt, F.M. Actin and serum response factor transduce physical cues from the microenvironment to regulate epidermal stem cell fate decisions. *Nat. Cell Biol.* **2010**, *12*, 711–718. [CrossRef]
25. Jiang, S.; Zhang, Q.; Su, Y.; Pan, L. Network-Based Differential Analysis to Identify Molecular Features of Tumorigenesis for Esophageal Squamous Carcinoma. *Molecules* **2018**, *23*, 88. [CrossRef]
26. Zhu, J.; Wu, G.; Li, Q.; Gong, H.; Song, J.; Cao, L.; Wu, S.; Song, L.; Jiang, L. Overexpression of Suprabasin is Associated with Proliferation and Tumorigenicity of Esophageal Squamous Cell Carcinoma. *Sci. Rep.* **2016**, *6*, 21549. [CrossRef]
27. Jeong, W.-J.; Ro, E.J.; Choi, K.-Y. Interaction between Wnt/β-catenin and RAS-ERK pathways and an anti-cancer strategy via degradations of β-catenin and RAS by targeting the Wnt/β-catenin pathway. *NPJ Precis. Oncol.* **2018**, *2*, 5. [CrossRef]
28. Lehn, S.; Tobin, N.P.; Berglund, P.; Nilsson, K.; Sims, A.H.; Jirström, K.; Härkönen, P.; Lamb, R.; Landberg, G. Down-regulation of the oncogene cyclin D1 increases migratory capacity in breast cancer and is linked to unfavorable prognostic features. *Am. J. Pathol.* **2010**, *177*, 2886–2897. [CrossRef]
29. Zhang, H.; Melamed, J.; Wei, P.; Cox, K.; Frankel, W.; Bahnson, R.R.; Robinson, N.; Pyka, R.; Liu, Y.; Zheng, P. Concordant down-regulation of proto-oncogene PML and major histocompatibility antigen HLA class I expression in high-grade prostate cancer. *Cancer Immun.* **2003**, *3*.
30. Harjunpää, H.; Asens, M.L.; Guenther, C.; Fagerholm, S.C. Cell Adhesion Molecules and Their Roles and Regulation in the Immune and Tumor Microenvironment. *Front. Immunol.* **2019**, *10*, 1078. [CrossRef]
31. Ashkenazi, E.; Baranovski, B.M.; Shahaf, G.; Lewis, E.C. Pancreatic islet xenograft survival in mice is extended by a combination of alpha-1-antitrypsin and single-dose anti-CD4/CD8 therapy. *PLoS ONE* **2013**, *8*, e63625. [CrossRef] [PubMed]

32. Hwa Kwon, C.; Ji Park, H.; Hwa Choi, J.; Rang Lee, J.; Kyung Kim, H.; Jo, H.J.; Sung Kim, H.; Oh, N.; Am Song, G.; Youn Park, D. Snail and serpinA1 promote tumor progression and predict prognosis in colorectal cancer. *Oncotarget* **2015**, *6*, 20312–20326. [CrossRef] [PubMed]
33. Fu, C.; Yu, Z.; He, Y.; Ding, J.; Wei, M. Down-Regulation of an Autophagy-Related Gene SERPINA1 as a Superior Prognosis Biomarker Associates with Relapse and Distant Metastasis in Colon Adenocarcinoma. *OncoTargets Ther.* **2021**, *14*, 3861–3872. [CrossRef] [PubMed]
34. Erez, A.; DeBerardinis, R.J. Metabolic dysregulation in monogenic disorders and cancer—Finding method in madness. *Nat. Rev. Cancer* **2015**, *15*, 440–448. [CrossRef] [PubMed]
35. Armstrong, P.B.; Quigley, J.P. Alpha2-macroglobulin: An evolutionarily conserved arm of the innate immune system. *Dev. Comp. Immunol.* **1999**, *23*, 375–390. [CrossRef]
36. Kurz, S.; Thieme, R.; Amberg, R.; Groth, M.; Jahnke, H.-G.; Pieroh, P.; Horn, L.-C.; Kolb, M.; Huse, K.; Platzer, M.; et al. The anti-tumorigenic activity of A2M—A lesson from the naked mole-rat. *PLoS ONE* **2017**, *12*, e0189514. [CrossRef] [PubMed]
37. Yamaguchi, H.; Wyckoff, J.; Condeelis, J. Cell migration in tumors. *Curr. Opin. Cell Biol.* **2005**, *17*, 559–564. [CrossRef]
38. Hayman, E.G.; Pierschbacher, M.D.; Ohgren, Y.; Ruoslahti, E. Serum spreading factor (vitronectin) is present at the cell surface and in tissues. *Proc. Natl. Acad. Sci. USA* **1983**, *80*, 4003–4007. [CrossRef]
39. Mendiola, M.; Pellinen, T.; Ramon-Patino, J.L.; Berjon, A.; Bruck, O.; Heredia-Soto, V.; Turkki, R.; Escudero, J.; Hemmes, A.; de la Calle, L.E.G.; et al. Prognostic implications of tumor-infiltrating T cells in early-stage endometrial cancer. *Mod. Pathol.* **2021**, *35*, 256–265. [CrossRef]
40. Krimbou, L.; Denis, M.; Haidar, B.; Carrier, M.; Marcil, M.; Genest, J., Jr. Molecular interactions between apoE and ABCA1: Impact on apoE lipidation. *J. Lipid Res.* **2004**, *45*, 839–848. [CrossRef]
41. Li, J.; Byrne, K.T.; Yan, F.; Yamazoe, T.; Chen, Z.; Baslan, T.; Richman, L.P.; Lin, J.H.; Sun, Y.H.; Rech, A.J.; et al. Tumor Cell-Intrinsic Factors Underlie Heterogeneity of Immune Cell Infiltration and Response to Immunotherapy. *Immunity* **2018**, *49*, 178–193.e7. [CrossRef] [PubMed]
42. Kemp, S.B.; Carpenter, E.S.; Steele, N.G.; Donahue, K.L.; Nwosu, Z.C.; Pacheco, A.; Velez-Delgado, A.; Menjivar, R.E.; Lima, F.; Espinoza, C.E.; et al. Apolipoprotein E Promotes Immune Suppression in Pancreatic Cancer through NF-κB-Mediated Production of CXCL1. *Cancer Res.* **2021**, *81*, 4305–4318. [CrossRef]
43. Zhang, T.; Wang, Q.; Wang, Y.; Wang, J.; Su, Y.; Wang, F.; Wang, G. AIBP and APOA-I synergistically inhibit intestinal tumor growth and metastasis by promoting cholesterol efflux. *J. Transl. Med.* **2019**, *17*, 161. [CrossRef] [PubMed]
44. Ura, B.; Scrimin, F.; Arrigoni, G.; Athanasakis, E.; Aloisio, M.; Monasta, L.; Ricci, G. Abnormal expression of leiomyoma cytoskeletal proteins involved in cell migration. *Oncol. Rep.* **2016**, *35*, 3094–3100. [CrossRef] [PubMed]
45. Ura, B.; Scrimin, F.; Franchin, C.; Arrigoni, G.; Licastro, D.; Monasta, L.; Ricci, G. Identification of proteins with different abundance associated with cell migration and proliferation in leiomyoma interstitial fluid by proteomics. *Oncol. Lett.* **2017**, *13*, 3912–3920. [CrossRef] [PubMed]
46. Ura, B.; Scrimin, F.; FZanconati, F.; Arrigoni, G.; Monasta, L.; Romano, A.; Banco, R.; Zweyer, M.; Milani, D.; Ricci, G. Two-dimensional gel electrophoresis analysis of the leiomyoma interstitial fluid reveals altered protein expression with a possible involvement in pathogenesis. *Oncol. Rep.* **2015**, *33*, 2219–2226. [CrossRef]
47. Ura, B.; Monasta, L.; Arrigoni, G.; Battisti, I.; Licastro, D.; Di Lorenzo, G.; Romano, F.; Aloisio, M.; Peterlunger, I.; Stabile, G.; et al. Phosphoproteins Involved in the Inhibition of Apoptosis and in Cell Survival in the Leiomyoma. *J. Clin. Med.* **2019**, *8*, 691. [CrossRef]

International Journal of *Molecular Sciences*

MDPI

Article

Palmitate Enhances the Efficacy of Cisplatin and Doxorubicin against Human Endometrial Carcinoma Cells

Zih-Syuan Wu [1], Shih-Ming Huang [1,2] and Yu-Chi Wang [1,3,*]

1 Graduate Institute of Life Sciences, National Defense Medical Center, Taipei City 114, Taiwan; g8401011@gapps.ndmctsgh.edu.tw (Z.-S.W.); shihming@ndmctsgh.edu.tw (S.-M.H.)
2 Department of Biochemistry, National Defense Medical Center, Taipei City 114, Taiwan
3 Department of Obstetrics and Gynecology, Tri-Service General Hospital, National Defense Medical Center, Taipei City 114, Taiwan
* Correspondence: yuchitsgh@mail.ndmctsgh.edu.tw

Abstract: Endometrial cancer is the most common gynecological cancer worldwide. At present there is no effective screening test for its early detection and no curative treatment for women with advanced-stage or recurrent disease. Overexpression of fatty acid synthase is a common molecular feature of a subgroup of sex steroid-related cancers associated with poor prognoses, including endometrial cancers. Disruption of this fatty acid synthesis leads to cell apoptosis, making it a potential therapeutic target. The saturated fatty acid palmitate reportedly induces lipotoxicity and cell death by inducing oxidative stress in many cell types. Here, we explored the effects of palmitate combined with doxorubicin or cisplatin in the HEC-1-A and RL95-2 human endometrial cancer cell lines. The results showed that physiological concentrations of exogenous palmitate significantly increased cell cycle arrest, DNA damage, autophagy, and apoptosis in both RL95-2 and HEC-1-A cells. It also increased the chemosensitivity of both cell types. Notably, we did not observe that palmitate lipotoxicity reflected increased levels of reactive oxygen species, suggesting palmitate acts via a different mechanism in endometrial cancer. This study thus provides a potential therapeutic strategy in which palmitate is used as an adjuvant in the treatment of endometrial cancer.

Keywords: lipotoxicity; palmitate; endometrial cancer; adjuvant chemotherapy

Citation: Wu, Z.-S.; Huang, S.-M.; Wang, Y.-C. Palmitate Enhances the Efficacy of Cisplatin and Doxorubicin against Human Endometrial Carcinoma Cells. *Int. J. Mol. Sci.* **2022**, *23*, 80. https://doi.org/10.3390/ijms23010080

Academic Editor: Laura Paleari

Received: 1 November 2021
Accepted: 17 December 2021
Published: 22 December 2021

1. Introduction

Cancer of the endometrium is the most frequently occurring gynecological cancer [1]. In recent years, both the incidence of endometrial cancer and its associated mortality have been increasing rapidly around the world. In Taiwan, for example, the incidence of endometrial cancer now exceeds that of both cervical and ovarian cancer [2]. Risk factors for endometrial cancer include endometrial hyperplasia, menopausal estrogen, obesity, nulliparity, polycystic ovary syndrome, high cumulative doses of tamoxifen, diabetes, and genetic factors, among others [3]. On the basis of their clinical pathogenic mechanisms, endometrial cancers have been classified into two major types: type I is estrogen-dependent and related to hormonal imbalances, while type II is non-estrogen-dependent [4]. Most patients with endometrial cancer are type I and associated with excessive use of estrogen, endometrial hyperplasia, or obesity. Type II tumors are usually more common in obese women who may have endocrine or metabolic disorders, and they are also related to atrophic endometrium [5]. There are a number of treatments for endometrial cancer, including surgery, chemotherapy, radiation, and hormone therapy. Stage, histology, and tumor characteristics are the most important determinants guiding therapeutic strategy [6]. Statistics indicate that while most patients with low-grade tumors can be cured through surgery and chemotherapy; the survival rate among patients with advanced or recurrent endometrial cancers is very low. The main reason is resistance of the cancer cells to chemotherapy and the limited treatment methods currently available. Further research is needed to improve existing therapies to address this challenge.

Palmitate is the most abundant saturated fatty acid, accounting for 70–80% of total plasma free fatty acids. It can be supplied in food or synthesized endogenously via de novo lipogenesis (DNL) [7], which is a tightly controlled process that converts carbohydrates into the fatty acids used to synthesize triglycerides or other lipid molecules for membrane biosynthesis and energy storage [8]. Fatty acid synthase (FAS) is the key rate-limiting enzyme in DNL and converts malonyl-CoA to palmitic acid, the primary fatty acid product in DNL. Palmitate is subsequently elongated and desaturated to produce complex fatty acids, including stearic acid, palmitoleic acid, and oleic acid. In normal tissues, palmitate content is regulated to within a range of concentrations, and ingestion does not significantly affect palmitate levels in tissues [9]. The mechanism by which tissue palmitate concentrations are strictly regulated may be primarily related to maintaining normal homeostasis with tissues, including maintenance of the physical properties of membranes and the biosynthesis of palmitoylethanolamide [10]. However, under certain physiological and pathological conditions and nutritional factors, DNL may be strongly induced, increasing tissue palmitate levels and disrupting its regulation [11].

Reactive oxygen species (ROS) are the intermediates produced during the metabolic processes of organelles, such as the endoplasmic reticulum, mitochondrial respiratory complex, peroxisomes, xenobiotic detoxification, and fatty acid oxidation [12]. Mitochondria are the primary source of intracellular ROS, including hydrogen peroxide and hydroxyl radicals, that act as second messengers in cell signal transduction for different biological processes as growth, differentiation, metabolism, and apoptosis [13,14]. In normal cells, the concentration of ROS is regulated by many antioxidant systems such as peroxiredoxins, glutathione peroxidases, and catalase, therefore keeping ROS at a basal non-toxic level [15]. In cancer cells, disruption of the redox balance has been proven to be one of the most important causes of cancer occurrence, progression, and metastasis. ROS activates the cancer cell survival signal cascade, involving MAPK/ERK1/2, p38, JNK, and PI3K/Akt to activate NF-κB, matrix metalloproteinases, and VEGF to initiate cancer angiogenesis, metastasis, and survival [16]. However, there are still many controversies regarding the definition of ROS as a tumor-promoter or tumor-suppressor [17]. Palmitate exerts adverse effects in part by inducing ROS generation, which leads to lipotoxicity associated with endoplasmic reticulum stress, mitochondrial dysfunction, and cell death in a number of cell types, including adipocytes [18], glomerular podocytes [19], pancreatic β cells [20], cardiomyocytes [21], endothelial cells [22], vascular smooth muscle cells [23], and hepatocytes [24]. However, contradictory findings from one study indicate that palmitate-induced pancreatic β cell death is not caused by ROS [25]. Consequently, the mechanism by which palmitate mediates ROS production remains unclear.

Additionally, recent studies into the role of palmitate in cancer has shown that compared to normal breast and some tumor cells, HER2/neu-positive breast cancer cells show significantly increased fatty acid synthesis and storage. Moreover, when physiological doses of exogenous palmitate are added, fatty acid synthesis is disrupted, leading to CHOP (C/EBP homologous protein)-dependent apoptosis [26]. Palmitate also induces reductions in the mitochondrial membrane potential (MMP) and release of cytochrome c into the cytosol in MDA-MB-231 cells [27]. Moreover, it has been observed in patients with endometrial cancer that, compared to benign endometrial tumors, malignant tumors are enriched in glycolytic and lipogenic metabolic pathways and depend on this metabolism for survival [28]. However, the effect of palmitate in endometrial cancer remains incompletely understood.

In light of these observations, we tried to explore the relationship between high dietary fat intake and cancer risk. Our work is based on the premise that assessing the effects of individual main dietary fatty acids on endometrial cancer cells would help us understand the mechanism by which palmitate may affect tumor cells. Here, we applied physiological concentrations (10–200 μM) of exogenous palmitate and used the RL95-2 and HEC-1-A endometrial cancer cell lines to investigate the mechanism(s) underlying palmitate cytotoxicity and its effect on the cells' responsiveness to cisplatin and doxorubicin. Our findings clarify the lipotoxicity of palmitate in endometrial cancer and have a synergistic effect with chemotherapy, providing a potential adjuvant treatment strategy.

2. Results

2.1. Palmitate Cytotoxicity in Human Endometrial Cancer Cell Lines

We first examined the metabolic activity of palmitate toward the RL95-2 (type I) and HEC-1-A (type II) endometrial cancer cell lines. The two cell lines were incubated for 24 or 48 h with increasing doses of palmitate, and metabolic activity was detected with MTT conversion assays (Figure 1A,B). Our preliminary data showed that the ED50 for palmitate toxicity was estimated to be 69.51 μM in RL95-2 treated for 24 h and 56.89 μM for HEC-1-A cells treated for 48 h.

To better understand the mechanism underlying palmitate cytotoxicity, we examined its effect on the levels of various proteins. Western blot analysis showed that levels of FAS, p53, cyclin D1 (a cell cycle G1 biomarker), and the ratio of phosphorylated to total ACC (fatty acid synthesis biomarker) were all significantly and dose-dependently decreased in both RL95-2 and HEC-1-A cells (Figure 1C). The H3P/H3 ratio, a cell cycle G2/M biomarker, was difficult to determine because H3 and H3P proteins were both dose-dependently elevated in RL95-2 cells. At higher doses, palmitate also induced expression of γH2A.x (a DNA damage biomarker) and dose-dependently increased p62 levels and the LC3B II/I ratio (two autophagy biomarkers). The effect of palmitate on levels of Nrf2 and HO-1 (anti-ROS stress biomarkers) differed between RL95-2 and HEC-1-A cells; although their levels were dose-dependently increased in RL95-2 cells, they were decreased in HEC-1-A cells. Overall, responsiveness to palmitate was more apparent in RL95-2 than HEC-1-A cells.

2.2. Synergistic Effects of Palmitate and Chemotherapeutic Drugs in Human Endometrial Cancer Cell Lines

To test whether palmitate has synergistic effects with conventional chemotherapeutic agents, we applied the Chou–Talalay method to calculate the CIs and dosage requirements of palmitate with cisplatin or doxorubicin in RL95-2 and HEC-1-A cells. When administered individually in RL95-2, the IC50 values for cisplatin or doxorubicin for 24 h were about 6.0 and 1.5 μM. In HEC-1-A, the IC50 values for cisplatin or doxorubicin for 48 h were about 53.4 and 0.6 μM (Figure 2A,B). In combination with palmitate and cisplatin, the CI was <1 in both RL95-2 and HEC-1-A cells, indicating synergistic effects (Figure 2C,D). On the other hand, for palmitate with doxorubicin, the CI was <1 only in HEC-1-A cells (Figure 2E,F). A significant synergistic effect was observed when palmitate was combined with cisplatin at concentrations ranging from 4.5 μM to 53.4 μM (Figure 2D).

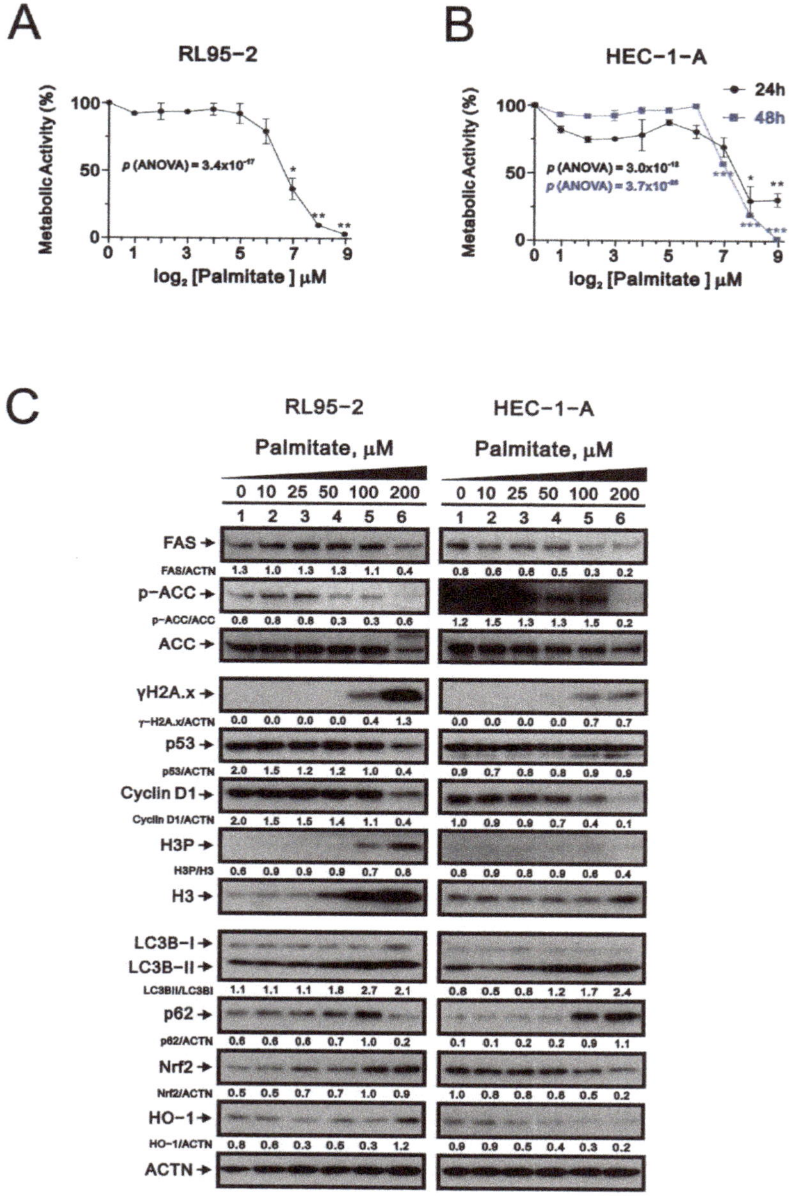

Figure 1. Responsiveness of human endometrial carcinoma cells to palmitate. (**A**,**B**) Metabolic activity measured using the MTT method. RL95-2, and HEC-1-A cells were treated for 24 h or 48 h with palmitate (0, 1.953125, 3.90625, 7.8125, 15.625, 31.25, 62.5, 125, 250, 500 μM). Symbols depict the mean ± SD of three independent experiments. * $p < 0.05$, ** $p < 0.01$, and *** $p < 0.001$ (Student's *t*-tests). (**C**) RL95-2, and HEC-1-A cells were treated for 24 h with indicated concentrations of palmitate. Cell lysates were subjected to Western blot analysis using antibodies against the indicated proteins. Alpha actinin (ACTN) was the loading control.

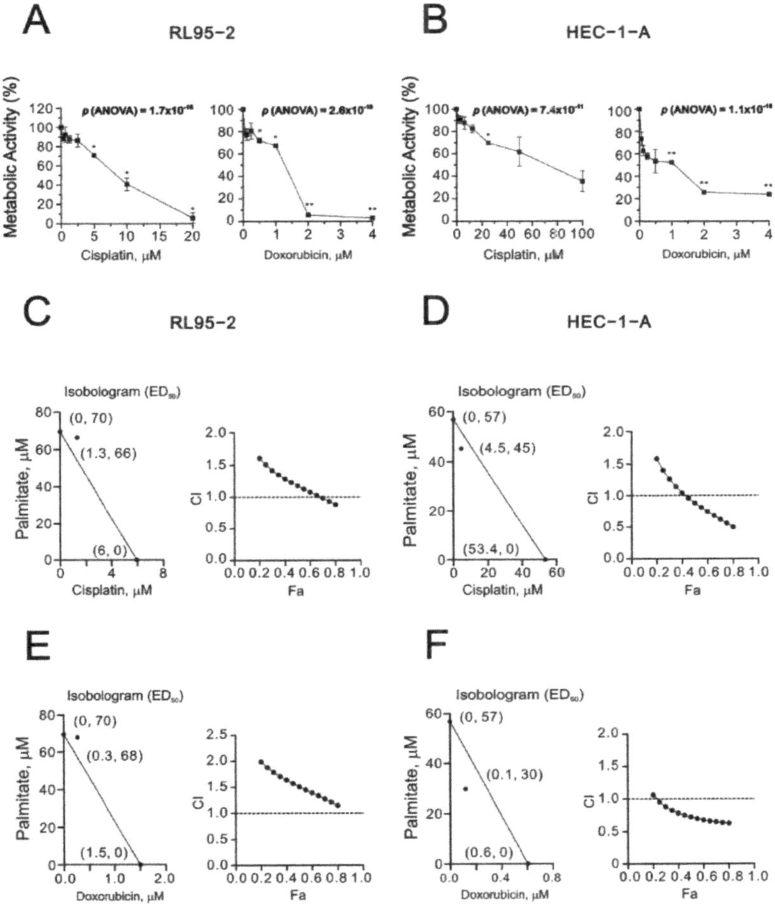

Figure 2. Combination indexes for palmitate with cisplatin or doxorubicin in RL95-2 and HEC-1-A cells. (**A**,**B**) Metabolic activity measured using the MTT method. RL95-2, and HEC-1-A cells were treated with doxorubicin (0, 0.0625, 0.125, 0.25, 0.5, 1, 2, 4 µM) or cisplatin (0, 0.3125, 0.625, 1.25, 2.5, 5, 10, 20 µM), (0, 1.5625, 3.125, 6.25, 12.5, 25, 50, 100 µM), respectively. Symbols depict the mean ± SD of three independent experiments. * $p < 0.05$, ** $p < 0.01$ (Student's *t*-tests). (**C**,**E**) RL95-2 cells were treated for 24 h with palmitate (0, 1.953125, 3.90625, 7.8125, 15.625, 31.25, 62.5, 125, 250, 500 µM) combined with cisplatin (0, 0.3125, 0.625, 1.25, 2.5, 5, 10, 20 µM) or doxorubicin (0, 0.0625, 0.125, 0.25, 0.5, 1, 2 µM). (**D**,**F**) HEC-1-A were treated for 48 h with palmitate (0, 1.953125, 3.90625, 7.8125, 15.625, 31.25, 62.5, 125, 250, 500 µM) combined with cisplatin (0, 1.5625, 3.125, 6.25, 12.5, 25, 50, 100 µM) or doxorubicin (0, 0.0625, 0.125, 0.25, 0.5, 1, 2 µM). Metabolic activity was measured using the MTT method. Isobolograms (ED_{50}) were calculated using CalcuSyn 2.0 software.

2.3. Molecular Mechanisms of Palmitate and Chemotherapeutic Drugs in Human Endometrial Cancer Cell Lines

We next investigated the effects of combination therapy on levels of various proteins in RL95-2 and HEC-1-A cells. In RL95-2 cells, cisplatin and doxorubicin individually increased γH2A.x, p53, and H3, and their abilities to increase levels of γH2A.x, p53, LC3B, CHOP, and cleaved PARP were all enhanced by palmitate (Figure 3A). Conversely, palmitate downregulated the effects of cisplatin and doxorubicin on FAS, HO-1, and the p-ACC/ACC and H3P/H3 ratios in RL95-2 cells. In HEC-1-A cells, palmitate downregulated the effects of cisplatin and doxorubicin on p53 and the p ACC/ACC and H3P/H3 ratios, whereas it enhanced their effects on LC3B, CHOP, and cleaved PARP (Figure 3B). These results indicate that the mechanisms by which palmitate acts in combination with chemotherapy drugs differ in RL95-2 and HEC-1-A cells.

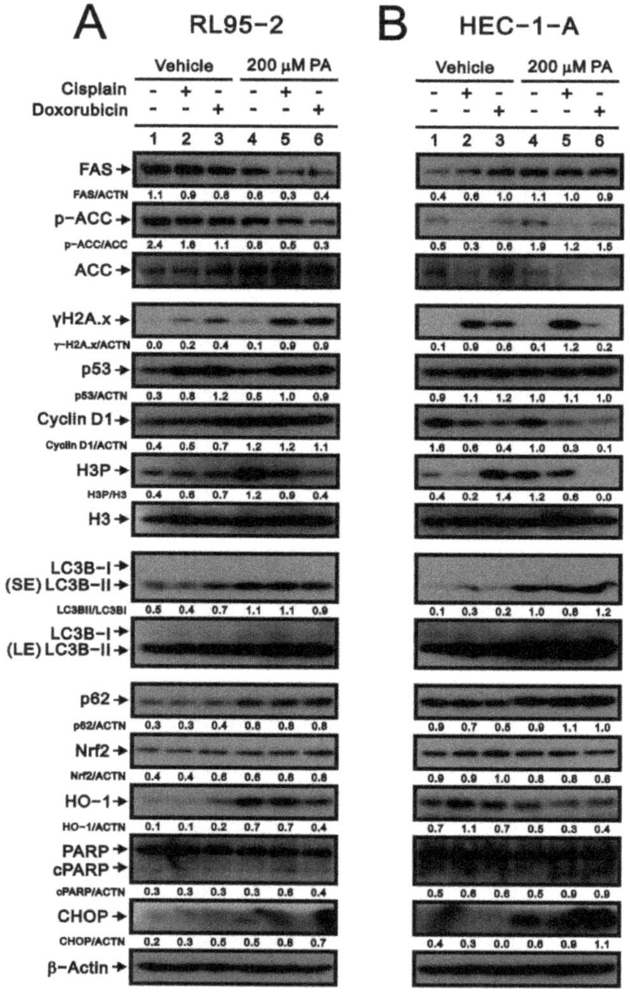

Figure 3. Effects of palmitate with cisplatin or doxorubicin on protein expression in RL95-2 and HEC-1-A cells. (**A**) RL95-2 cells were incubated for 24 h with 200 μM palmitate plus 5 μM cisplatin or 0.5 μM doxorubicin. (**B**) HEC-1-A cells were incubated for 24 h with 200 μM palmitate plus 50 μM cisplatin or 0.5 μM doxorubicin. Cell lysates were subjected to Western blot analysis using antibodies against the indicated proteins. Beta-actin was the loading control.

2.4. Effects of Palmitate and Chemotherapeutic Drugs on the Cell Cycle Profile, Cellular Proliferation, and Apoptosis in Human Endometrial Cancer Cell Lines

Because palmitate induced downregulation of cyclin D1, a cell cycle-related protein, in both RL95-2 and HEC-1-A cells and also significantly induced H3 and H3P (a cell cycle

G2/M biomarker) in RL95-2 cells (Figure 1B), we used PI staining to test the effect of palmitate on cell cycle profiles in RL95-2 and HEC-1-A cells. In both RL95-2 and HEC-1-A cells, palmitate significantly induced cell cycle arrest in the subG1 and G2/M phases and inhibited the S phase (Figure 4A,B). Moreover, combining cisplatin or doxorubicin with palmitate enabled us to verify the cell cycle changes they reportedly induce in cancer cells (Figure 4C,D for cisplatin; Figure 4E,F for doxorubicin). It is well known that cisplatin works on G1 populations [29], while doxorubicin works on the G2/M population [30]. We found that in RL95-2 cells, combined treatment with palmitate and cisplatin increased the population at sub-G1 phase, whereas the combined treatment with doxorubicin decreased the sub-G1 phase population. In HEC-1-A cells, for palmitate in combination with cisplatin or doxorubicin, the sub-G1 phase tended to increase, which is consistent with the results summarized in Figures 2 and 3 and points to a synergistic effect of the combined therapy in HEC-1-A cells.

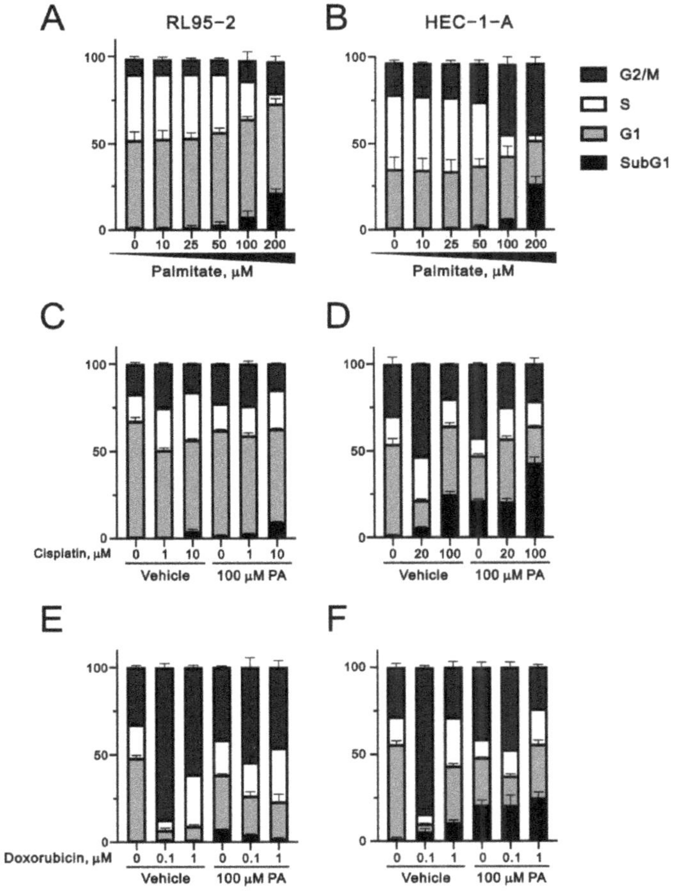

Figure 4. Effect of palmitate alone and in combination with cisplatin or doxorubicin on the cell cycle profiles in human endometrial cancer cells. (**A**,**C**,**E**) RL95-2 and (**B**,**D**,**F**) HEC-1-A cells were incubated for 24 h, after which they were treated with the indicated concentrations of doxorubicin or cisplatin plus 100 μM palmitate for 24 or 48 h. Cell cycle profiles were then analyzed using flow cytometry. Bars depict the mean ± SD of three independent experiments.

Given the finding that palmitate significantly reduced the numbers of cells at S-phase (Figure 4A,B), we used BrdU staining to assess palmitate effects on cell proliferation. Our results showed that palmitate significantly and dose-dependently reduced cell proliferation capability in RL95-2 and HEC-1-A cells (Figure 5A,B).

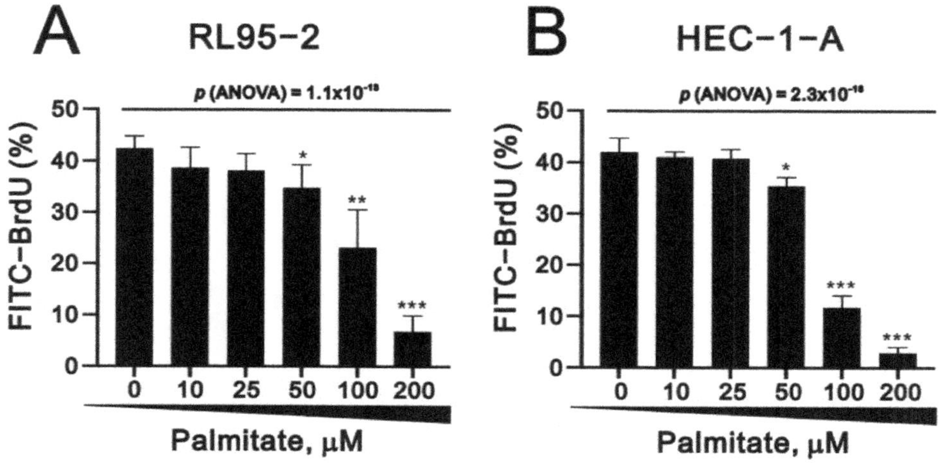

Figure 5. Effect of palmitate on proliferation of human endometrial cancer cells. (**A**) RL95-2 and (**B**) HEC-1-A cells were incubated for 24 h, after which they were treated for 24 or 48 h with the indicated concentrations of palmitate. Cell proliferation indicated by BrdU incorporation was analyzed using flow cytometry. Bars depict the mean ± SD of three independent experiments. * $p < 0.05$, ** $p < 0.01$, and *** $p < 0.001$ (Student's *t*-tests).

To determine whether palmitate-induced cytotoxicity and the elevation in sub-G1 populations leads to increased apoptosis among RL95-2 and HEC-1-A cells, we used Annexin V-PE and 7-AAD labelling to quantitatively assess cellular apoptosis. After palmitate treatment, the early and late apoptotic cell populations were increased significantly among both RL95-2 and HEC-1-A cells (Figure 6A,B). Cisplatin and doxorubicin each increased the early and late apoptotic populations among RL95-2 cells, but cisplatin only increased the early and late apoptotic populations in HEC-1-A cells (Figure 6C,D). Synergistic effects on total (early plus late) apoptotic populations were observed when RL95-2 and HEC-1-A cells were treated with palmitate plus cisplatin or doxorubicin. In RL95-2 cells, the synergistic effect with cisplatin was in the early apoptotic population, while the synergistic effect with doxorubicin was in both the early and late apoptotic populations. In HEC-1-A cells, the synergistic effect with cisplatin was in the late apoptotic population, while the synergistic effect with doxorubicin was in the early and late apoptotic populations.

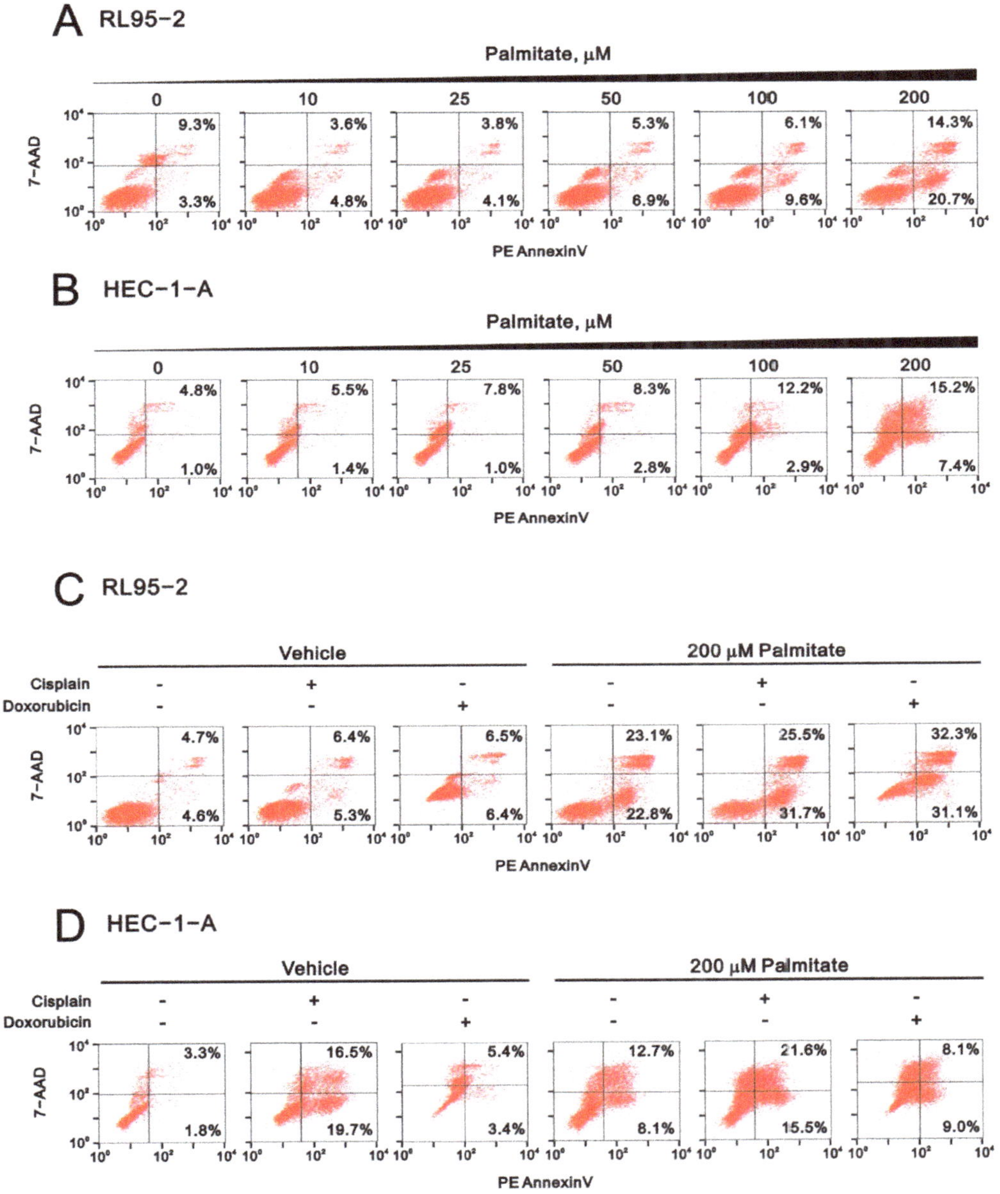

Figure 6. Effect of palmitate alone and in combination with cisplatin or doxorubicin on apoptosis among human endometrial cancer cells. (**A**) RL95-2 and (**B**) HEC-1-A cells were incubated for 24 h, after which they were treated for 24 or 48 h with palmitate (0, 10, 25, 50, 100, 200 μM). (**C**) RL95-2 cells treated for 24 h with 5 μM cisplatin or 0.5 μM doxorubicin plus 200 μM palmitate. (**D**) HEC-1-A cells treated for 48 h with 50 μM cisplatin or 0.5 μM doxorubicin plus 200 μM palmitate. Apoptosis markers labeled by PE-Annexin V and 7-AAD were analyzed using flow cytometry.

2.5. Effects of Palmitate and Chemotherapeutic Drugs on Mitochondrial Function in Human Endometrial Cancer Cell Lines

Mitochondria play key roles in cellular survival, ROS generation, and stress-induced programmed cell death. Moreover, recent studies suggest ROS are involved in palmitate-induced apoptosis [31–33]. We therefore assessed ROS levels in palmitate-treated RL95-2 and HEC-1-A cells. We found that whether using DCFH-DA to measure overall cell ROS levels or MitoSox to measure mitochondrial superoxide, ROS levels declined as the palmitate concentration increased in both RL95-2 and HEC-1-A cells (Figure 7A–D). H_2O_2, which served as a positive control, significantly increased ROS levels in our two assays.

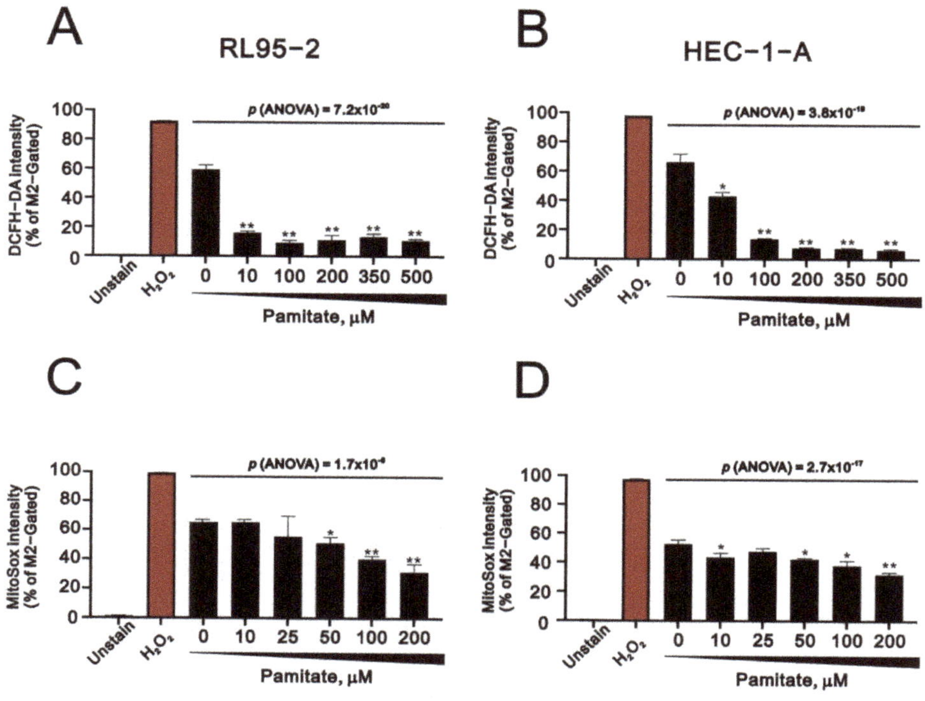

Figure 7. Effects of palmitate on ROS levels in human endometrial cancer cells. (**A**) RL95-2 and (**B**) HEC-1-A cells were incubated for 24 h, after which they were treated for 1.5 h with the indicated concentration of palmitate. Cellular ROS levels were monitored using 10 μM DCFH-DA with flow cytometry. (**C**) RL95-2 and (**D**) HEC-1-A cells were incubated for 24 h, after which they were treated for 1.5 h with the indicated concentration of palmitate. Mitochondrial ROS levels were monitored using 5 μM MitoSOX with flow cytometry. H_2O_2 serves as a positive control in all panels. Bars depict the mean ± SD of three independent experiments. * $p < 0.05$, ** $p < 0.01$ (Student's *t*-tests).

Because loss of MMP is a hallmark of apoptosis activation [34], we used JC-1 dye to measure changes in MMP after palmitate treatment in RL95-2 and HEC-1-A cells (Figure 8A,B). The results revealed that palmitate dose-dependently reduced MMP in both RL95-2 and HEC-1-A cells.

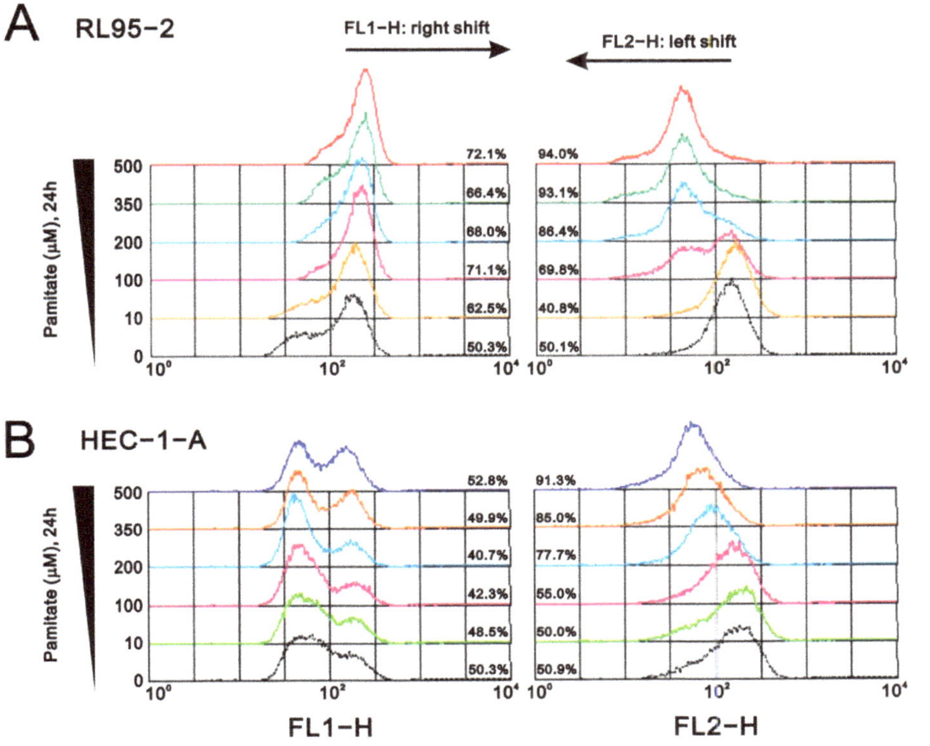

Figure 8. Effects of palmitate on mitochondrial membrane potential in human endometrial cancer cells. (**A**) RL95-2 and (**B**) HEC-1-A cells were incubated for 24 h, treated for 24 h with the indicated concentration of palmitate, and stained for 15 min with JC-1 dye. Mitochondrial membrane potential was detected using flow cytometry. Traces shown are representative of three independent experiments.

Mitochondrial morphology is dynamic, as the organelles continually undergo fission and fusion in response to their environmental conditions [35]. When cells sense mild stress, mitochondria form an elongated and interconnected network to increase ATP production. Under severe cellular stress, however, mitochondria divide into fragments for mitophagy or apoptosis. MitoView ^{TM} Green is a green fluorescent mitochondrial dye, the signal from which is based on mitochondrial mass rather than mitochondrial membrane potential. We used immunofluorescent staining with TOM20 and MitoView^{TM} Green to evaluate the effect of palmitate on mitochondrial morphology and mass in RL95-2 and HEC-1-A cells (Figure 9A,B). The results showed that palmitate not only triggered mitochondrial fragmentation in both RL95-2 and HEC-1-A cells, but they also decreased mitochondrial mass. These findings suggest that palmitate causes mitochondrial damage in RL95-2 and HEC-1-A cells.

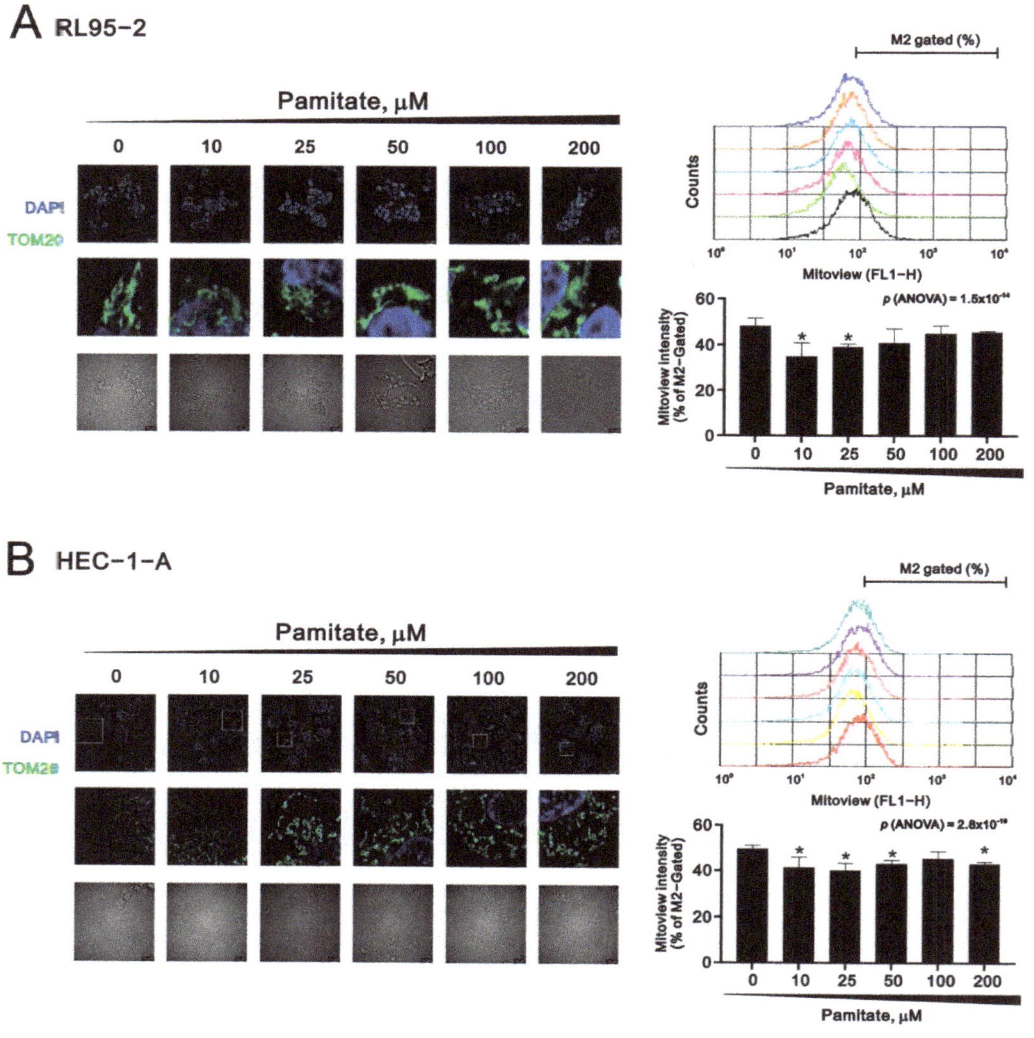

Figure 9. Effects of palmitate on mitochondrial morphology in human endometrial cancer cells. (**A**) RL95-2 and (**B**) HEC-1-A cells were treated for 1.5 h with the indicated concentrations of palmitate, after which they immunostained for TOM20 (mitochondria, green). Nuclei were stained with 4′,6-diamidino-2-phenylindole (DAPI, blue). Images were obtained using a Leica THUNDER Imager microscope (100x oil-immersion objective). Scale bar: 10 μm. Mitochondrial mass was assayed using MitoView^{TM} Green with flow cytometry. Traces are representative of three independent experiments. Bars depict the mean ± SD of three independent experiments. * $p < 0.05$ (Student's *t*-tests).

3. Discussion

In this work, we found that palmitate may increase the sensitivity of endometrial cancer cells to chemotherapy drugs. Our findings suggest it significantly increased cell cycle arrest, DNA damage, autophagy, and apoptosis in RL95-2 and HEC-1-A cells, with RL95-2 cells being more sensitive to palmitate than HEC-1-A cells. An earlier study of the metabolic profiles of seven endometrial cancer cell lines revealed that RL95-2 and HEC-1-A cells depend on different metabolic pathways [28]. The extracellular acidification rate;

oxidation of glucose, glutamine, and palmitate; and DNL from glucose, glutamine, and acetate were all higher in RL95-2 than HEC-1-A cells. Notably, RL95-2 cells were the most dependent on DNL pathways among the seven endometrial cancer cell lines tested. We suggest that this may explain why RL95-2 cells are more sensitive to palmitate than are HEC-1-A cells.

The overall purpose of this study was to investigate the impact and underlying mechanisms of palmitate in two types of endometrial cancer. Known as the "triple endometrial cancer syndrome," obesity, diabetes, and hypertension often co-exist in patients with endometrial cancer [36]. Recent studies indicate that palmitate levels are increased in cerebrospinal fluid from overweight and obese individuals and correlate positively with body mass index and abdominal circumference [37]. Elevation of palmitate in cerebrospinal fluid was also seen in overweight people with diabetes, dyslipidemia, and/or hypertension. One study reported that enzymes catalyzing glycolysis and DNL, including ACC1, ACC2, and FAS, are upregulated in most endometrial tumor tissues as compared to adjacent nonmalignant tissues [28]. Overexpression of FAS has been detected during the early stages of cancer development, is more pronounced in more advanced tumors, and is typically associated with a poor prognosis [38]. Recent findings have also shown that FAS blockade decreases cell proliferation and viability by stimulating apoptosis in endometrial carcinoma cells [39]. In addition to FAS, our present findings demonstrate that palmitate directly downregulates ACC and H3 protein levels. Details of the mechanisms remain to be investigated in the future.

Results from several studies have led to differing conclusions about the properties of palmitate. Several studies have reported that palmitate exhibits potential tumorigenic properties [40–42], although they also reported that it exhibits anticancer activity. In terms of promoting cancer, evidence suggests that palmitate may increase carcinogenesis by regulating DNA damage and inflammation [43], induce invasion of pancreatic cancer cells through TLR4/ROS/NF-κB/MMP-9 pathway [44], and increase colorectal cancer cell proliferation in a β2-adrenergic receptor-dependent manner [45]. Regarding anticancer activity, palmitate reduced cell membrane fluidity and limited glucose metabolism to enhance the anticancer effect of methylseleninic acid in hepatocellular carcinoma cells [46] and induced cell cycle G2/M arrest and promoted apoptosis in human neuroblastoma cells and breast cancer cells [47,48]. Moreover, in breast cancer, it has been observed that palmitate induced a different transcription program, reducing expression of HER2 and HER3, thereby sensitizing the cells to trastuzumab [26]. Palmitic acid isolated from the red alga Amphiroa zonata has anti-tumor activity both in vitro and in vivo. Amazingly, palmitic acid showed selective toxicity and induced apoptosis in leukemia cell lines but showed very low cytotoxicity to normal cell lines used as controls in the study [49]. Furthermore, in one study, palmitic acid is expected to become a novel anticancer agent for the treatment of prostate cancer. Their research showed that palmitic acid may inhibit the tumor metastasis regulator protein in human prostate cancer cells by inhibiting the PI3K/Akt pathway to induce cell cycle G1 phase arrest and anti-metastatic efficiency [50]. Due to the diversity of cancer phenotypes, involving factors are complicated in terms of defining palmitate as a tumor-promoter or tumor-suppressor. The working model of endometrial cancer in this study is one of the cancers that has been reported to overexpress FAS, which have developed peculiar metabolic pathways to gain a greater glucose uptake and to synthetize endogenous fatty acids. Our results indicated that palmitate plays a role in the inhibition of endometrial cancer cells. Furthermore, we used palmitate as an adjuvant to increase the sensitivity of endometrial cancer cells to chemotherapy drugs, cisplatin and doxorubicin, for the development a potential therapeutic strategy of endometrial cancer treatment.

Many literatures pointed out that the lipotoxicity induced by palmitate is characterized by the accumulation of intracellular ROS, which leads to an increase in oxidative stress and ultimately to cell apoptosis [7,51–53]. The effect of lipo-apoptosis caused by ROS seems to be cell type-dependent, including adipocytes, glomerular podocytes, pancreatic β cells, cardiomyocytes, endothelial cells, vascular smooth muscle cells, and hepatocytes. In contrast, apoptosis of palmitate-treated neonatal cardiomyocytes is independent of oxidative stress, and ROS is not the primary cause of palmitate-induced pancreatic β cell death [25]. It has been reported that redox homeostasis appears to be a key factor in the normal function of mitochondria and organisms. Both high levels of ROS (oxidative stress) and too low levels of ROS (reductive stress) are harmful and apparently play a pathogenic role. It has been shown that palmitate stimulated the function of UCP 1 in mitochondria, which can reduce the generation of ROS by mild uncoupling [54,55]. Moreover, the amphiphile nature of free fatty acids promotes their incorporation into the inner mitochondrial membrane, which leads to changes in membrane fluidity [56]. Consequently, it is necessary to verify whether palmitate-induced superoxide directly activates the UCP, which leads to negative feedback controlling both ROS production and their levels.

Autophagy also contributes to the reduction of ROS levels under various stress conditions [57]. The level of ROS may play an important role in regulating the formation of autophagy through various signaling pathways, including FOXO3–LC3/BNIP3 and NRF2–P62 pathways [58]. ROS produced by damaged mitochondria may induce mitophagy and eliminate damaged organelles to reduce ROS levels. The loss of mitochondrial membrane potential was initially considered to be a clue to mitophagy [59]. While palmitate was not observed to increase ROS in our study, palmitate treatment demonstrated the loss of mitochondrial membrane potential, triggered mitochondrial fragmentation, and decreased mitochondrial mass in both RL95-2 and HEC-1-A cells. Either alone or in combination, palmitate has been showed to increase the protein levels of p62 and LC3B II. These findings suggest that mitophagy might be the cause of the decline in ROS levels in RL95-2 and HEC-1-A cells. The detailed mechanism remains to be investigated in the future.

4. Materials and Methods

4.1. Cell Culture and Reagents

The RL95-2 (ATCC®CRL-1671™) and HEC-1-A (ATCC®HTB-112™) human endometrial carcinoma cell lines were purchased from the American Type Culture Collection (Manassas, VA, USA). RL95-2 cells were cultured in Dulbecco's modified Eagle's medium nutrient mixture F-12 (DMEM/F12) supplemented with 10% fetal bovine serum (FBS), 0.005 mg/mL insulin, and 1% penicillin–streptomycin (Thermo Fisher Scientific, Waltham, MA, USA). HEC-1-A cells were cultured in McCoy's 5A medium supplemented with 10% FBS and 1% penicillin–streptomycin. Doxorubicin, cisplatin, sodium palmitate, 2′,7-dichlorofluorescein diacetate (DCFH-DA), propidium iodide (PI), and thiazolyl blue tetrazlium bromide (MTT) were obtained from Sigma Aldrich (St. Louis, MO, USA).

4.2. Analysis of Cell Metabolic Activity

RL95-2 (8×10^3) and HEC-1-A (5×10^3) cells were seeded into 96-well plates and incubated in their respective media. The next day, they were exposed to the indicated drugs in fresh DMEM/F12 or McCoy's 5A medium for the indicated periods. MTT solution (0.5 mg/mL in phosphate-buffered saline (PBS)) was then added to each well, and the cells were incubated for 4 h at 37 °C. After removing the supernatants, we added dimethyl sulfoxide (DMSO; 100 µL) to dissolve the precipitate, and the absorbances at 570 nm and 650 nm were then measured using an enzyme-linked immunosorbent assay plate reader (Multiskan EX, Thermo Fisher Scientific). The relative metabolic activity was calculated on the basis of the absorbance ratio between cells cultured with the indicated drugs and the untreated controls, which were assigned a value of 100. A combination index (CI) was also calculated using CalcuSyn 2.0 software (Biosoft, Cambridge, United Kingdom) to produce an isobologram where a CI < 1 indicates a synergistic combination effect and a CI > 1 indicates an antagonistic combination effect [60].

4.3. Western Blot Analysis

RL95-2 and HEC-1-A cells were lysed in radio-immunoprecipitation assay buffer (100 mM Tris-HCl (pH 8.0), 150 mM NaCl, 0.1% SDS, and 1% Triton 100) at 4 °C. Proteins in the resultant lysates were separated by sodium dodecyl sulfate polyacrylamide gel electrophoresis, after which the resolve proteins were immunoblotted with antibodies against β-actin, p53, p62, FAS, nuclear factor-erythroid factor 2-related factor 2 (Nrf2) (Santa Cruz Biotechnology, Santa Cruz, CA, USA), phospho-histone H3 (H3P; serine phosphorylation at residue 10), histone H3 (H3), microtubule-associated proteins 1A/1B light chain 3B (LC3B), phospho-acetyl-CoA carboxylase (p-ACC; serine phosphorylation at residue 79), ACC, cleaved poly-ADP-ribose polymerase (cPARP), CHOP (Cell Signaling, Danvers, MA, USA), phospho-histone H2A.X (γH2A.X; serine phosphorylation at residue 139), Cyclin D1 (Abcam, Cambridge, United Kingdom), and heme oxygenase 1 (HO-1) (Enzo Life Sciences, Farmingdale, NY, USA).

4.4. Cell Cycle Profiles and Cellular Proliferation Analysis

RL95-2 (6×10^5) and HEC-1-A (3×10^5) cells were seeded into 6-well plates in their respective media. The next day, the indicated drugs were added in fresh medium, and the cells were incubated for an additional 24 or 48 h. For cell cycle analysis, the cells were fixed in 70% ice-cold ethanol and stored at −20 °C overnight. The fixed cells were then centrifuged (1000 rpm, 5 min), washed twice with ice-cold PBS supplemented with 1% FBS, and stained with PI solution (5 μg/mL PI in PBS, 0.5% Triton X-100, and 0.5 μg/mL RNase A) for 30 min at 37 °C in the dark. For each condition, 10,000 cells were analyzed using a BD FACScalibur™ flow cytometer and Cell Quest Pro software (version 5.1) (BD Biosciences, Franklin Lakes, NJ, USA).

For cell proliferation assays, following the incubation protocol described above, the cells were incubated for an additional 1 h with 10 μM BrdU (BD Pharmingen BrdU Flow Kit, San Diego, CA, USA). The medium was then discarded, and the cells were fixed at room temperature for 30 min and treated with FITC-conjugated anti-BrdU antibody (BD Pharmingen). After washing, the cells were incubated with 7-AAD and analyzed using a BD FACSCalibur™ flow cytometer and CellQuest Pro software (BD Biosciences).

4.5. Apoptosis Analysis

For apoptosis assays, the cells were stained with Annexin V-PE and 7-AAD and then detected with flow cytometry using the manufacturer's protocol (BD PharMingen, San Diego, CA, USA). Briefly, after treatment with the indicated drugs, cells were washed twice with ice-cold PBS and stained with 5 μL of Annexin V-PE and 10 μL of 7-AAD (5 μg/mL) in 1 mL of binding buffer for 15 min at room temperature in the dark. Apoptotic cells were then counted using a BD FACSCalibur™ flow cytometer and Cell Quest Pro software (BD Biosciences, Franklin Lakes, NJ, USA).

4.6. ROS Analysis

The DCFH-DA fluorescent marker was used to identify intracellular ROS levels. In addition, MitoSOX™ Red (Invitrogen, Carlsbad, CA, USA) is a fluorescent dye that reacts selectively with mitochondrial superoxide in live cells. Cells were incubated for 1.5 h with selected doses of palmitate and then stained with DCFH-DA (10 μM) or MitoSOX™ Red (10 μM) in serum-free medium for 30 min at 37 °C and harvested. Samples were then evaluated using a FACSCalibur flow cytometer and Cell Quest Pro software (BD Biosciences, Franklin Lakes, NJ, USA).

4.7. Mitochondrial Membrane Potential Analysis

RL95-2 (6×10^5) and HEC-1-A (3×10^5) cells were seeded into 6-well plates in their respective media. The next day, selected doses of palmitate were added in fresh medium, and the cells were incubated for an additional 24 or 48 h. All dead and viable cells were then harvested, washed with PBS, and incubated with 1× binding buffer containing the MMP-sensitive fluorescent dye JC-1 for 30 min at 37 °C in the dark. The cells were then washed twice with PBS, resuspended in 500 μL of 1× binding buffer, and analyzed using a FACSCalibur flow cytometer and Cell Quest Pro software (BD Biosciences, Franklin Lakes, NJ, USA). Mitochondrial depolarization was measured on the basis of a decrease in the red/green fluorescence intensity ratio.

4.8. Immunocytochemistry

Immunocytochemical analysis was carried out with cells adhering to cover slips in 24-well plates. After treatment with selected doses of palmitate for 1.5 h, the cells were fixed for 10 min in 4% formaldehyde, incubated for 10 min in 0.1% Triton X-100 solution, washed 3 times in PBS, and treated for 1 h at room temperature with 1% BSA. Thereafter, the cells were incubated with anti-TOM20 antibody at 4 °C overnight. The next day, the cells were washed with PBS and incubated for 1 h with FITC. Cell nuclei were stained by DAPI. Mitochondrial morphology was observed using a THUNDER Imager microscope equipped with a 100× objective (Leica, Wetzlar, Germany).

4.9. Mitochondrial Mass Assay

RL95-2 (6×10^5) and HEC-1-A (3×10^5) cells were seeded into 6-well plates in their respective media. The next day, the cells were treated with selected doses of palmitate for 1.5 h and stained with 20 μM MitoView™ Green (Biotium, CA, USA) for 30 min at 37 °C in the dark. Samples were then evaluated using a FACSCalibur flow cytometer and Cell Quest Pro software (BD Biosciences, Franklin Lakes, NJ, USA).

4.10. Statistical Analysis

Values were expressed as the mean ± SD from at least three independent experiments. All comparisons between groups were made using Student's *t*-tests. Statistical significance: n.s., not significant, * $p < 0.05$; ** $p < 0.01$; *** $p < 0.001$. The comparison between multiple groups was conducted using analysis of variance (ANOVA). Statistical significance was set at $p < 0.05$.

5. Conclusions

Our findings indicate that palmitate exhibits potential anti-endometrial cancer activity, especially in HEC-1-A cells, which are type II endometrial cancer cells and therefore less sensitive to chemotherapy. Palmitate administered as an adjuvant treatment significantly increased cancer cell sensitivity to cisplatin and doxorubicin. This study thus provides a potential therapeutic strategy for the treatment of endometrial cancers otherwise resistant to treatment.

Author Contributions: Z.-S.W.: data curation, formal analysis, investigation, validation, writing—original draft. S.-M.H.: conceptualization, formal analysis, methodology, project administration, supervision, writing—original draft. Y.-C.W.: conceptualization, formal analysis, funding acquisition, supervision, writing—review & editing. All authors have read and agreed to the published version of the manuscript.

Funding: This work was supported by grants from the Ministry of Science and Technology (MOST-109–2221-E-016-002-MY3 to Y-C WANG), Taiwan, Republic of China.

Institutional Review Board Statement: Not applicable.

Informed Consent Statement: Not applicable.

Acknowledgments: We gratefully acknowledge all of the funding sources.

Conflicts of Interest: The authors declare no conflict of interest.

Abbreviations

ACC	acetyl-CoA carboxylase
ACLY	ATP-citrate lyase
ANOVA	analysis of variance
CHOP	C/EBP homologous protein
CI	combination index
cPARP	cleaved poly-ADP-ribose polymerase
DCFH-DA	2′,7-dichlorofluorescein diacetate
DMSO	dimethyl sulfoxide
DNL	de novo lipogenesis
ED50	median effective dose
ER	endoplasmic reticulum
FAS	fatty acid synthase
FBS	fetal bovine serum
H3P	phospho-histone H3
H3	histone H3
HO-1	heme oxygenase 1
LC3B	microtubule-associated proteins 1A/1B light chain 3B
MMP	mitochondrial membrane potential
MTT	thiazolyl blue tetrazlium bromide
Nrf2	nuclear factor-erythroid factor 2-related factor 2
PBS	phosphate-buffered saline
PI	propidium iodide
ROS	reactive oxygen species
TCA	tricarboxylic acid

References

1. Torre, L.A.; Islami, F.; Siegel, R.L.; Ward, E.M.; Jemal, A. Global Cancer in Women: Burden and Trends. *Cancer Epidemiol. Biomark. Prev.* **2017**, *26*, 444–457. [CrossRef]
2. Tseng, C.J.; Lu, C.J.; Chang, C.C.; Chen, G.D.; Cheewakriangkrai, C. Integration of data mining classification techniques and ensemble learning to identify risk factors and diagnose ovarian cancer recurrence. *Artif. Intell. Med.* **2017**, *78*, 47–54. [CrossRef]
3. Felix, A.S.; Brinton, L.A. Cancer Progress and Priorities: Uterine Cancer. *Cancer Epidemiol. Biomark. Prev.* **2018**, *27*, 985–994. [CrossRef] [PubMed]
4. Sherman, M.E. Theories of endometrial carcinogenesis: A multidisciplinary approach. *Mod. Pathol.* **2000**, *13*, 295–308. [CrossRef] [PubMed]
5. Murali, R.; Soslow, R.A.; Weigelt, B. Classification of endometrial carcinoma: More than two types. *Lancet Oncol.* **2014**, *15*, e268–e278. [CrossRef]
6. Cosgrove, C.M.; Barrington, D.; Backes, F.J. Impact of Molecular Classification on Treatment Paradigms in Uterine Cancers. *Curr. Oncol. Rep.* **2021**, *23*, 75. [CrossRef]
7. Ly, L.D.; Xu, S.; Choi, S.K.; Ha, C.M.; Thoudam, T.; Cha, S.K.; Wiederkehr, A.; Wollheim, C.B.; Lee, I.K.; Park, K.S. Oxidative stress and calcium dysregulation by palmitate in type 2 diabetes. *Exp. Mol. Med.* **2017**, *49*, e291. [CrossRef] [PubMed]
8. Song, Z.; Xiaoli, A.M.; Yang, F. Regulation and Metabolic Significance of De Novo Lipogenesis in Adipose Tissues. *Nutrients* **2018**, *10*, 1383. [CrossRef] [PubMed]
9. Song, X.; Huang, Y.; Neuhouser, M.L.; Tinker, L.F.; Vitolins, M.Z.; Prentice, R.L.; Lampe, J.W. Dietary long-chain fatty acids and carbohydrate biomarker evaluation in a controlled feeding study in participants from the Women's Health Initiative cohort. *Am. J. Clin. Nutr.* **2017**, *105*, 1272–1282. [CrossRef] [PubMed]
10. Carta, G.; Murru, E.; Banni, S.; Manca, C. Palmitic Acid: Physiological Role, Metabolism and Nutritional Implications. *Front. Physiol.* **2017**, *8*, 902. [CrossRef]
11. Wilke, M.S.; French, M.A.; Goh, Y.K.; Ryan, E.A.; Jones, P.J.; Clandinin, M.T. Synthesis of specific fatty acids contributes to VLDL-triacylglycerol composition in humans with and without type 2 diabetes. *Diabetologia* **2009**, *52*, 1628–1637. [CrossRef]
12. Murphy, M.P. How mitochondria produce reactive oxygen species. *Biochem. J.* **2009**, *417*, 1–13. [CrossRef] [PubMed]
13. Chio, I.I.C.; Tuveson, D.A. ROS in cancer: The burning question. *Trends Mol. Med.* **2017**, *23*, 411–429. [CrossRef]

14. Zou, Z.; Chang, H.; Li, H.; Wang, S. Induction of reactive oxygen species: An emerging approach for cancer therapy. *Apoptosis* **2017**, *22*, 1321–1335. [CrossRef] [PubMed]
15. Saikolappan, S.; Kumar, B.; Shishodia, G.; Koul, S.; Koul, H.K. Reactive oxygen species and cancer: A complex interaction. *Cancer Lett.* **2019**, *452*, 132–143. [CrossRef] [PubMed]
16. Aggarwal, V.; Tuli, H.S.; Varol, A.; Thakral, F.; Yerer, M.B.; Sak, K.; Varol, M.; Jain, A.; Khan, M.A.; Sethi, G. Role of Reactive Oxygen Species in Cancer Progression: Molecular Mechanisms and Recent Advancements. *Biomolecules* **2019**, *9*, 735. [CrossRef]
17. Wang, J.; Yi, J. Cancer cell killing via ROS: To increase or decrease, that is the question. *Cancer Biol. Ther.* **2008**, *7*, 1875–1884. [CrossRef]
18. Davis, J.E.; Gabler, N.K.; Walker-Daniels, J.; Spurlock, M.E. The c-Jun N-terminal kinase mediates the induction of oxidative stress and insulin resistance by palmitate and toll-like receptor 2 and 4 ligands in 3T3-L1 adipocytes. *Horm. Metab. Res.* **2009**, *41*, 523–530. [CrossRef] [PubMed]
19. Xu, S.; Nam, S.M.; Kim, J.H.; Das, R.; Choi, S.K.; Nguyen, T.T.; Quan, X.; Choi, S.J.; Chung, C.H.; Lee, E.Y.; et al. Palmitate induces ER calcium depletion and apoptosis in mouse podocytes subsequent to mitochondrial oxidative stress. *Cell Death Dis.* **2015**, *6*, e1976. [CrossRef] [PubMed]
20. Sato, Y.; Fujimoto, S.; Mukai, E.; Sato, H.; Tahara, Y.; Ogura, K.; Yamano, G.; Ogura, M.; Nagashima, K.; Inagaki, N. Palmitate induces reactive oxygen species production and β-cell dysfunction by activating nicotinamide adenine dinucleotide phosphate oxidase through Src signaling. *J. Diabetes Investig.* **2014**, *5*, 19–26. [CrossRef]
21. Joseph, L.C.; Barca, E.; Subramanyam, P.; Komrowski, M.; Pajvani, U.; Colecraft, H.M.; Hirano, M.; Morrow, J.P. Inhibition of NADPH Oxidase 2 (NOX2) Prevents Oxidative Stress and Mitochondrial Abnormalities Caused by Saturated Fat in Cardiomyocytes. *PLoS ONE* **2016**, *11*, e0145750. [CrossRef] [PubMed]
22. Yamagishi, S.; Okamoto, T.; Amano, S.; Inagaki, Y.; Koga, K.; Koga, M.; Choei, H.; Sasaki, N.; Kikuchi, S.; Takeuchi, M.; et al. Palmitate-induced apoptosis of microvascular endothelial cells and pericytes. *Mol. Med.* **2002**, *8*, 179–184. [CrossRef]
23. Brodeur, M.R.; Bouvet, C.; Barrette, M.; Moreau, P. Palmitic acid increases medial calcification by inducing oxidative stress. *J. Vasc. Res.* **2013**, *50*, 430–441. [CrossRef] [PubMed]
24. Meyn, R.E.; Jenkins, W.T.; Murray, D. Radiation damage to DNA in various animal tissues: A comparison of yields and repair in vivo and in vitro. *Basic Life Sci.* **1986**, *38*, 151–158. [PubMed]
25. Choi, S.E.; Jung, I.R.; Lee, Y.J.; Lee, S.J.; Lee, J.H.; Kim, Y.; Jun, H.S.; Lee, K.W.; Park, C.B.; Kang, Y. Stimulation of lipogenesis as well as fatty acid oxidation protects against palmitate-induced INS-1 beta-cell death. *Endocrinology* **2011**, *152*, 816–827. [CrossRef] [PubMed]
26. Baumann, J.; Wong, J.; Sun, Y.; Conklin, D.S. Palmitate-induced ER stress increases trastuzumab sensitivity in HER2/neu-positive breast cancer cells. *BMC Cancer* **2016**, *16*, 551. [CrossRef] [PubMed]
27. Hardy, S.; El-Assaad, W.; Przybytkowski, E.; Joly, E.; Prentki, M.; Langelier, Y. Saturated fatty acid-induced apoptosis in MDA-MB-231 breast cancer cells. A role for cardiolipin. *J. Biol. Chem.* **2003**, *278*, 31861–31870. [CrossRef] [PubMed]
28. Byrne, F.L.; Poon, I.K.; Modesitt, S.C.; Tomsig, J.L.; Chow, J.D.; Healy, M.E.; Baker, W.D.; Atkins, K.A.; Lancaster, J.M.; Marchion, D.C.; et al. Metabolic vulnerabilities in endometrial cancer. *Cancer Res.* **2014**, *74*, 5832–5845. [CrossRef] [PubMed]
29. Wagner, J.M.; Karnitz, L.M. Cisplatin-induced DNA damage activates replication checkpoint signaling components that differentially affect tumor cell survival. *Mol. Pharmacol.* **2009**, *76*, 208–214. [CrossRef] [PubMed]
30. Ling, Y.H.; El-Naggar, A.K.; Priebe, W.; Perez-Soler, R. Cell cycle-dependent cytotoxicity, G2/M phase arrest, and disruption of p34cdc2/cyclin B1 activity induced by doxorubicin in synchronized P388 cells. *Mol Pharm.* **1996**, *49*, 832–841.
31. Zhang, Y.; Xia, G.; Zhang, Y.; Liu, J.; Liu, X.; Li, W.; Lv, Y.; Wei, S.; Liu, J.; Quan, J.J. Palmitate induces VSMC apoptosis via toll like receptor (TLR) 4/ROS/p53 pathway. *Atherosclerosis* **2017**, *263*, 74–81. [CrossRef] [PubMed]
32. Wei, C.D.; Li, Y.; Zheng, H.Y.; Tong, Y.Q.; Dai, W.J.M. Palmitate induces H9c2 cell apoptosis by increasing reactive oxygen species generation and activation of the ERK1/2 signaling pathway. *Mol. Med. Rep.* **2013**, *7*, 855–861. [CrossRef] [PubMed]
33. Liu, J.; Chang, F.; Li, F.; Fu, H.; Wang, J.; Zhang, S.; Zhao, J.; Yin, D.J.B. Palmitate promotes autophagy and apoptosis through ROS-dependent JNK and p38 MAPK. *Biochem. Biophys. Res. Commun.* **2015**, *463*, 262–267. [CrossRef]
34. Henry-Mowatt, J.; Dive, C.; Martinou, J.C.; James, D. Role of mitochondrial membrane permeabilization in apoptosis and cancer. *Oncogene* **2004**, *23*, 2850–2860. [CrossRef] [PubMed]
35. Youle, R.J.; van der Bliek, A.M. Mitochondrial fission, fusion, and stress. *Science* **2012**, *337*, 1062–1065. [CrossRef]
36. Yang, X.; Wang, J. The Role of Metabolic Syndrome in Endometrial Cancer: A Review. *Front. Oncol.* **2019**, *9*, 744. [CrossRef] [PubMed]
37. Melo, H.M.; Seixas da Silva, G.D.S.; Sant'Ana, M.R.; Teixeira, C.V.L.; Clarke, J.R.; Coreixas, V.S.M.; de Melo, B.C.; Fortuna, J.T.S.; Forny-Germano, L.; Ledo, J.H.; et al. Palmitate Is Increased in the Cerebrospinal Fluid of Humans with Obesity and Induces Memory Impairment in Mice via Pro-inflammatory TNF-α. *Cell Rep.* **2020**, *30*, 2180–2194.e8. [CrossRef] [PubMed]
38. Lupu, R.; Menendez, J.A. Targeting fatty acid synthase in breast and endometrial cancer: An alternative to selective estrogen receptor modulators? *Endocrinology* **2006**, *147*, 4056–4066. [CrossRef] [PubMed]
39. Menendez, J.A.; Oza, B.P.; Atlas, E.; Verma, V.A.; Mehmi, I.; Lupu, R.J.O. Inhibition of tumor-associated fatty acid synthase activity antagonizes estradiol-and tamoxifen-induced agonist transactivation of estrogen receptor (ER) in human endometrial adenocarcinoma cells. *Oncogene* **2004**, *23*, 4945–4958. [CrossRef]

40. Nath, A.; Oak, A.; Chen, K.Y.; Li, I.; Splichal, R.C.; Portis, J.; Foster, S.; Walton, S.P.; Chan, C. Palmitate-Induced IRE1-XBP1-ZEB Signaling Represses Desmoplakin Expression and Promotes Cancer Cell Migration. *Mol. Cancer Res.* **2021**, *19*, 240–248. [CrossRef]
41. Pan, J.; Dai, Q.; Zhang, T.; Li, C. Palmitate acid promotes gastric cancer metastasis via FABP5/SP1/UCA1 pathway. *Cancer Cell Int.* **2019**, *19*, 69. [CrossRef] [PubMed]
42. Kim, S.; Yang, X.; Yin, A.; Zha, J.; Beharry, Z.; Bai, A.; Bielawska, A.; Bartlett, M.G.; Yin, H.; Cai, H. Dietary palmitate cooperates with Src kinase to promote prostate tumor progression. *Prostate* **2019**, *79*, 896–908. [CrossRef] [PubMed]
43. Zeng, L.; Wu, G.Z.; Goh, K.J.; Lee, Y.M.; Ng, C.C.; You, A.B.; Wang, J.; Jia, D.; Hao, A.; Yu, Q.; et al. Saturated fatty acids modulate cell response to DNA damage: Implication for their role in tumorigenesis. *PLoS ONE* **2008**, *3*, e2329. [CrossRef]
44. Binker-Cosen, M.J.; Richards, D.; Oliver, B.; Gaisano, H.Y.; Binker, M.G.; Cosen-Binker, L.I. Palmitic acid increases invasiveness of pancreatic cancer cells AsPC-1 through TLR4/ROS/NF-κB/MMP-9 signaling pathway. *Biochem. Biophys. Res. Commun.* **2017**, *484*, 152–158. [CrossRef] [PubMed]
45. Fatima, S.; Hu, X.; Huang, C.; Zhang, W.; Cai, J.; Huang, M.; Gong, R.-H.; Chen, M.; Ho, A.H.; Su, T. High-fat diet feeding and palmitic acid increase CRC growth in β2AR-dependent manner. *Cell Death Dis.* **2019**, *10*, 711. [CrossRef] [PubMed]
46. Lin, L.; Ding, Y.; Wang, Y.; Wang, Z.; Yin, X.; Yan, G.; Zhang, L.; Yang, P.; Shen, H. Functional lipidomics: Palmitic acid impairs hepatocellular carcinoma development by modulating membrane fluidity and glucose metabolism. *Hepatology* **2017**, *66*, 432–448. [CrossRef] [PubMed]
47. Hsiao, Y.-H.; Lin, C.-I.; Liao, H.; Chen, Y.-H.; Lin, S.-H. Palmitic acid-induced neuron cell cycle G2/M arrest and endoplasmic reticular stress through protein palmitoylation in SH-SY5Y human neuroblastoma cells. *Int. J. Mol. Sci.* **2014**, *15*, 20876–20899. [CrossRef]
48. Hardy, S.; Langelier, Y.; Prentki, M. Oleate activates phosphatidylinositol 3-kinase and promotes proliferation and reduces apoptosis of MDA-MB-231 breast cancer cells, whereas palmitate has opposite effects1. *Cancer Res.* **2000**, *60*, 6353–6358.
49. Harada, H.; Yamashita, U.; Kurihara, H.; Fukushi, E.; Kawabata, J.; Kamei, Y. Antitumor activity of palmitic acid found as a selective cytotoxic substance in a marine red alga. *Anticancer Res.* **2002**, *22*, 2587–2590. [PubMed]
50. Zhu, S.; Jiao, W.; Xu, Y.; Hou, L.; Li, H.; Shao, J.; Zhang, X.; Wang, R.; Kong, D. Palmitic acid inhibits prostate cancer cell proliferation and metastasis by suppressing the PI3K/Akt pathway. *Life Sci.* **2021**, *286*, 120046. [CrossRef] [PubMed]
51. Nakamura, S.; Takamura, T.; Matsuzawa-Nagata, N.; Takayama, H.; Misu, H.; Noda, H.; Nabemoto, S.; Kurita, S.; Ota, T.; Ando, H.; et al. Palmitate induces insulin resistance in H4IIEC3 hepatocytes through reactive oxygen species produced by mitochondria. *J. Biol. Chem.* **2009**, *284*, 14809–14818. [CrossRef]
52. Geng, Y.; Hernández Villanueva, A.; Oun, A.; Buist-Homan, M.; Blokzijl, H.; Faber, K.N.; Dolga, A.; Moshage, H. Protective effect of metformin against palmitate-induced hepatic cell death. *Biochim. Biophys. Acta (BBA) Mol. Basis Dis.* **2020**, *1866*, 165621. [CrossRef] [PubMed]
53. Zorov, D.B.; Juhaszova, M.; Sollott, S.J. Mitochondrial reactive oxygen species (ROS) and ROS-induced ROS release. *Physiol. Rev.* **2014**, *94*, 909–950. [CrossRef] [PubMed]
54. Echtay, K.S.; Roussel, D.; St-Pierre, J.; Jekabsons, M.B.; Cadenas, S.; Stuart, J.A.; Harper, J.A.; Roebuck, S.J.; Morrison, A.; Pickering, S. Superoxide activates mitochondrial uncoupling proteins. *Nature* **2002**, *415*, 96–99. [CrossRef] [PubMed]
55. Breen, E.P.; Gouin, S.G.; Murphy, A.F.; Haines, L.R.; Jackson, A.M.; Pearson, T.W.; Murphy, P.V.; Porter, R.K. On the Mechanism of Mitochondrial Uncoupling Protein 1 Function. *J. Biol. Chem.* **2006**, *281*, 2114–2119. [CrossRef] [PubMed]
56. Schönfeld, P.; Wojtczak, L. Fatty acids as modulators of the cellular production of reactive oxygen species. *Free Radic. Biol. Med.* **2008**, *45*, 231–241. [CrossRef]
57. Li, L.; Tan, J.; Miao, Y.; Lei, P.; Zhang, Q. ROS and Autophagy: Interactions and Molecular Regulatory Mechanisms. *Cell Mol. Neurobiol.* **2015**, *35*, 615–621. [CrossRef]
58. Mahalingaiah, P.K.; Singh, K.P. Chronic oxidative stress increases growth and tumorigenic potential of MCF-7 breast cancer cells. *PLoS ONE* **2014**, *9*, e87371. [CrossRef]
59. Kim, I.; Rodriguez-Enriquez, S.; Lemasters, J.J. Selective degradation of mitochondria by mitophagy. *Arch. Biochem. Biophys.* **2007**, *462*, 245–253. [CrossRef]
60. Chou, T.C. Theoretical basis, experimental design, and computerized simulation of synergism and antagonism in drug combination studies. *Pharmacol. Rev.* **2006**, *58*, 621–681. [CrossRef] [PubMed]

International Journal of
Molecular Sciences

MDPI

Article

TET3- and OGT-Dependent Expression of Genes Involved in Epithelial-Mesenchymal Transition in Endometrial Cancer

Piotr Ciesielski, Paweł Jóźwiak, Ewa Forma and Anna Krześlak *

Department of Cytobiochemistry, Faculty of Biology and Environmental Protection, University of Lodz, 90-236 Łódź, Poland; piotr.ciesielski@biol.uni.lodz.pl (P.C.); pawel.jozwiak@biol.uni.lodz.pl (P.J.); ewa.forma@biol.uni.lodz.pl (E.F.)
* Correspondence: anna.krzeslak@biol.uni.lodz.pl

Abstract: TET3 is a member of the TET (ten-eleven translocation) proteins family that catalyzes the conversion of the 5-methylcytosine into 5-hydroxymethylcytosine. TET proteins can also affect chromatin modifications and gene expression independently of their enzymatic activity via interactions with other proteins. O-GlcNAc transferase (OGT), the enzyme responsible for modification of proteins via binding of *N*-acetylglucosamine residues, is one of the proteins whose action may be dependent on TET3. Here, we demonstrated that in endometrial cancer cells both TET3 and OGT affected the expression of genes involved in epithelial to mesenchymal transition (EMT), i.e., *FOXC1*, *TWIST1*, and *ZEB1*. OGT overexpression was caused by an increase in *TWIST1* and *ZEB1* levels in HEC-1A and Ishikawa cells, which was associated with increased O-GlcNAcylation of histone H2B and trimethylation of H3K4. The TET3 had the opposite effect on gene expressions and histone modifications. OGT and TET3 differently affected *FOXC1* expression and the migratory potential of HEC-1A and Ishikawa cells. Analysis of gene expressions in cancer tissue samples from endometrial cancer patients confirmed the association between OGT or TET3 and EMT genes. Our results contribute to the knowledge of the role of the TET3/OGT relationship in the complex mechanism supporting endometrial cancer progression.

Keywords: endometrial cancer; TET proteins; O-GlcNAc transferase; *TWIST1*; *ZEB1*; *FOXC1*; migration; invasion

Citation: Ciesielski, P.; Jóźwiak, P.; Forma, E.; Krześlak, A. TET3- and OGT-Dependent Expression of Genes Involved in Epithelial-Mesenchymal Transition in Endometrial Cancer. *Int. J. Mol. Sci.* **2021**, *22*, 13239. https://doi.org/10.3390/ijms222413239

Academic Editor: Laura Paleari

Received: 2 November 2021
Accepted: 3 December 2021
Published: 8 December 2021

1. Introduction

TET family proteins play a significant role in changing the pattern of DNA methylation through participation in DNA demethylation. In humans, three TET proteins named TET1, TET2, and TET3 have been identified. These proteins are iron(II)- and 2-ketoglutarate-dependent dioxygenases, which catalyze the conversion of the 5-methylcytosine (5-mC) into 5-hydroxymethylcytosine (5-hmC) and further into 5-formylcytosine (5-fC) and 5-carboxycytosine (5-caC), which initiates the process of DNA demethylation [1–3]. Besides being involved in DNA demethylation, TET proteins can also affect the epigenetic modifications, regardless of their enzymatic activity by interacting with other proteins that are involved in the modification of chromatin, for example, O-GlcNAc transferase (OGT) [4–6]. OGT is an enzyme responsible for the modification (O-GlcNAcylation) of cellular proteins by linking the single *N*-acetylglucosamine moieties to serine or threonine residues by the O-glycosidic bond. This dynamic modification affects the activity and stability of a great number of proteins, including metabolic enzymes, kinases, phosphatases, transcription factors, and many others. Increased expression of OGT and hyper-O-GlcNAcylation are the hallmarks of many tumors [7–9]. It has been found that the O-GlcNAcylation may be a part of histone code, and OGT modifies H2A, H2B, H3, and H4 histones [10]. Studies using chromatin immunoprecipitation (ChIP-Seq) showed that TET1, TET2, TET3, and OGT can colocalize in H3K4me3-rich sites of chromatin in promoters of transcriptionally active genes [10,11].

It is suggested that TET proteins play an essential role in the recruitment of OGT to chromatin so that it can modify histones [4,5,12]. Chen et al. [4] demonstrated that in murine embryonic stem cells (mESC), Tet2 protein is necessary for OGT to modify histone H2B at serine 112 (H2BS112GlcNAc). Glycosylation of this serine residue facilitates the further modification of histone H2B, the monoubiquitination of lysine 120 (H2BK120Ub). Studies of Ito et al. [5] indicated that TET3 plays the main role in the recruitment of O-GlcNAc transferase to chromatin and regulates its stability.

The epithelial to mesenchymal transition (EMT) is a biological process in which epithelial cells experience profound changes in motility and their ability to invade surrounding tissues [13]. Major players in the regulation of EMT are transcription factors, designated as EMT-TFs, which include, among others, Zinc finger E-box-binding homeobox 1 and 2 (ZEB1 and ZEB2), and Twist-related protein 1 and 2 (TWIST1 and TWIST2). These factors take part in a complex network that activates a specific molecular program aimed at repressing epithelial markers (e.g., E-cadherin, *CDH1*) and activating mesenchymal markers (e.g., vimentin, *VIM*) [13].

The pioneer factors, a special class of transcription factors that can associate with compacted chromatin to facilitate the binding of additional transcription factors are also involved in cancer progression. The function of pioneer factors was originally described during development. More recently, they have been implicated, especially FOXA1, in hormone-dependent cancers, such as estrogen receptor-positive breast cancer and androgen receptor-positive prostate cancer [14]. The role of FOXA1 in human malignancy remains incompletely defined, as both pro- and antitumorigenic functions have been uncovered. FOXA1 is strongly associated with metastatic disease in prostatic adenocarcinoma [15]. In breast cancer, high FOXA1 expression positively correlates with the outcome, but the potential impact of its expression depends on the ERα status and tumor molecular subtype [15]. The other important factor whose role in cancer development and progression has begun to emerge is FOXC1 [16]. Although FOXC1 has not been formally confirmed as a pioneer factor it seems probable it is one since it exhibits conservation of the critical amino acids which confer pioneer activity in FOXA1. Overexpression of FOXC1 has been reported in at least 16 types of cancer, often in association with a poor prognosis [16].

Here, we demonstrate that OGT and TET3 affect the expression of genes associated with epithelial-mesenchymal transition via changes of histones modifications and the ability of endometrial cancer cells for migration and invasion.

2. Results

2.1. *Effect of OGT and TET3 on the Expression of Genes Involved in EMT*

The impact of changes in OGT and TET3 amounts on the expression of genes involved in epithelial-mesenchymal transition (*FOXA1, FOXC1, TWIST, ZEB1*) in endometrial cancer cells has been analyzed. Expressions of TET3 and OGT were changed by treating HEC-1A and Ishikawa cells with plasmid vectors or siRNA (Figure 1A–D). In endometrial cancer cells with TET3 and OGT dysregulation, the expressions of *FOXA1, FOXC1, TWIST, ZEB1* were analyzed and the results are shown for HEC-1A and Ishikawa cells in Figure 1E,F, respectively. The expressions were analyzed in cells with unchanged OGT and TET3 (control), cells with overexpression of TET3 (pTET3), cells with overexpression OGT (pOGT), and cells with overexpression both OGT and TET3 (pTET3/pOGT), and cells cotransfected with pOGT and siTET3. Changes in expression of TET3 and OGT affected the expression of *FOXC1, TWIST1,* and *ZEB1* but did not significantly influence the expression of *FOXA1.* Overexpression of TET3 and OGT caused decreased expression of *FOXC1* in HEC1A cells and increased expression of this gene in Ishikawa cells. TET3 and OGT seem to have the opposite effect on *TWIST1* expressions both in HEC1A and Ishikawa cells, i.e., TET3 causes decreased expression and OGT causes increased expression. Expression of *ZEB1* was affected mostly by OGT, especially in Ishikawa cells. When TET3 and OGT were coexpressed, the most interesting results were found for *TWIST1*. TET3 seems to counteract the *TWIST1* expression increase caused by OGT. In the case of *ZEB1*, co-overexpression of OGT and

TET3 give the same results as overexpression of OGT alone. To confirm the results of the TET3 and OGT impact on these genes, we also analyzed the expression of genes in cells treated with siRNAs specific for TET3 or OGT. The results are shown in Supplementary Figure S1. These results support the findings from overexpression experiments.

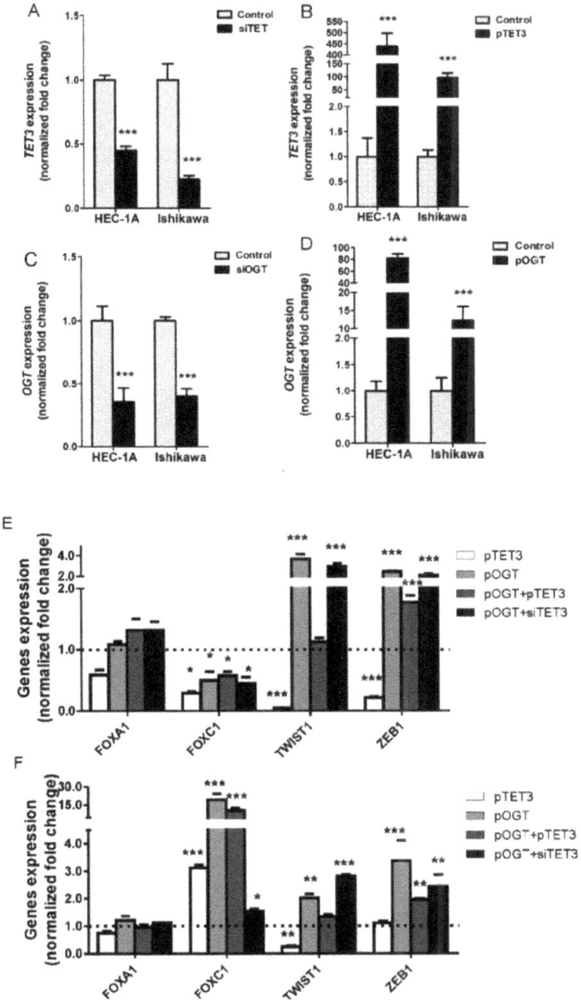

Figure 1. Gene expressions in cells with TET3 and OGT dysregulation. In HEC-1A and Ishikawa cells, TET3 (**A**) and OGT (**C**) expression were decreased by transfection of cells with specific siRNA or increased via transfection with plasmid DNA (**B**,**D**) respectively). The mRNA levels were analyzed by real-time PCR methods (for details see Methods). The mRNA expression of *FOXA1*, *FOXC1*, *TWIST1*, and *ZEB1* were evaluated in HEC-1A (**E**) and Ishikawa (**F**) cells with increased expression of TET3 (pTET3), OGT (pOGT), co-expression of OGT and TET3 (pOGT + pTET3), and increased expression of OGT and reduced expression of TET3 (pOGT + siTET3). The expressions of genes in each kind of sample were compared to appropriate controls, i.e., cells treated with empty vectors or nonsilent siRNA duplexes in which expression was assumed to be 1. Data show mean ± SE (n = 5), * $p < 0.01$, ** $p < 0.001$, *** $p < 0.0001$.

2.2. Amount and Localization of OGT in Endometrial Cancer Cells with TET3 Overexpression

It has been suggested that interactions between TET3 and OGT may impact the stability and intracellular localization of OGT [5]. Thus, the impact of TET3 dysregulation on the expression and localization of OGT in chromatin fraction in endometrial cells was analyzed. In HEC-1A and Ishikawa cells treated with TET3 siRNA or plasmid DNA, the OGT expression was analyzed. The mRNA and protein levels of OGT in cells with TET3 down- and up-regulation did not change (Figure 2A,B). TET3 overexpression did not affect the localization of OGT as well. There was no difference in the global cytoplasmic and nuclear amount of OGT between control cells and cells overexpressing TET3 (Figure 2C). Control cells and cells treated with transcription vector were also fractionated to obtain the chromatin fraction. The results showed enrichment of OGT in chromatin fraction, but TET3 overexpression did not impact OGT localization in chromatin fraction (Figure 2D).

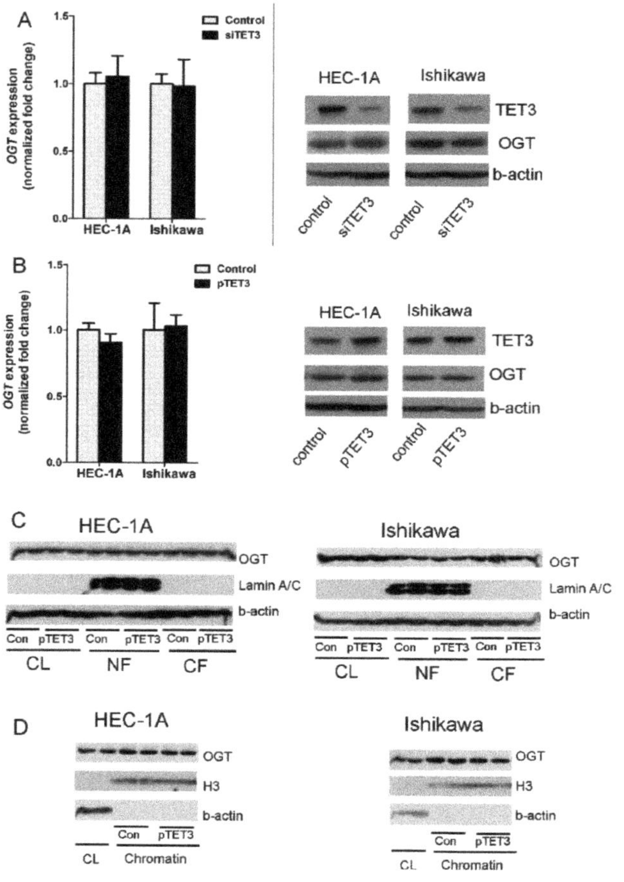

Figure 2. OGT expression and localization in cells with TET3 deregulation. OGT expression was analyzed at mRNA and protein levels in HEC-1A and Ishikawa cells with TET3 reduced (**A**) or increased (**B**) expression using the real-time PCR method or Western blot method. (**C**,**D**) are the results of OGT identification by Western blot method in cellular fractions in control cells and cells with TET3 overexpression. CL—whole cell lysate; NF—nuclear fraction; CF—cytoplasmic fraction.

2.3. Impact of TET3 and OGT on Each Other Binding to Chromatin and Histones Modifications

Since both TET3 and OGT deregulation had a significant impact on *FOXC1*, *TWIST*, and *ZEB1* expressions, the binding of TET3 and OGT proteins to chromatin associated with these genes were analyzed. The results showed that both TET3 and OGT bound to chromatin in regions of EMT genes in HEC-1A and Ishikawa cells (Figure 3A,B, respectively). Overexpression of TET3 or OGT in both cells caused increased and siRNA decreased binding to chromatin compared to appropriate control cells. The effect of overexpression or downregulation of OGT and TET3 on each other's amounts in chromatin associated with *FOXC1*, *TWIST1*, and *ZEB1* was analyzed. In HEC-1A cells, the OGT changed significantly the anti-TET3 antibody binding only in the case of *ZEB1*, but TET3 overexpression reduced the anti-OGT antibody binding to chromatin in regions of all three genes. TET3 overexpression in Ishikawa cells was associated with increased binding of the anti-OGT antibody to the region of *FOXC1* and *TWIST1*.

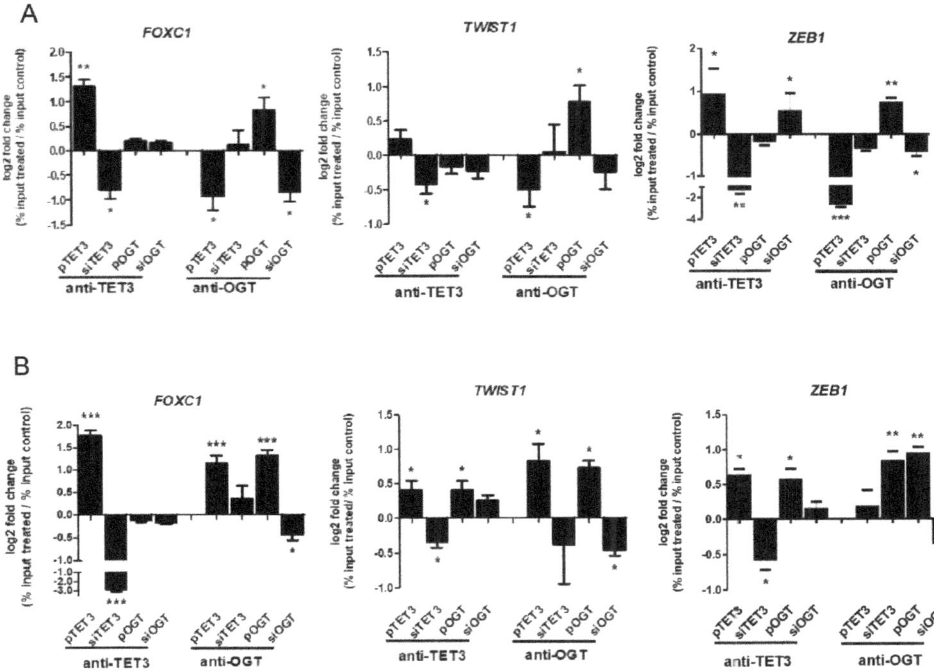

Figure 3. OGT and TET3 binding to chromatin in *FOXC1*, *TWIST1*, and *ZEB1* regions. Chromatin immunoprecipitation (ChIP) analysis was performed for HEC-1A (**A**) and Ishikawa (**B**) endometrial cancer cells treated with plasmid DNA or siRNA with specific antibodies. Differential chromatin enrichment was quantified using real-time quantitative PCR. The results were calculated as % input (% input = % input of specific antibodies probes—% input mock IgG) and are presented as a log2fold change (Fold change = % input of treated cells sample/% input of control cells sample). Data show mean ± SE, ($n = 4$) * $p < 0.01$, ** $p < 0.001$, *** $p < 0.0001$.

The modifications of histones were analyzed in cells with overexpression of OGT or TET3 (Figure 4). Generally, overexpression of OGT caused increased O-GlcNAcylation of histone H2B and trimethylation of lysine 4 of histone H3 (Figure 4A,D). The only exception was the *FOXC1* region where methylation was decreased after OGT overexpression. The overexpression of TET3 caused generally decreased O-GlcNAcylation and ubiquitination of histone H2B and methylation of histone H3 in the case of all analyzed genes (Figure 4B,F). However, there was also an exception. In Ishikawa cells overexpression of TET3 caused increased O-GlcNAcylation, methylation, and ubiquitination of histones in chromatin in the *FOXC1* region. Thus, OGT and TET3 had a similar effect on the methylation of H3K4 in HEC-1A and Ishikawa cells. Increased methylation of H3K4 was associated with increased expression of *FOXC1*.

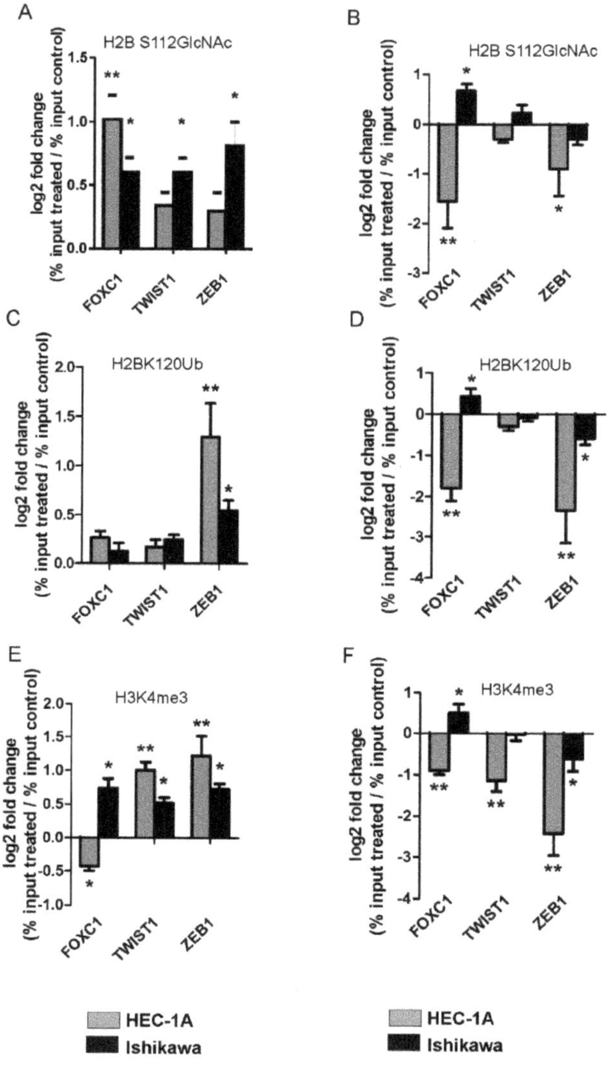

Figure 4. OGT and TET3 dysregulation affect histone modifications. The O-GlcNAcylation of serine 112 of histone H2B (H2BS112GlcNAc), ubiquitination of lysine 120 of histone H2B (H2BK120Ub), and trimethylation of lysine 4 of histone H3 (H3K4me3) have been analyzed using the ChiP method in cells with OGT (**A**,**C**,**E**) or TET3 (**B**,**D**,**F**) overexpression. Differential chromatin enrichment was quantified using Real-time quantitative PCR. The results were calculated as % input (% input = % input of specific antibodies probes—% input mock IgG) and are presented as a log2fold change (Fold change = % input of treated cells sample/% input of control cells sample). Data show mean ± SE, (n = 5) * $p < 0.01$, ** $p < 0.001$.

2.4. OGT and TET3 Affect Migration and Invasion of Endometrial Cancer Cells

Wound healing and Boyden chamber assays were conducted to evaluate the migration and invasion capacity of endometrial cancer cells with overexpression of OGT and TET3 (Figure 5). Interestingly, the effect of TET3 and OGT overexpression was opposite in HEC-1A and Ishikawa cells. In HEC-1A cells, the overexpression of TET3 and OGT caused decreased migration and invasion. The migration potential of Ishikawa control cells was very small, and it was slightly increased after OGT or TET3 expression. The results of Boyden chamber assays showed that both migration and invasion of cells were significantly increased after coexpression of OGT and TET3.

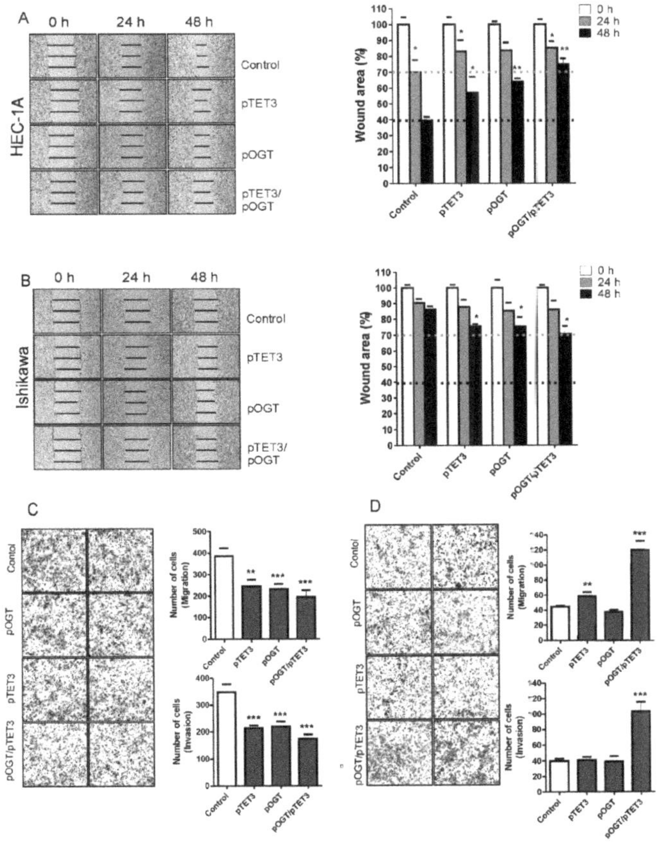

Figure 5. Effect of OGT and TET3 on migration and invasion of cells. The wound-healing assay was performed to examine the migration rate of HEC-1A (**A**) and Ishikawa cells (**B**) transduced with OGT and TET3 plasmid vectors. Photographs were taken at 0, 24 h, and 48 h following the initial scratch. Migration rates were quantified by measuring three different wound areas. Three separate experiments were performed. Cell migration assays using Transwell chambers and invasion assays using Transwell chambers with Matrigel were performed for HEC-1A (**C**) and Ishikawa cells (**D**). Representative images of migrating cells stained with Giemsa are displayed (left) for HEC-1A (**C**) and Ishikawa cells (**D**). Quantitative data of migration and invasion assay are expressed relative to the migration and invasion abilities of control cells. Plots show average counts from three independent testings * $p < 0.01$; ** $p < 0.001$; *** $p < 0.0001$.

2.5. Correlations between Expression of EMT Genes and OGT or TET3 in Endometrial Cancer

To further investigate whether OGT and TET3 were associated with the expression of EMT genes in endometrial cancer, we analyzed the correlations between these gene expressions in 131 samples of endometrial cancer tissues using quantitative PCR. The results showed a significant moderate correlation between OGT expression and *TWIST1* (r = 0.427) or *ZEB1* (r = 0.479) (Figure 6). There was also a weak correlation between TET3 and *FOXC1* (r = 0.229) and *TWIST1* (r = 0.247) (Figure 6).

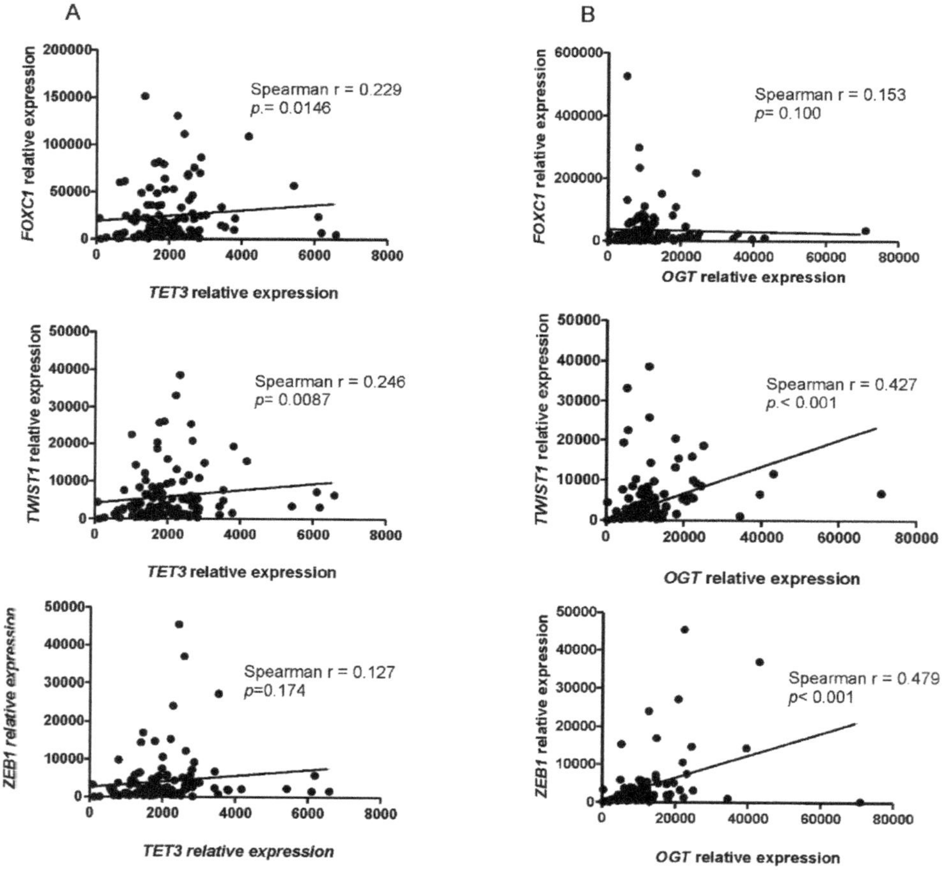

Figure 6 Correlations between gene expressions in endometrial cancer samples. Expression levels of TET3, *OGT*, *FOXC1*, *TWIST1*, and *ZEB1* in samples of endometrial cancers were evaluated by real-time quantitative PCR analysis with the *HPRT1* gene applied as a reference. Spearman rank correlation analysis was performed to analyze the association of TET3 (**A**) or OGT (**B**) expression with *FOXC1*, *TWIST1*, and *ZEB1*. The correlations were considered significant when $p < 0.05$.

3. Discussion

Most earlier studies of TET proteins focused on their ability to facilitate DNA demethylation through the production of 5-hmC. It has only recently been recognized that TET proteins can also affect chromatin modifications and gene expressions independently of their enzymatic activity. It is suggested that individual TET proteins via interactions with other proteins may indirectly change the expressions of specific genes [17]. Moreover, the individual TET proteins may have a different impact on cancer onset and progression.

TET1 decreased expression and low 5-hmC levels are frequently observed in many different types of cancers, including gastric, prostate, liver, lung, and breast cancer as well as glioblastoma and melanoma [18]. The TET2 gene is subjected to frequent somatic mutations in an extensive range of hematopoietic cancers, including myeloid and lymphoid cancers and several solid cancers [19]. In contrast to decreased expression of TET1/2 in cancers, it was found that TET3 expression was upregulated in ovarian cancer, and its high expression was correlated with poor clinicopathological features [20]. Our previous studies showed that TET1 and TET2 messenger RNA expression was lower and TET3 expression was higher in endometrial cancers compared to normal tissues [21]. A positive correlation between 5-hmC and the relative expression of TET1 and TET2 was found, but no correlation was observed in the case of TET3 [21]. Thus, the role of TET3 in endometrial cancer seems to be different than TET1 or TET2, and this protein may be involved in modifications of histones by targeting other proteins to chromatin. It has been suggested that O-GlcNAc transferase is one of the TET3 partners [5].

In this study, we analyzed the impact of TET3 and OGT on migration and invasion of endometrial cancer cells via regulation of EMT genes expression. Our results show that both TET3 and OGT affect cell migration and invasion of endometrial cancer cells; however, unexpectedly, their effects are different in HEC-1A and Ishikawa cells. Ishikawa cells are well-differentiated and their migratory potential is low. TET3 and OGT overexpression caused an increase in the migratory potential of Ishikawa cells, which was especially seen when both proteins were coexpressed. Increased expression of TET3 and OGT caused a significant increase of *FOXC1* expression. Interestingly, increased expression of TET3 caused increased binding of OGT to chromatin in the region of *FOXC1* and increased histone H2B O-GlcNAcylation and H3K4 trimethylation. This may suggest that TET3 plays a role in the targeting of OGT to chromatin in Ishikawa cells. Although our results did not show general enrichment of OGT in chromatin fraction after TET3 overexpression, that did not exclude the possibility that TET3 is involved in the targeting of OGT to specific chromatin sites. HEC-1A cells, contrary to Ishikawa, showed decreased migratory potential after TET3 and OGT overexpression. Interestingly, in these cells, *FOXC1* expression was decreased after TET3 or OGT overexpression. We cannot explain the reason for different TET and OGT impacts on *FOXC1* in the two cell lines. Both cell types represent type I endometrial cancer. Traditionally, endometrial cancer has been divided into two subtypes with distinct clinical, pathological, histological, and molecular behavior. Type I endometrial carcinomas account for 85% of all endometrial cancers, and they are mainly low grade, estrogen-dependent, hormone-receptor-positive adenocarcinomas with endometrioid morphology [22]. However, despite many similarities, these two types of cells differ in many ways. For example, Ishikawa cells do not express *PTEN*, and HEC-1A cells are *PTEN*-wild-type. On the other hand, Ishikawa cells expressed all estrogen receptors, and HEC-1A cells lack expression of *ESR1*. Ishikawa cells form more estradiol from estrone than HEC-1A cells [23]. The relationship between ER α and the expression of EMT genes is well established for breast cancer [24]. Thus, we think that the different molecular background lies behind the different effects of TET3 and OGT on *FOXC1*. Future studies are necessary to identify all TET and OGT partners involved in FOXC1 regulation. Although the impact of TET3 and OGT on *FOXC1* expression in HEC-1A and Ishikawa are opposite, it seems that endometrial cell migration and invasion are correlated with this factor expression. Several studies have linked *FOXC1* activity to the aggressive phenotype in cancer cells, especially in basal-like breast cancer and hepatocellular carcinoma [25]. Although studies of FOXC1 in endometrial cancer are not advanced, it is suggested that *FOXC1* may be a potential oncogene also in endometrial carcinoma [25]. In endometrial cancer, the downregulation of *FOXC1* by miRNA—specifically miRNA 204 and miRNA 495—caused inhibition of cancer cell growth and migration [26,27]. The results of our research seem to confirm the significance of *FOXC1* in the aggressive phenotype of endometrial cancer cells. However, the analysis of mRNA expression of FOXC1 in endometrial cancer tissues did not show any association of *FOXC1* expression with clinicopathological characteristics (Supplemental Figure S2).

The findings of this study suggest that OGT and TET have the opposite effect on *TWIST1* and *ZEB1* expressions and histone modifications, both in HEC-1A and Ishikawa cells. OGT increases of O-GlcNAcylation of H4S112 and methylation of H3K4 in the ZEB1 region, which results in the increased expression of *ZEB1*. On the contrary, TET3 overexpression was associated with decreased O-GlcNAcylation and methylation. OGT overexpression also significantly increased H3K4 methylation and expression of *TWIST1*, while TET3 reduced both methylation and expression. However, the migratory potential of HEC-1A cells was not correlated with the increased expression of *ZEB1* or *TWIST1*. The previous studies of Feng et al. [28] showed that *ZEB1* was related to the metastasis of endometrial cancer. *ZEB1* expression was significantly associated with subtype, grade, myometrial invasion, and lymph node metastases in endometrial cancer [28]. Similarly, Shen et al. [29] also showed significantly higher *TWIST1* expression in patients with type I endometrial cancer compared to normal endometrium, and aberrant *TWIST1* expression was significantly associated with clinical parameters, indicating poor prognosis and shorter patient survival. However, Sadłecki et al. [30] did not find significant associations between the clinicopathological characteristics of endometrial cancer patients and the expressions of TWIST1, TWIST2, ZEB1, and *SNAIL*. Our results also did not show any correlation between TWIST1 expression and clinicopathological parameters and *ZEB1* expression was even lower in more advanced cancers compared to less aggressive cancers (Supplementary Figures S3 and S4).

Thus, the function of EMT inducers in endometrial cancers needs further explanation. The expression and function of EMT inducers may vary considerably across different cancer types or even cell types depending on the molecular context [31]. For example, *TWIST1* was found to promote metastasis in the breast cancer model but was shown to be dispensable for metastasis in a pancreatic cancer model. Sometimes, factors may have even opposite effects in different tumors. *ZEB2* is associated with metastasis in ovarian, gastric and pancreatic tumors but reduces aggressiveness in melanoma [31].

In conclusion, our results showed that both TET3 and OGT are involved in the regulation of *FOXC1*, *TWIST1*, and *ZEB1* in endometrial cancer and affect cancer cell migration and invasion. These results contribute to the knowledge of the complex mechanism supporting endometrial cancer progression; however, further studies are needed to elucidate the significance of particular EMT inducers in endometrial cancer progression.

4. Materials and Methods

4.1. Patients and Tissue Samples

Samples of endometrial cancer were obtained from 131 patients who underwent surgery in the Department of Gynecological Oncology Copernicus Memorial Hospital (Łódź, Poland). Tissue samples after tumor resection were immediately placed in RNAlater (Ambion®, Carlsbad, CA, USA). Samples were subsequently stored at −80 °C until RNA and DNA extraction. All cancer samples were characterized in terms of tumor stage according to the International Federation of Gynecology and Obstetrics (FIGO) criteria, histological grade, and type, according to WHO classification, and the ability of cancer cells to metastasize to lymph nodes. The investigations were approved by the Bioethical Commissions of the University of Lodz (6/KBBN-UŁ/III/2014).

4.2. Cell Culture and Treatment

Endometrial cancer cell line HEC-1A was obtained from the American Type Culture Collection (Manassas, VA, USA) and Ishikawa cells were obtained from the European Collection of Authenticated Cell Cultures (Wiltshire, UK). Cells were cultured in DMEM: F12 media (Lonza, Basel, Switzerland) containing 10% (HEC-1A) or 5% (Ishikawa) (*v/v*) FBS at 37 °C and 5% CO_2.

Overexpression of TET3 and OGT was established by transfection of FH-TET3-pEF (#49446, Addgene, Watertown, MA, USA) or pCMV6-OGT-myc (#RC224481, OriGene, Rockville, MD, USA) into cells with Lipofectamine™2000 (Invitrogen™, ThermoFisher

Scientific, Grand Island, NY, USA). For controls, empty vectors were used. Knockdown experiments were performed using Silencer Select siRNA (ID: s47238 ID: s16093) (Ambion®, Carlsbad, CA, USA). To knockdown TET3 or OGT siRNA targeting both genes were complexed to Lipofectamine RNAiMAX (Invitrogen™, ThermoFisher Scientific, Grand Island, NY, USA) following the manufacturer's specifications. The siRNAs were used at a concentration of 30 nM. Cotransfection was performed with Lipofectamine™ 2000 (Invitrogen™, ThermoFisher Scientific, Grand Island, NY, USA). Similar to single transfection in co-transfection experiments, the concentration of siRNA was 30 nM and the ratio of DNA to Lipofectamine was 1:2. The effect was analyzed 24 h after transfection.

4.3. RNA Isolation and RT-PCR

Total RNA from cancer tissue samples was isolated using Trizol® Reagent (Sigma Aldrich, Saint Louis, MO, USA) and from cells using the ExtractMe Total RNA Kit (Blirt, Gdańsk, Poland) according to the manufacturer's instructions. First-strand cDNAs were obtained by reverse transcription of 2 µg of total RNA using High Capacity cDNA Reverse Transcription Kit (ThermoFisher Scientific, Waltham, MA, USA) following the manufacturer's protocol. Real-time amplification of the cDNA was performed using TaqMan® Gene Expression Assay (ThermoFisher Scientific, Waltham, MA, USA) according to the manufacturer's instructions. The fluorogenic, FAM-labeled probes, and the sequence-specific primers for TET3, OGT, FOXA1, FOXC1, TWIST1, and ZEB1, and the internal control HPRT1 were obtained as inventoried assays: Hs00379125_m1, Hs01023894_m1, Hs00559473_s1, Hs01379963_m1, Hs01379963_m1, and Hs02800695_m1 (Applied Biosystems, ThermoFisher Scientific, Waltham, MA, USA). PCR reactions were carried out using the Mastercycler ep realplex (Eppendorf, Hamburg, Germany). The equation $1000*2^{-\Delta Ct}$ was applied to calculate the expression of studied genes in tissue samples, where ΔCt = Ct of the target gene − Ct the reference gene (HPRT1). Results are expressed as a number of target gene mRNA copies per 1000 copies of HPRT1 mRNA. Fold differences in genes expression in cells normalized to HPRT1 levels were calculated using the formula $2^{\Delta\Delta Ct}$.

4.4. Isolation of Cytoplasmic, Nucleoplasmic, and Chromatin Fractions

Cytoplasmic, nucleoplasmic, and chromatin fractions were prepared from a pellet of cultured cells according to Yu et al. [32]. Cells were resuspended in lysing buffer (10 mM HEPES (ang. (4-(2-hydroxyethyl)-1-piperazineethanesulfonic acid)) pH 7.4, 10 mM KCl, 0.05% Triton X-100). After centrifugation at 14,000 rpm (4C), 10 min supernatant containing cytoplasmic proteins was moved to new Eppendorf probes and pellet containing nuclear proteins after washing with lysing buffer was resuspended in a low salt buffer (10 mM TrisHCl pH 7.4, 0.2 mM $MgCl_2$, 1% Triton X100). After centrifugation at 14,000 rpm for 10 min, supernatant containing nucleoplasmic proteins was moved to new Eppendorf probes and pellet with chromatin proteins was resuspended in 0.2 M HCl. Supernatant obtained after centrifugation contained chromatin proteins.

4.5. Chromatin Immunoprecipitation

Chromatin immunoprecipitation (ChIP) analysis was performed for HEC-1A and Ishikawa endometrial cancer cells treated with plasmid DNA or siRNA. Following treatment, cells were cross-linked for 10 min with formaldehyde at room temperature; the cross-linking was stopped by adding glycine at a final concentration of 125 mM. Cells were washed and lysed in lysing buffer (10 mM HEPES, 85 mM KCL, 0.5% Triton X-100, 1 mM PMSF). After centrifugation (700 rpm, 4 °C, 2 min), nuclei in pellet were resuspended in high salt buffer and sonicated on ice using Vibra Cell TM model VCX-130 (Sonics & Materials Inc. Newtown, CT, USA). The chromatin fragments were then immunoprecipitated with specific antibodies. The following antibodies were used: anti-OGT (#5368; Cell Signaling Technology), anti-TET3 [C3] C-term (GTX121453, GeneTex Irvine, CA, USA), anti-Histone H2B (glcnac S112) (ab130951; Abcam, Cambridge, UK), anti-Ubiquityl-Histone H2B (Lys120) (D11) XP® (#5546, Cell Signaling Technology, Danvers, MA, USA), and anti-Tri-Methyl-Histone H3 (Lys4) (C42D8) (#9751, Cell Signaling Technology, Danvers, MA, USA). A control immunoprecipitation using IgG was set up in parallel to distinguish nonspecific precipitation. The anti-Normal Rabbit IgG (#2729, Cell Signaling Technology, Danvers, MA, USA) antibody was used. Protein A/G Plus agarose beads were used to bind protein–antibody complexes. Differential chromatin enrichment was quantified using real-time quantitative PCR. The primer pairs were: for FOXC1 forward ATGGC-GATTTGATTACAGAC and reverse ATTACTGCTTAAGTGTTGCC, for TWIST1 forward CTAGATGTCATTGTTTCCAGAG and reverse CCCTGTTTCTTTGAATTTGG, for ZEB1 forward AAAGATGATGAATGCGAGTC and reverse TCCATTTTCATCATGACCAC. The results were calculated as % input (% input = % input of specific antibodies probes − % input mock IgG) and are presented as a fold change (FC = % input of treated cells sample/% input of control cells sample).

4.6. Western Blotting

Endometrial cancer cells were lysed in a RIPA buffer (50 mM Tris HCl pH 8, 150 mM NaCl, 1% Nonidet P-40, 0.5% sodium deoxycholate, 0.1% SDS, 1 mM EDTA, 1 mM PMSF). Concentrations of protein were determined using the Lowry method. Proteins of the cell lysates were resolved by 8% SDS-PAGE and transferred to Immobilon P membranes. The blots were incubated for two hours at room temperature with the following primary antibodies: anti-OGT (#5368) (diluted 1:2000, Cell Signaling Technology, Danvers, MA, USA), anti-lamin A/C (sc-376248) (diluted 1:2000; Santa Cruz Biotechnology, Dallas, TX, USA), anti-Histone H3 (ab1791) (diluted 1:2500; Abcam), and anti-β-actin (sc-4778) (diluted 1:5000; Santa Cruz Biotechnology, Dallas, TX, USA). After washing with TBST (Tris buffered saline with Tween-20), immunoblots were incubated 1h at room temperature with goat anti-mouse or anti-rabbit secondary antibodies conjugated with horseradish peroxidase (diluted 1:5000, Cell Signaling Technology, Danvers, MA, USA).

4.7. Migration/Invasion Assay

The Transwell assay was performed to assess the rate of migration or invasion of HEC-1A and Ishikawa cells after treatment. Cell culture inserts Millicell™ (polyethylene terephthalate PET membranes with 8 μm pores) (Merck Millipore, Burlington, MA, USA) were used. Cells were plated in serum-free medium 24 h after treatment and placed in the upper chamber. The lower chamber was filled with serum-containing medium. Cells were cultured for 24 h. After that, cells in the upper chamber were removed, and migrated cells at the bottom of the inserts were fixed in 4% paraformaldehyde and stained with Giemsa. In case of invasion, assay chambers were coated with Matrigel® Matrix Basement Membrane (Corning, New York, NY, USA).

4.8. Statistical Analysis

Differences between the expression levels of genes among the studied tissue sample groups were analyzed using the nonparametric Kruskal-Wallis test with the post-hoc Dunn test. Correlations between different gene expressions were analyzed using the Spearman test. The Student's paired *t*-test was used to compare the differences between treated and control cells. A p-value < 0.05 was considered to indicate a statistically significant difference.

Supplementary Materials: The following are available online at https://www.mdpi.com/article/10.3390/ijms222413239/s1.

Author Contributions: Conceptualization, A.K. and P.C.; methodology, P.C., P.J. and E.F.; investigation, P.C., P.J. and E.F.; writing—original draft preparation, P.C. and A.K.; writing—review and editing, A.K. All authors have read and agreed to the published version of the manuscript.

Funding: This research was funded by a grant from the National Science Centre of Poland UMO-2015/19/N/NZ3/01311.

Institutional Review Board Statement: The investigations were approved by the Bioethical Commissions of University of Lodz (6/KBBN-UŁ/III/2014).

Informed Consent Statement: Informed consent was obtained from all subjects involved in the study.

Conflicts of Interest: The authors declare no conflict of interest.

References

1. Guo, J.U.; Su, Y.; Zhong, C.; Ming, G.L.; Song, H. Hydroxylation of 5-methylcytosine by TET1 promotes active DNA demethylation in the adult brain. *Cell* **2011**, *145*, 423–434. [CrossRef]
2. Ito, S.; Shen, L.; Dai, Q.; Wu, S.C.; Collins, L.B.; Swenberg, J.A.; He, C.; Zhang, Y. Tet proteins can convert 5-methylcytosine to 5-formylcytosine and 5-carboxylcytosine. *Science* **2011**, *333*, 1300–1303. [CrossRef] [PubMed]
3. Kroeze, L.I.; van der Reijden, B.A.; Jansen, J.H. 5-Hydroxymethylcytosine: An epigenetic mark frequently deregulated in cancer. *Biochim. Biophys. Acta* **2015**, *1855*, 144–154. [CrossRef]
4. Chen, Q.; Chen, Y.; Bian, C.; Fujiki, R.; Yu, X. TET2 promotes histone O-GlcNAcylation during gene transcription. *Nature* **2013**, *493*, 561–564. [CrossRef] [PubMed]
5. Ito, R.; Katsura, S.; Shimada, H.; Tsuchiya, H.; Hada, M.; Okumura, T.; Sugawara, A.; Yokoyama, A. TET3-OGT interaction increases the stability and the presence of OGT in chromatin. *Genes Cells* **2014**, *19*, 52–65. [CrossRef] [PubMed]
6. Zhang, Q.; Liu, X.; Gao, W.; Li, P.; Hou, J.; Li, J.; Wong, J. Differential regulation of the ten-eleven translocation(TET) family of dioxygenases by O-linked β-N-acetylglucosamine transferase (OGT). *J. Biol. Chem.* **2014**, *289*, 5986–5996. [CrossRef]
7. Hart, G.W.; Slawson, C.; Ramirez-Correa, G.; Lagerlof, O. Cross talk between O-GlcNAcylation and phosphorylation: Roles in signaling, transcription, and chronic disease. *Annu. Rev. Biochem.* **2011**, *80*, 825–858. [CrossRef] [PubMed]
8. Hanover, J.A.; Chen, W.; Bond, M.R. O-GlcNAc in cancer: An Oncometabolism-fueled vicious cycle. *J. Bioenerg. Biomembr.* **2018**, *50*, 155–173. [CrossRef] [PubMed]
9. de Queiroz, R.M.; Carvalho, E.; Dias, W.B. O-GlcNAcylation: The Sweet Side of the Cancer. *Front. Oncol.* **2014**, *4*, 132. [CrossRef] [PubMed]
10. Sakabe, K.; Wang, Z.; Hart, G.W. Beta-N-acetylglucosamine (O-GlcNAc) is part of the histone code. *Proc. Natl. Acad. Sci. USA* **2010**, *107*, 19915–19920. [CrossRef]
11. Deplus, R.; Delatte, B.; Schwinn, M.K.; Defrance, M.; Méndez, J.; Murphy, N.; Dawson, M.A.; Volkmar, M.; Putmans, P.; Calonne, E.; et al. TET2 and TET3 regulate GlcNAcylation and H3K4 methylation through OGT and SET1/COMPASS. *EMBO J.* **2013**, *32*, 645–655. [CrossRef]
12. Vella, P.; Scelfo, A.; Jammula, S.; Chiacchiera, F.; Williams, K.; Cuomo, A.; Roberto, A.; Christensen, J.; Bonaldi, T.; Helin, K.; et al. Tet proteins connect the O-linked N-acetylglucosamine transferase Ogt to chromatin in embryonic stem cells. *Mol. Cell* **2013**, *49*, 645–656. [CrossRef] [PubMed]
13. Tania, M.; Khan, M.A.; Fu, J. Epithelial to mesenchymal transition inducing transcription factors and metastatic cancer. *Tumor Biol.* **2014**, *35*, 7335–7342. [CrossRef]
14. Jozwik, K.M.; Carroll, J.S. Pioneer factors in hormone-dependent cancers. *Nat. Rev. Cancer* **2012**, *12*, 381–385. [CrossRef] [PubMed]
15. Augello, M.A.; Hickey, T.E.; Knudsen, K.E. FOXA1: Master of steroid receptor function in cancer. *EMBO J.* **2011**, *30*, 3885–3894. [CrossRef] [PubMed]
16. Gilding, L.N.; Somervaille, T.C.P. The Diverse Consequences of FOXC1 Deregulation in Cancer. *Cancers* **2019**, *11*, 184. [CrossRef]
17. Lian, H.; Li, W.B.; Jin, W.L. The emerging insights into catalytic or non-catalytic roles of TET proteins in tumors and neural development. *Oncotarget* **2016**, *7*, 64512–64525. [CrossRef] [PubMed]
18. Lio, C.J.; Yue, X.; Lopez-Moyado, I.F.; Tahiliani, M.; Aravind, L.; Rao, A. TET methylcytosine oxidases: New insights from a decade of research. *J. Biosci.* **2020**, *45*, 21. [CrossRef] [PubMed]
19. Jiang, S. Tet2 at the interface between cancer and immunity. *Commun. Biol.* **2020**, *3*, 667. [CrossRef] [PubMed]
20. Cao, T.; Pan, W.; Sun, X.; Shen, H. Increased expression of TET3 predicts unfavorable prognosis in patients with ovarian cancer-a bioinformatics integrative analysis. *J. Ovarian Res.* **2019**, *12*, 101. [CrossRef]
21. Ciesielski, P.; Jóźwiak, P.; Wójcik-Krowiranda, K.; Forma, E.; Cwonda, Ł.; Szczepaniec, S.; Bieńkiewicz, A.; Bryś, M.; Krześlak, A. Differential expression of ten-eleven translocation genes in endometrial cancers. *Tumour Biol.* **2017**, *39*, 1010428317695017. [CrossRef]
22. Chang, Z.; Talukdar, S.; Mullany, S.A.; Winterhoff, B. Molecular characterization of endometrial cancer and therapeutic implications. *Curr. Opin. Obstet. Gynecol.* **2019**, *31*, 24–30. [CrossRef] [PubMed]
23. Hevir-Kene, N.; Rižner, T.L. The endometrial cancer cell lines Ishikawa and HEC-1A, and the control cell line HIEEC, differ in expression of estrogen biosynthetic and metabolic genes, and in androstenedione and estrone-sulfate metabolism. *Chem. Biol. Interact* **2015**, *234*, 309–319. [CrossRef]
24. Voutsadakis, I.A. Epithelial-Mesenchymal Transition (EMT) and Regulation of EMT Factors by Steroid Nuclear Receptors in Breast Cancer: A Review and in Silico Investigation. *J. Clin. Med.* **2016**, *5*, 11. [CrossRef] [PubMed]
25. Elian, F.A.; Yan, E.; Walter, M.A. FOXC1, the new player in the cancer sandbox. *Oncotarget* **2017**, *9*, 8165–8178. [CrossRef] [PubMed]
26. Chung, T.K.; Lau, T.S.; Cheung, T.H.; Yim, S.F.; Lo, K.W.; Siu, N.S.; Chan, L.K.; Yu, M.Y.; Kwong, J.; Doran, G.; et al. Dysregulation of microRNA-204 mediates migration and invasion of endometrial cancer by regulating FOXC1. *Int. J. Cancer* **2012**, *130*, 1036–1045. [CrossRef] [PubMed]

27. Xu, Y.Y.; Tian, J.; Hao, Q.; Yin, L.R. MicroRNA-495 downregulates FOXC1 expression to suppress cell growth and migration in endometrial cancer. *Tumour Biol.* **2016**, *37*, 239–251. [CrossRef]
28. Feng, G.; Wang, X.; Cao, X.; Shen, L.; Zhu, J. ZEB1 expression in endometrial biopsy predicts lymph node metastases in patient with endometrial cancer. *Dis. Markers* **2014**, *2014*, 680361. [CrossRef]
29. Shen, J.; Chen, Q.; Li, N.; Bai, X.; Wang, F.; Li, B. TWIST1 expression and clinical significance in type I endometrial cancer and premalignant lesions: A retrospective clinical study. *Medicine* **2020**, *99*, e23397. [CrossRef]
30. Sadłecki, P.; Jóźwicki, J.; Antosik, P.; Walentowicz-Sadłecka, M. Expression of Selected Epithelial-Mesenchymal Transition Transcription Factors in Endometrial Cancer. *Biomed Res. Int.* **2020**, *2020*, 4584250. [CrossRef]
31. Lu, W.; Kang, Y. Epithelial-Mesenchymal Plasticity in Cancer Progression and Metastasis. *Dev. Cell* **2019**, *49*, 361–374. [CrossRef] [PubMed]
32. Yu, Z.; Huang, Z.; Lung, M.L. Subcellular Fractionation of Cultured Human Cell Lines. *Bio-Protocol* **2013**, *3*, e754. [CrossRef]

MDPI
St. Alban-Anlage 66
4052 Basel
Switzerland
Tel. +41 61 683 77 34
Fax +41 61 302 89 18
www.mdpi.com

International Journal of Molecular Sciences Editorial Office
E-mail: ijms@mdpi.com
www.mdpi.com/journal/ijms

www.ingramcontent.com/pod-product-compliance
Lightning Source LLC
Chambersburg PA
CBHW061613210926
51850CB00009B/2304
9783036579153